Circuit Design for Electronic Instrumentation

ANALOG AND DIGITAL DEVICES FROM SENSOR TO DISPLAY

Darold Wobschall

Associate Professor, Electrical Engineering
State University of New York at Buffalo

McGRAW-HILL BOOK COMPANY
New York St. Louis San Francisco Auckland Bogotá
Düsseldorf Johannesburg London Madrid
Mexico Montreal New Delhi Panama
Paris São Paulo Singapore
Sydney Tokyo Toronto

Library of Congress Cataloging in Publication Data

Wobschall, Darold.
  Circuit design for electronic instrumentation.

  Includes index.
  1. Electronic instruments—Design and
construction.   I. Title.
TK7878.4.W62   621.3815'3   79-9084
ISBN 0-07-071230-1

234567890   KPKP   86543210

*The editors for this book were Tyler G. Hicks and
Joseph Williams, the designer was Naomi Auerbach,
and the production supervisors were Sally Fliess
and Paul A. Malchow.
It was set in Baskerville by University Graphics, Inc.*

Printed and bound by The Kingsport Press.

Contents

Part Two   Transducers

Part Three   Signal Amplification and Processing

16. Noise and Noise Reduction   261

Part Four   Data Switching, Control, and Readout

17. Pulse Timers and Counters   277

18. Multiplexing   294

19. Data Transmission and Recording   306

Part Five   Power Circuits

Preface

Electronic instrument design is now a widely practiced craft. Its increase in popularity in recent years with both professional and amateur engineers can be attributed to the wide variety of integrated circuits which reduce the cost and complexity of implementing a design approach.

With the proliferation of electronic devices there has been a corresponding expansion of electronic literature and a subdivision into areas of specialization which treat only certain aspects of instrument design. As a consequence, students and non-specialists may spend more effort in finding electronic design information than in designing it, or may not even become aware of standard techniques. I have attempted to simplify the design process by gathering and describing standard devices and techniques in a single, reasonably compact book. The designer can then go to a single source for ideas and approaches to an electronic design problem.

The objective of this book is to provide guidance in the design of complete electronic instruments, from input to output or sensor to readout. Therefore, interfacing and interrelation of circuits is emphasized. All the parts of a digital thermometer, for example, which include a thermosensor, an analog amplifier, an A/D converter, and digital displays, are described.

Sections of the text cover temperature, electrooptical, and displacement transducers or sensors, analog and digital devices, analog and digital signal processing, digital displays, analog-to-digital conversion, data transmission, and data recording. Many or most instruments in use today utilize several of these circuits or devices.

While microprocessor systems are not covered in full detail, many microprocessor-based instruments employ the circuits described here as peripheral devices. The discussions of multiplexing and data transmission are especially appropriate. The chapter on microprocessors expands on the background by describing the interfacing of microprocessors to external devices.

Coverage of a wide variety of topics in compact form requires brevity and careful selection. Preference is given here to more basic and useful circuits. Sufficient explanation is included to allow the reader to understand circuit operation, but extensive mathematical treatment and detailed descriptions of specialized topics are avoided. Should an interest in understanding the details of specific circuits be kindled, I expect that most readers will prefer, as I do, to consult specialized texts or to analyze the circuit themselves.

Students of electronics engineering design will find this book a useful reference text. In one sense, design is not taught by books but through practice and laboratory experience. It must be done by the student. A text can only complement the process by providing guidelines and the basic circuits which are the building blocks of electronic instruments. This text is intended to do so while not relieving the student of exercise by leaving no detail unsaid. Students in a senior-level electronic design course, for which this book was developed, design, analyze, construct, and test instruments on an individual basis as a part of the course requirements. Such projects, even in the academic environment, provide experience akin to that encountered industrially, which is widely recognized as a valuable facet of an engineer's education. Several examples of how the various circuits are combined to form complete instruments are given in the text. Design problems are also included to provide further design exercise for students.

I am indebted to many students who have used earlier drafts of this book and whose suggestions and encouragement have resulted in the present version. I also wish to thank the typists who worked on the various drafts, particularly Kay Ward and my daughter, Tina. Finally, I thank my wife, Katrina, for proofreading when she would rather be planting tulips.

Darold Wobschall

Part One

Semiconductor Devices and Basic Circuits

Chapter One

An Approach to Electronic-Instrument Design

The design of an electronic instrument consists largely of the creative arrangement of proved circuits and devices to achieve a chosen performance goal. Of course, technical requirements must be met in the choice and interconnection of subunits. A textbook cannot teach creativity, but it can describe the devices and circuits which are the ingredients of the design process. It also can provide examples. Actually experience is the best teacher. This chapter has been written as an aid to the inexperienced designer, who may lack confidence or general guidelines to instrument design.

1-1 The Parts of an Electronic Instrument

The purpose of an electronic instrument is typically the measurement and display of a physical parameter. Generally an instrument can be divided into the sections indicated by the block diagram of Fig. 1-1. An input transducer is a device with a measurable electrical parameter which is a function of, and often linearly proportional to, the physical parameter being measured. For example, the resistance change (electrical parameter) of a thermistor (transducer) is proportional to the temperature (physical parameter) at least over a limited range. Usually some type of read-in circuit is present, which converts the change in the electrical parameter of the transducer into an electrical signal which can be easily measured. Signal voltage or frequency, in particular, can be measured accurately. A transducer read-in or converter circuit may be quite simple. For the

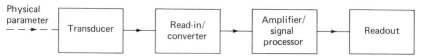

Fig. 1-1 General block diagram of an electronic instrument.

thermistor a dc voltage source and a resistor are all that is needed to convert the resistance change into a proportional voltage change (signal).

Amplification and perhaps further processing of the signal are often required before it is suitable for readout. Signal processing can be elaborate, involving separation into frequency components, timing of various segments, or conversion from ac into dc. Signal readout may consist of a panel meter, a digital display, or perhaps conversion into a form suitable for transfer to a digital computer. In addition to these parts, an instrument generally requires a power supply, which may be so standard that it needs no separate discussion, and control circuits, which may be an integral part of the blocks already discussed.

1-2   Circuit Design and Refinement

Instrument design consists in many cases simply of choosing an appropriate circuit for each block from a catalog of available circuits with due consideration to the matching or interfacing of these blocks. It is assumed, of course, that the function, performance characteristics, or specifications are known. A wide variety of transducers and circuits are available. More often than not the problem facing the designer is how to choose compatible circuits from the wide variety available rather than finding a circuit which performs a particular function. Compatibility between sections must be stressed since without it the required interfacing circuits can increase the complexity of the instrument and may require more effort than a careful selection of basic circuit blocks.

Circuits to perform a particular function can be obtained from a number of sources, e.g., a text like this, journals or magazines concerned with electronic instrumentation, electronic manufacturer's application notes, and local consultants or friends interested in this subject. An experienced designer also builds up a repertoire of circuits or bag of tricks on which to draw, together with a feeling for which circuits, singly or in combination, are easy to implement.

Sometimes a complete and detailed circuit is found to perform the task required, in which case no design is required; all that remains is construction and test of the instrument. More often than not the circuit found will be incomplete, have too many or too few functions, or otherwise not have quite the right specifications. The designer's task is then to edit, modify, or fill in missing parts of the circuit. Circuits found in textbooks or application notes generally have one or more components whose values are unspecified, but a formula or procedure is given so that appropriate values to fit a specific application can be chosen. For example, the gain A of a feedback amplifier might be determined by the ratio of two resistors ($A = R_b/R_a$), and the designer is free to set, within limits, the gain by the resistor selection.

Considerable design freedom is usually available, and the student who has been narrowly schooled in the analytical methods of circuit analysis is bothered by this extra freedom at first. A gain of 10 in the example above can be obtained with the combination $R_a = 100 \text{ k}\Omega$ and $R_b = 10 \text{ k}\Omega$, or $R_a = 20 \text{ k}\Omega$ and $R_b = 2 \text{ k}\Omega$, or an infinite number of other choices. Many analytically inclined students would like to see only one solution. They must learn to live with these extra degrees of freedom and even appreciate them.

Once a tentative circuit has been chosen, it should be analyzed, refined, and tested. This is an iterative process, as the flowchart of Fig. 1-2

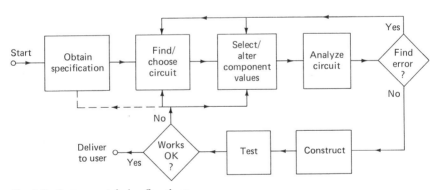

Fig. 1-2   Instrument-design flowchart.

suggests. In the analysis step, the voltage, current, gain, frequency response, or other appropriate parameters are examined quantitatively. This may only consist of a check on the compatibility of voltages between stages or may involve a more thorough mathematical analysis of the system response. A computer-aided design procedure may be used, especially in complicated high-frequency circuits. If errors or deviations from the given specifications are uncovered, the circuit component values are appropriately modified or if necessary a better circuit is found. These steps are repeated until no errors are apparent. In other words, the circuit should look good on paper before one proceeds to the next stages, the constructing and testing of the instrument. If problems are uncovered in the testing stage, as is usually the case, the component values may have to be modified, or in the worst case the circuit will have to be scrapped and a new one selected (back to the drawing board).

If the instrument does not meet the specifications provided, the possibility of changing the specifications to fit the actual performance should be considered. Often the specifications provided initially are unnecessarily stringent or unrealistic in some respects, but until the design is complete, the problem is not recognized. Quite possibly a particular requirement was set rather arbitrarily or with a very generous margin of safety which upon closer examination might be modified. Drawing up a complete and realistic set of specifications, in other words, is often part of the design problem and the iterative loop rather than an unalterable input.

1-3   Integrated-Circuit Advantages

Instrument design has been enormously simplified by the wide availability of integrated circuits (ICs). Circuits which might take hours to design using discrete transistors take only minutes with ICs. Further, the instrument with ICs will probably work much better, cost much less, and be much easier to construct. A beginner today can in some cases design and construct an amplifier in an hour which will outperform an amplifier requiring days to make by an electrical engineer one or two decades ago. With many IC configurations, the choice of components is simple, even cookbooklike. Novices are not likely to run into many problems with these configurations, and the more experienced designers also appreciate them because they take so little time. Of course, not all circuits involving ICs are easy to design or even understand; there is still an adequate challenge to the experienced engineer. The point is that the easy end of the circuit spectrum is easy indeed and should not be feared by the inexperienced.

An important advantage of most ICs is that they approach ideal behavior closer than the typical discrete circuit. They are more linear and insensitive to power-supply fluctuations, for example. Actually these desirable characteristics are achieved through external negative-feedback stabilization and internal regulation and stabilization circuits on the IC chip itself, which can be duplicated with discrete transistors. However an IC may have the equivalent of 10, 50, or even more transistors at less than the cost of a single transistor (not to mention the cost of wiring the equivalent circuit), and therefore the implementation of the equivalent circuit with discrete transistors is usually impractical.

It would be a mistake to leave the impression that all electronic functions can be carried out best with ICs. An IC may not be available to perform the specialized function the designer has in mind. Some functions are simple enough to be done quite adequately by a single transistor or diode, and an IC is an unnecessary complication. Furthermore voltage, power, and frequency limitations are generally greater in ICs than in discrete transistors. It will be noted that a number of the circuits described in this text and in commercial instruments involve discrete components.

1-4 Interfacing and Matching Sections

Circuit blocks which work well individually may not work when connected. Compatibility is a problem in electronic-instrument design, as it is in marriage. Primarily the interfacing procedure between stages, sections, or blocks involves (1) matching the voltage levels between stages and (2) avoiding excessive loading of a stage by the following stage. In Fig. 1-3 an

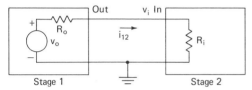

Fig. 1-3 Stage input-output equivalent circuit.

input-output equivalent circuit for two stages is diagramed. Of interest here is the matching of the output of stage 1 to the input of stage 2, where stage 1 might represent any section of an instrument, including a transducer, and stage 2 the following section. The output of stage 1 is represented by a voltage generator v_o with a series resistance R_o, referred to as the stage output resistance or impedance. This equivalent-circuit model for a circuit is rather general and known as the *Thevenin equivalent*. The input resistance of stage 2 is represented by R_i. When stage 2 is connected to stage 1, the load on stage 1 is R_i and a current i_{12} flows from 1 to 2.

At the output terminals of stage 1 the voltage v_i, which is equal to the voltage input of stage 2, is

$$v_i = v_o - R_o i_{12} = \frac{v_o}{1 + R_o/R_i} \tag{1-1}$$

Usually R_o and R_i are not constant or accurately known. Precise predictions of circuit behavior cannot be made unless the effect of these resistances is made negligible. As Eq. (1-1) indicates, this can be done by making the output impedance R_o small and/or the input impedance R_i high, that is, $R_o \ll R_i$, in which case $v_i = v_o$ under all conditions. Circuit behavior is often described and derivations made under the implicit assumption that this condition is met. Should this not be the case, an additional high-input-impedance amplifier such as the noninverting operational amplifier (op-amp) configuration (perhaps unity gain) can be added, as described later. Digital circuits present a special problem which will be discussed in Chap. 5.

1-5   To Build or to Buy?

The question often arises whether it is less costly to design and build an instrument from components than to purchase a complete unit from a commercial source. The production of electronic circuits even in modest quantities results in a much lower unit construction cost. Equally or more important, the cost of design and debugging the instrument can be spread over a large number of units. Of course if no commercial instrument can be found, users have no choice but to build it themselves or have it custom-built. It might appear that the cost advantage lies heavily on the side of the commercial instrument, but this is often not the case. A commercial company must write service manuals, maintain a service group, advertise, and carry administrative overhead. The instrument must also be as versatile as possible to appeal to a reasonably wide range of users. The increased cost of a higher-performance unit is only partly compensated for by the economics of greater production. If the required instrument versatility and performance are much more limited than a commercially available unit, it is quite possible that the user-built unit will be cheaper.

Consider the audio-frequency oscillator as an example. Wide-range, variable-frequency, and amplitude-test oscillators are available from many manufacturers and are competitively priced. Low distortion, good stability, and accuracy are standard. An individual user building an oscillator of similar versatility would find that the cost is much higher than that of the commercial unit. On the other hand, if a fixed-frequency fixed-amplitude oscillator is all that is required, the user can build a unit at a much lower cost with an IC. It has the further advantage of small size and easy incorporation into a system, with less risk that connecting wires will come off or that passing knob twiddlers will misadjust the unit.

If a general guideline can be stated, it is that popular, versatile, multi-control complex instruments are less costly obtained from commercial services† while special-purpose, fixed- or limited-range simple units can be assembled by the user at lower cost.

†The prime example is the oscilloscope. The cost advantage is so heavily on the side of the commercial manufacturer that very few oscilloscope circuits are published although it is one of the most common instruments in use. No one is foolish enough to make his own oscilloscope, at least with the thought of saving money.

1-6 Analog or Digital?

Unmistakably the trend in instrument design is to increase the utilization of digital circuits. The advantages of digital techniques are widely recognized—reliability, flexibility of data reduction, and easy, accurate readout. Extreme enthusiasts of digital systems often feel that instruments should be as close to 100 percent digital as possible and that the incorporation of any analog circuits represents a deviation from perfection. However, this is an analog world; few physical parameters vary in a discrete fashion. Analog signals are often best processed by analog devices, and it is not obvious that a conversion to digital form is necessarily an improvement. In any case conversion increases instrument cost, complexity, and response time. As new analog and digital circuits are developed, the best circuit for a particular application changes and it is unwise to assume without a detailed examination that either the analog or digital approach is necessarily superior.

A few generalizations can be made. In instruments or systems with multiple inputs and/or extensive data-processing requirements, the digital approach is usually best. Conversion of analog signals from sensors to standard digital form close to the input of a system simplifies processing the data. Transmission of data to a remote point is usually more reliable and accurate in digital form. Readout of a voltage or other variable to an arbitrary high of precision is possible with a multiple-digit display. On the other hand, with a small instrument requiring both analog input and output, the overhead of conversion from analog to digital and the reverse may be costly and unnecessary. Furthermore digital readouts are not always easiest to read. Maximization of a tuned-circuit voltage, for example, is definitely easier with a standard (analog) voltmeter than with a digital voltmeter. Finally since most sensors are analog, some degree of analog signal processing is ordinarily required at the input even in a basically digital instrument.

A competent instrument designer, in short, must be familiar with both analog and digital methods and must consider the alternatives fairly before deciding on a specific approach.

1-7 The Role of Circuit Analysis

Mathematical analysis of circuits is strongly advocated by some designers and just as strongly disliked by others. As pointed out above, analysis is only one step in the design process. Its purpose is to determine whether a proposed circuit will work as expected and/or to determine the proper value of components. If there is any creative part to the design process, it is usually in thinking up the circuit initially and rarely in the analysis step.

Furthermore, analysis is often time-consuming and can produce either obvious or irrelevant results, particularly if nonlinear processes are involved. Experienced, successful practical designers, in fact, often ignore mathematical circuit analysis. Academicians, on the other hand, usually stress analysis, perhaps because it is easier to teach a clean-cut mathematical process than a fuzzy creative approach or because of a desire to give students more general methods which are not quickly outdated as discussions of specific circuits are destined to be. Numerous examples might be cited of seat-of-the-pants approaches which failed or consumed large quantities of time and money in the testing and redesign process when a straightforward circuit analysis would have foreseen the error and allowed correction at an early stage.

What then is the role of mathematical circuit analysis? Let us assume that the objective is to design an electronic instrument with a given set of specifications efficiently, i.e., with minimum design, construction, and testing time. A proposed circuit is at hand, and a judgment on the depth of analysis must be made before construction and test. Of course, no strict rules can be applied, but the following guidelines are suggested:

1. Analyze the circuit by sections. Check the interfacing (voltage levels and impedance matching) between sections. Determine whether the gains or other needed characteristics of a section are immediately known or easily calculated. If so, they should be listed for each section.

2. Feedback involving more than one section calls for a detailed analysis of the sections involved. At least linear feedback and stability analysis should be applied.

3. Sections whose performance are uncertain should be singled out for careful analysis (and perhaps separately tested).

4. Logic circuits should be analyzed by appropriate methods such as timing diagrams and truth tables.

5. Complex and/or nonlinear analog circuits should be analyzed by computer with a program such as SPICE, SEPTRE, or ECAP.

Bear in mind that an analysis which takes more time than circuit construction and testing is probably inefficient (unless it leads to an understanding of circuit operation which might be helpful in later work). Overkill on the analysis side is thus to be avoided. Likewise, blind trial and error is inefficient, often leading nowhere. Obviously the middle road is best.

Chapter Two

Discrete-Device Characteristics

Discrete semiconductor devices, in particular diodes and transistors, are important components of electronic instruments and are the basic functional units of ICs. The discussion here will be brief and emphasize devices and characteristics of interest to this text.

2-1 Basic Diode Characteristics

An ideal diode passes current in only one direction. If a positive voltage is applied with the polarity convention shown in Fig. 2-1, the diode is considered forward-biased. It then exhibits no resistance, and the current flow (in the direction of the arrow) is limited only by an external resistance. When the battery polarity is reversed, the diode is reverse-biased and no current flows. The conventional current-flow direction (not electron flow) is assumed.

A real diode only approximates the ideal. When it is forward-biased, the current increases exponentially with voltage and the resistance remains finite. For most practical purposes, an appreciable current does not flow until 0.5 to 0.7 V forward bias (silicon diode) is applied, but at and above this voltage, the current rises rapidly. In effect there is a conduction threshold V_T at about 0.6 V. This fit to the actual exponential rise is referred to as the *piecewise approximation*. Note that the voltage across a forward-biased diode is always approximately the same (0.6 V for silicon) and almost independent of current.

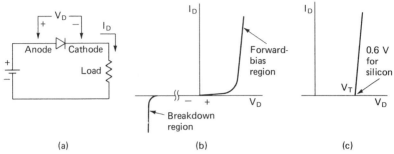

(a) (b) (c)

Fig. 2-1 Diode characteristics: (*a*) polarity convention (forward bias shown); (*b*) real diode current-voltage curve; (*c*) piecewise approximation.

Germanium diodes have a less well-defined threshold at about 0.2 V, while GaAs diodes (light-emitting diodes) have a threshold at 1.6 V.

Temperature is also an important parameter. The threshold V_T shifts to lower values at higher temperatures (Fig. 2-2) at a rate of 2.1 mV/°C.

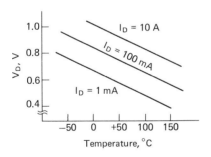

Fig. 2-2 Silicon-diode voltage-threshold temperature dependence at constant current.

The current in a reverse-biased diode is small (typically 10^{-8} A) and approximately independent of voltage until the breakdown region at high reverse voltages is reached (Fig. 2-1*b*). At this point, referred to as the *avalanche* or *Zener breakdown region,* the current rises rapidly with increasing (reverse) voltage. Voltage-regulator diodes which utilize this effect are discussed below.

For most applications, the only important parameters which distinguish various diodes are the maximum permissible operating voltage and current. The maximum voltage refers to the reverse direction and specifies the minimum voltage at which breakdown will occur for a specific device. Since breakdown can occur even for brief pulses or peaks, a diode should be chosen with an inverse voltage comfortably higher than the highest transient voltage which might occur. Breakdown voltages less than 50 V are uncommon, and units which can be operated above 1 kV are available. Similarly the maximum peak current for a diode is specified, and the actual current surge in a circuit should be kept less than this value by a substantial margin. Both an instantaneous (surge) and a time-average (few seconds) maximum current are specified for diodes.

Numerous types of diodes are available commercially, but except for special-purpose diodes the differences for most applications are not very significant. A circuit might work equally well if any diode is selected out of a list of hundreds. For critical applications, the diode speed and reverse-current leakage should be considered in addition to the maximum voltage and current, but otherwise the choice is arbitrary.

2-2   Special-Purpose Diodes

Rectifier (High-Current) Diodes

Diodes intended for power-supply rectifier applications have a relatively high maximum current capacity and usually a high inverse voltage as well. A typical small rectifier diode would be able to handle 1 A average (10 A surge) and have an inverse voltage of 100 V or more. Units rated well over 100 A or 1 kV are available. Large units have a stud (bolt) type of case intended for mounting on a heat sink to dissipate the power produced by internal heating. The case is connected to the cathode except in units designated R (reverse), in which case the anode is connected to the case.

Since the current leakage (reverse bias) can be rather high and the response time comparatively slow in rectifier diodes, they may not be suitable for low-level switching or signal applications.

Computer Diodes

A diode intended for high-speed-switching applications is referred to as a *computer diode*. Diodes do not respond instantly because a finite time is required to input or remove charge near the junction to produce the junction potential. The charge is referred to as *storage charge*, and the time to remove the charge is the *storage time*. It is also referred to as the *off time* τ_{off} since this process is responsible for the delay in the current or conductance drop when the potential across the diode is suddenly changed from forward to reverse bias. The *on time* τ_{on} when the bias is changed from reverse to forward is due to charge injection and is often shorter than τ_{off}. Switching times (both τ_{off} and τ_{on}) for a computer diode are typically 4 ns. Other high-speed diodes are the PIN type, especially useful as high-frequency detectors (rectifiers).

Voltage-Regulator Diodes

Devices designed to be operated in the reverse breakdown region are referred to as *voltage-regulator, avalanche,* or *Zener diodes.* A good regulator diode will exhibit a sharp knee and a high slope on a current-voltage plot (Fig. 2-3*a*) in the breakdown region. A regulator diode is operated

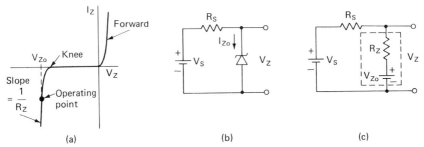

(a)   (b)   (c)

Fig. 2-3   Zener diode characteristics and equivalent circuits: (*a*) current-voltage characteristics; (*b*) measurement circuit; and (*c*) small-signal equivalent circuit.

through a resistor from a voltage source V_S. Always V_S is chosen substantially higher than the operating voltage (V_{Zo}). Current is adjusted by a series limiting resistance R_S.

At a specified operating point (V_{Zo}, I_{Zo}) a small-signal Zener resistance R_Z can be defined as the inverse slope at that point. For example, a 10-V V_{Zo} diode operated at a current I_{Zo} of 20 mA might have a Zener resistance R_Z of 10 Ω (the dc or large-signal resistance is, of course, 500 Ω from Ohm's law). An equivalent circuit for the breakdown diode in the reverse direction is shown in Fig. 2-3*c*.

In a practical circuit R_S is much greater than R_Z, and therefore

$$I_Z = (V_S - V_{Zo})/R_S \tag{2-1}$$

The output voltage, or voltage across the diode V_Z, is

$$V_Z = V_{Zo} - R_Z I_Z \tag{2-2}$$

Since R_Z is small, the voltage V_Z is almost constant and rises only slightly as the current increases. This is the basis of its voltage-regulator action (see Chap. 21).

Care must be taken to keep the power dissipation ($P_D = I_Z V_Z$) less than the specified maximum under all conditions of operation. Typically the maximum power dissipation is 0.4 to 1 W. Diodes with breakdown voltages from below 3 up to 200 V are available in steps of about 20 percent.

Voltage-Variable Capacitor (Varactor)

A voltage-variable capacitor is actually a diode operated in the reverse-bias region. Applications include automatic tuning and frequency modulation of oscillators. As mentioned above, a diode under reverse-biased conditions can be thought of as a capacitor with the depletion region as the insulator. Increasing the bias voltage increases the width of the depletion layer and thus decreases the capacitance. A diode acts as a capacitor only at higher frequencies. At low frequencies the resistive component may dominate. Capacitance varies approximately as the inverse square root of voltage. A typical diode might have a capacitance of 50 pF at 4 V and vary ±50 percent in the range of 2 to 8 V.

PIN Diodes

A PIN diode differs from a standard diode by the addition of a narrow insulating region between the p-type and n-type semiconductor region (the I in PIN signifies an insulator between the pn junction). Because the region is very thin, the normal depletion region at the junction is extended only slightly and the dc diode characteristics are only slightly different from those of a standard diode; the high-frequency characteristics are better than those of standard high-speed (computer) diodes by one or two orders of magnitude, so that operation above 100 MHz or even 1 GHz is possible. The PIN diode is an excellent rectifier or demodulator at high frequencies.

When it is biased into the forward-conduction region, the PIN diode can be used as a current-controlled resistor for small signals. Its ac resistance can be varied from well above 10 kΩ to below 1 Ω. Attenuators for high-frequency operation often incorporate PIN diodes for this purpose.

Other Diodes

Photodiodes and light-emitting diodes are discussed in Chap. 7. The trigger diode (diac) is mentioned in Chap. 22.

2-3 Bipolar Junction Transistors (BJT)

The input and output characteristics of an npn transistor are shown in graph form in Fig. 2-4. The various nonlinearities have been exaggerated for the purposes of illustration. The usual common-emitter configuration is assumed.

It should be noted that the current-voltage characteristic of the input (I_B-V_B) is just that of the forward-bias diode (Fig. 2-1), as expected since the base-emitter junction is in fact a diode. A single curve is shown since I_B is little affected by the output V_C. Note that the input resistance of a transistor is highly nonlinear and in fact exponential with V_B. If a transistor is biased to a particular operating point, a small-signal input resistance R_i can be defined as the inverse slope or derivative of the $I_B V_B$ curve at that point.

Commonly R_i is in the range of 100 to 1000 Ω. Detailed discussions of transistor biasing and input independence in the linear operating regions can be found in many texts. This topic is not discussed here since the use of ICs eliminates or greatly reduces situations where discrete-transistor biasing techniques into a linear region are required.

From the output characteristics (Fig. 2-4) it is apparent that collector

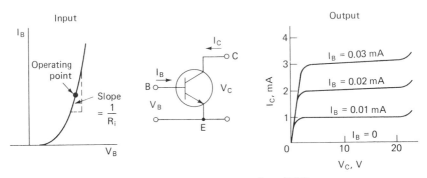

Fig. 2-4   Input-output characteristics for an *npn* transistor (BJT).

current I_C is nearly independent of collector voltage V_C except at very low voltage (the saturation region) or at very high voltages (voltage break-down). Actually there is a slight upward slope, which will be ignored here. For each value of input (base) current, a proportionally higher output current is obtained. This is just the current gain (usually designated as β or h_{fe}) defined as

$$\beta = \frac{I_C}{I_B} \tag{2-3}$$

Strictly speaking β varies slightly with temperature, V_C, and I_C but for many purposes can be considered constant for a given transistor. It is perhaps the most important transistor parameter and typically is of the order of 100.

Maximum values of transistor operating voltage, current, and power are specified for a particular type of transistor. Maximum (breakdown) voltages for transistors in most cases are in the 15- to 100-V range. Maximum (peak) currents typically are 50 to 500 mA but may be above 10 A for certain power transistors. Maximum power dissipation ranges from 0.1 W (small plastic case) to 150 W (large, stud-mounted metal cases). Care must be taken to cool any transistors operated near the power limit, and it is good practice to keep the average power dissipation several times below the limit for long transistor life. Higher-power types must be mounted on heat sinks.

Two transistor models or equivalent circuits are shown in Fig. 2-5. The first model shows simply that the transistor output acts as a current generator I_C proportional to the input current I_B. An exponential I-V input relation is suggested by the diode, and when appreciable current flows (transistor on, or active), V_B is approximately 0.6 V.

The second model is the conventional small-signal or hybrid-parameter model slightly simplified. In addition to the output current generator βI_B

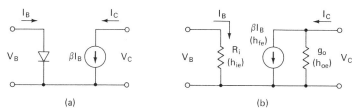

Fig. 2-5   BJT models: (*a*) diode input (*npn*) and (*b*) linear hybrid model.

a small output conductance g_o which models for the slope in the output $I_C V_C$ curve is present. The input is represented by a constant resistance R_i, which depends on the bias point. In the hybrid-parameter notation h_{fe} = β, h_{ie} = R_i, and h_{oe} = g_o. If these parameters are known, either by direct measurements or from manufacturer's data, it is possible to calculate the small-signal performance of a circuit with the transistor included by standard analysis techniques.

Operation of the *pnp* transistor is similar to that of the *npn* except that the polarities of the charge carriers (and therefore the bias voltages) are reversed. Proper bias-voltage polarity and output characteristics for the *pnp* and *npn* in the normal operating region are indicated in Fig. 2-6.

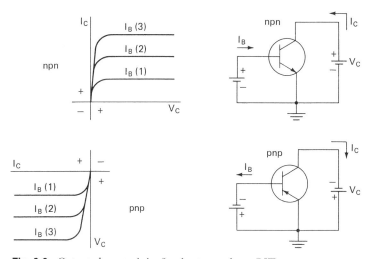

Fig. 2-6   Output characteristics for the *pnp* and *npn* BJTs.

2-4   Field-Effect-Transistor (FET) Characteristics

An FET is basically a voltage-variable resistor. Application of a voltage between gate and source V_{GS} varies the resistance between source and drain, resulting in the characteristics shown in Fig. 2-7. At lower drain voltages the drain current I_D, which is also always equal to the source current, increases linearity with the drain voltage V_{DS} for a fixed gate voltage V_{GS}. The slope, equal to the inverse of the FET source-to-drain resistance R_S, decreases as the gate voltage is made more negative (for an n-channel FET). When a sufficiently negative gate voltage is applied, no current flows and the transistor is in the cutoff region (turned off). More positive gate voltages increase the drain current somewhat but only to a certain limit, at which point the transistor is turned fully on and R_S has its minimum value, R_{on}. At high drain voltages the current levels off; that is, I_D becomes almost independent of V_D (for small changes). The point above which I_D is almost constant is roughly equal to the pinchoff voltage.

When the FET is operated as a linear amplifier (common-source configuration), it is desirable to have I_D independent of V_D and therefore normally the FET is operated beyond pinchoff (higher voltages). However, for applications as a voltage-variable signal attenuator (Chap. 15) linear operation in the low-voltage region well below pinchoff is required.

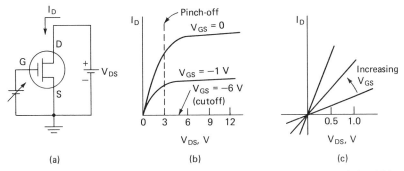

(a)   (b)   (c)

Fig. 2-7 Typical FET output characteristics (n-channel): (a) FET symbol (MOSFET); (b) normal operating region; (c) low-drain-voltage region (below pinchoff).

An important parameter of the FET is its gain or transconductance g_m, defined as

$$g_m = \frac{I_D}{V_{GS}} \qquad (2\text{-}4)$$

at or near a particular operating point (I_D, V_D). A typical value of g_m is 5 mS† at I_D = 2 mA and V_{DS} = 10 V. Unfortunately g_m varies considerably with I_D; in other words, the FET is a rather nonlinear device.

A distinction is made between the FET with an insulated gate (IGFET), usually referred to as a MOSFET (metal-oxide-semiconductor field-effect transistor, named after the process of forming the insulating layer) and the FET gate, formed by an *np*-type diode junction (JFET). The JFET gate is reverse-biased in switching applications provided the gate current is limited to a small value, usually by a large series resistor. A MOSFET may be biased with either polarity, but care must be taken not to exceed the gate breakdown voltage. Damage to the gate by excess voltage such as static electricity when picking up the transistor by its leads is a problem with the MOSFET.

Proper biasing of the FET requires that experimental data on the current I_D as a function of gate voltage V_{GS} be available. This may be obtained from the output characteristic (family of curves for various V_{GS}) such as Fig. 2-7 or from the input-output characteristics indicated in Fig. 2-8. The normal operating and cutoff regions vary substantially with FET

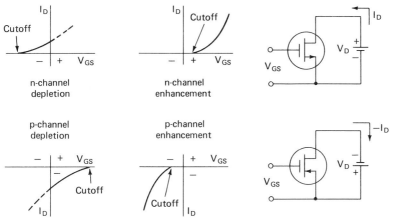

Fig. 2-8 Transfer characteristics for several FET types.

type and can be classified into two general bias types, *depletion* and *enhancement* (named for the physical processes involved).

Both *p*-channel and *n*-channel versions of each bias type are available, where the *p*-channel FET, like the *pnp* transistor, normally uses a negative voltage supply on the drain side. Actually only MOSFETs are available in both bias types; the JFET cannot be operated in the enhancement mode because the gate would be forward-biased and draw current. Note that

---

†S stands for siemens, the SI unit replacing the mho.

while a *sufficiently* positive bias (say +10 V) will in general turn on an *n* channel and a sufficiently negative bias (say −10 V) will turn it off, the effect of voltage not far from zero (+2 to −2 V) is not known without reference to the characteristics of a specific type.

Several common FET symbols are explained in Fig. 2-9; note that (1)

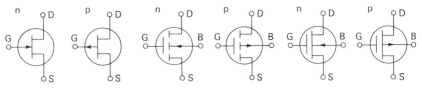

Fig. 2-9 Symbols for several FET types.

the direction of the gate or substrate arrow distinguishes between *n* and *p* channel, (2) a separated gate (line) distinguishes a MOSFET from a JFET, and (3) a broken line between source and drain indicates an enhancement type. The substrate *B* of a MOSFET may be brought out on a separate terminal or internally connected to the source. Where a separate substrate lead is available, to avoid current flow it must be biased so that it is more negative than the source (or drain) for an *n* channel or more positive than the source (or drain) for a *p* channel. Also it should not be left open for reliable operation. Care must be taken to keep the gate voltage of an *n*-channel JFET negative (or positive for a *p*-channel) to avoid gate current. Actually a slight positive (0.4-V) gate is permissible if a gate-limiting resistance (typically 100 kΩ to 10 MΩ) is included.

For many FETs the source and drain are interchangeable (except for the substrate connection). They are actually labeled according to circuit function (drain more positive for an *n* channel).

2-5 The SCR and Triac

A silicon-controlled rectifier (SCR) is a diode which will conduct in the usual forward direction only if turned on by a current pulse applied to a gate. Reverse current does not flow under any condition (short of break-down). Actually an SCR is a four-layer (*pnpn*) device, but only three leads are brought out. Since most SCRs are intended for power-switching application, they have a high current capacity, ranging from 1 to 100 A or more, and are mounted in cases which can dissipate the necessary power. The SCR is turned off by reversing the anode voltage (negative with respect to the cathode) or at a minimum by reducing the anode voltage (and current) below a threshold value termed the *holding voltage* (or *current*). Application of gate current can only turn on the standard SCR;

once on, it loses control, and the SCR cannot be turned off by the gate, as indicated in Fig. 2-10.

Standard SCR gate-current requirements are fairly high, typically 20 to 100 mA (or more for higher-power devices). The gate current need be applied only briefly (~10 μs), however, and therefore SCR gate drivers

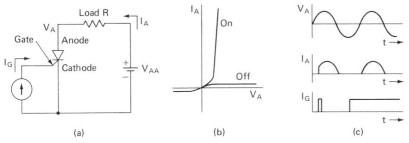

Fig. 2-10   The silicon-controlled rectifier: (*a*) symbol and polarities; (*b*) output characteristics; (*c*) example of control by gate current.

are frequently designed to deliver a current pulse. Control circuits for the SCR are discussed in Chap. 22. Sensitive SCRs require much less gate current for turn-on (typically 0.1 to 10 mA). Voltage drop across the SCR in the conducting state is typically 1.5 V.

One problem with the SCR is inadvertent triggering by large fast-rising voltage pulses on the anode (*dv/dt* effect). This occurs because the internal capacitance between the anode and gate allows a high-frequency current to pass from the anode to the gate, where it can turn on the SCR. Applications where inductive transients or radio-frequency (RF) interference are present may require a filter on the SCR anode.

Triacs are similar to SCRs except that they conduct current in either direction. They are intended for ac applications such as switching a light or motor connected to a 110-V (60-Hz) line. It is convenient to think of a triac as two SCRs back to back, although the device is actually more complex because the control voltage is applied between the gate and only one anode (T_1).

Permissible gate current and polarity of triacs differ substantially. Some classes of triacs work only if the gate and anode polarity are the same; i.e., the gate current must be ac as well as the anode voltage. In terms of Fig. 2-11*c* this means that operation only in quadrants I and III is allowable. Other types of triacs will work in all quadrants, so that a dc current can be used to switch the triac on. These devices are sometimes termed *logic triacs* because the current available from a logic IC (TTL or CMOS) is sufficient to turn the triac on, usually in quadrants I and IV. Triac drive circuitry is discussed in Chap. 22.

2-6   Programmable Unijunction Transistors (PUT)

Both the standard unijunction transistors (UJT) and the programmable unijunction transistor (PUT) are devices which switch to a high-conductance, or *on,* state when an upper threshold voltage is reached and switch

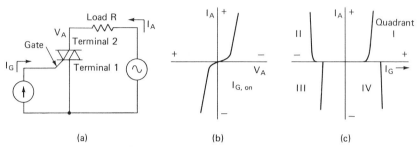

(a)   (b)   (c)

Fig. 2-11   Triacs: (*a*) symbol and polarity; (*b*) output characteristics; (*c*) gate-control quadrants.

back to a lower conductance, or *off,* state when brought back to a lower current. The main application of both is as a relaxation oscillator or pulse generator, in which a capacitor charged through a resistor is rapidly discharged by the transistor when threshold is reached. The relatively high pulse current during discharge is useful for turning on higher-current devices like the SCR.

A UJT consists of an *n*-type semiconductor strip with a *p*-type region attached along (e.g., halfway) the strip (Fig. 2-12*a*). The junction between the *n* and *p* material forms a reverse-biased emitter if the V_E is less than (more negative than) the voltage V_T at the point of attachment. When the UJT is reverse-biased, I_E is zero and V_T is determined by the ratio of the slab resistance from V_T to B_1 and B_2 as well as V_{BB}. In effect the slab acts as a resistive voltage divider. If V_E is made sufficiently positive, the junction will become forward-biased and current I_E will flow into the *n* region. As the charge carriers (holes) enter the *n* region, the conductivity will rise, predominantly at the negative B_2 end, where the charges are attracted. As the conduction increases, the junction will become more forward-biased in a regenerative manner. Thus a capacitor attached to the emitter will rapidly discharge once V_E reaches a threshold. The threshold for the UJT is fixed and depends on the transistor construction. It is specified by the *intrinsic stand-off ratio,* which is the fraction of V_{BB} at which forward conduction occurs. If the emitter current falls sufficiently, the conduction will fall and the junction will return to the normal reverse-biased state.

A programmable unijunction transistor (PUT) is a four-layer device related to the SCR and is so named because it is functionally similar to the

UJT (even though it has three junctions). Aside from the symbol, the main difference between the UJT and PUT is that the threshold voltage V_T is adjustable (programmable) by external resistors (R_1 and R_2 of Fig. 2-12c). The PUT anode takes the place of the UJT emitter.

The device will turn on when a threshold voltage (V_E for the UJT, V_A

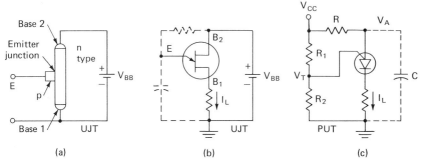

(a)   (b)   (c)

Fig. 2-12 The unijunction transistor (UJT) and programmable unijunction transistor (PUT): (a) principle of the UJT; (b) UJT symbol showing operating-voltage polarity; and (c) PUT showing operating-voltage polarity.

for the PUT) is reached, but turn-off is specified by a minimum current (I_E or I_A). The designer should make sure that the emitter or anode resistor is sufficiently high to reduce the direct current below the threshold value if a return to the off state is desired.

2-7 Lamp and Relay Drivers

A discrete transistor (BJT) is an excellent switch or driver for higher-current devices, e.g., incandescent lamps, light-emitting diodes, relays, and even small heaters, because of its good current- and power-handling capacity. Several circuits are shown in Fig. 2-13. It is assumed that a positive dc control signal V_S at least several volts in magnitude is available. Digital devices (TTL and CMOS) as well as op-amps provide adequate drive.

In the first circuits shown (Fig. 2-13a–c) the transistors act as switches, i.e., are either in saturation (full on) or in cutoff (full off). Many transistors, especially those designed for switching operation, have dc drops of only 0.5 V at saturation. Cutoff currents are typically under 1 to 10 μA but can be much higher if the transistor is hot. In saturation the base current I_B exceeds that given by the linear-transistor current-gain relation $I_B = I_C/\beta$, where $I_C = I_L$ the collector (load) current. It is good practice to exceed the base current given by the linear relation (using the minimum value of β if a range is specified) by a factor of 2 to 10. A resistor R_B limits the base current I_B, the voltage of which can be calculated from the equivalent circuit (Fig. 2-13d).

Relay drivers (Fig. 2-13c) are the same as lamp drivers except that a diode across the load is practically essential to limit the inductive voltage transient when the relay is switched off. Miniature reed relays with coil

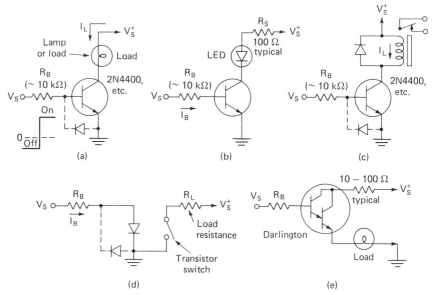

Fig. 2-13 Transistor drivers: (a) transistor switch with collector-load-type driver; (b) switch with LED load; (c) switch with relay load; (d) equivalent circuit; (e) emitter follower.

voltage ratings between 5 and 24 V (current 10 to 100 mA) are convenient for many applications.

Transistor selection is not critical since the main requirement is that the maximum transistor current and voltage exceed that of the load. Power dissipation of the transistor is usually not a problem because of the low voltage drop when turned on. Input voltage V_S may be negative when driven by an op-amp, but the maximum collector-to-base voltage V_{CBO} should not be exceeded. Alternatively the input diode, which keeps the base from becoming more than 0.6 V negative, may be added as indicated.

An emitter-follower configuration is also a good relay or lamp driver. It operates in the linear region ($I_B = I_L/\beta$) unless the input voltage V_S exceeds the supply voltage. Load voltage is 0.6 V (single transistor) or 1.2 V (Darlington) less than the input voltage V_S in saturation. A base-current-limit resistor protects the transistor from destruction by limiting current if the load is accidently shorted. An advantage of the emitter-follower driver is low input current, which is still lower if the high-gain Darlington ($\beta \approx 1000$ typically) or dual emitter-follower configuration is used. Note that the voltage difference from input-base to output-emitter is two diode drops (1.2 V). Actually the Darlington transistor can be used with all the circuits of Fig. 2-13.

Chapter Three
Operational-Amplifier Characteristics

Characteristics and limitations of op-amps are mainly of concern when the devices are operated near their limits. The novice designer may wish to read only the first part of this chapter and return to it if problems are encountered.

3-1 Ideal Op-amps

The voltage output v_o of an ideal differential-input op-amp, symbolized in Fig. 3-1, is proportional to the difference in voltage of two signal sources (v_+ and v_-), as indicated by

$$v_o = A_o(v_+ - v_-) \tag{3-1}$$

where A_o is the open-loop gain. Usually A_o is quite high (10^4 to 10^5), and applications as a linear amplifier will require a feedback circuit, to be discussed at length. Either polarity can be obtained from the output, and the amplifier can be used for dc as well as ac applications. The inputs labeled (+) and (−) are the noninverting and inverting inputs, respectively. Here the notation (+) and (−) refers to the sign of the gain [Eq. (3-1)] and not to the polarity of the inputs (v_+ or v_-) with respect to ground.

A dual power supply (typically ±15 V) to furnish the voltages $+V_S$ and $-V_S$ is ordinarily required. Note that amplifier output v_o is given with respect to the power-supply common or ground connection. Most circuit

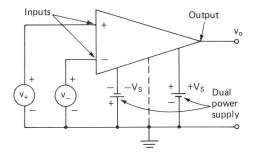

Fig. 3-1 Ideal op-amp.

diagrams, including those in this text, do not indicate the dual power supply explicitly, but its presence is assumed. Operation with a single-ended supply is possible under conditions discussed in Chap. 21.

Novices should be clear on the following points:

1. Either input may have a positive or negative applied voltage; it is the sign of the difference voltage which determines the sign of the output voltage.

2. A dual (three-terminal) power supply is required, even when not shown explicitly, with the common or center-tap terminal connected to ground.

3. Op-amps are intended to be used with feedback (or with saturated output) because without feedback small deviations from ideality, magnified by the high gain, can result in extremely unstable behavior.

3-2   Inside the Voltage-controlled Op-amp

Few users of op-amps have more than a casual interest in the details of the circuit which constitutes the op-amp, nor is such information necessary for its proper use. All needed information is given by the manufacturer in the electrical-characteristics portion of the specification sheets. However, a brief examination provides insight into the limitations and deviations from ideality of real circuits.

A simplified op-amp (Fig. 3-2) consists of two stages of dc amplification. The first is an emitter-coupled differential-input amplifier with a single-ended output. Resistor R_1 provides a constant-current source. If the inputs (base voltages of Q_1 and Q_2) are equal, the current from the constant-current source I_0 will divide equally between Q_1 and Q_2 and the output voltage from the first stage (collector of Q_2) will be $+v_s - R_2 i_0/2$. Without the constant-current source Q_3, the current through Q_1 and Q_2, will vary with the level of v_+ and v_- (not just their difference). The second stage Q_4 acts as a level translator as well as an amplifier, and the output

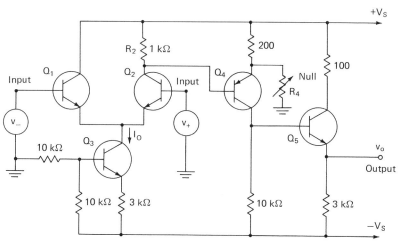

Fig. 3-2   Internal circuit of a simple op-amp.

voltage can be adjusted to zero voltage through an adjustment of R_4. An emitter follower Q_5 provides low output impedance. Note that the output of Q_5 can swing negative as well as positive even though the output of the first stage cannot. If v_+ is made more positive than v_-, the current through Q_2 (and thus Q_4) will increase, and as a result output will become more positive. Here v_+ functions as a noninverting input. If v_- is made more positive than v_+, the output will swing negative, as expected from an inverting input.

Several characteristics of op-amps are evident from this diagram. One is that both a positive and negative (dual) power supply is required. Another is that both the input and output voltage is limited to a span less than the power-supply voltage, possibly much less. If the input voltage is too great, Q_4 will saturate, or cut off, and cause the output to swing to a positive or negative limit, i.e., saturation. Another characteristic is the bias current (dc) which flows through the input terminals into the base of the transistors (Q_1 and Q_2). Practical circuits often have Darlington inputs to reduce the bias current to low but not always negligible levels. FET input transistors can be used if near zero bias current is required.

In the circuit shown, a potentiometer is varied to set the output to zero input. Amplifiers without a potentiometer, however, can be made if the input transistors and the other components are carefully matched. Matching to within a few millivolts (referred to inputs) is about the limit of a typical IC.

More elaborate and practical circuits than Fig. 3-2 have more gain features to maximize the input- and output-voltage range and provide output current limit. Unlike the circuit shown, many op-amps draw equal currents from the negative and positive supply, and these do not require a connection to power-supply common or ground (although the output must still be returned to common).

3-3   Nonideal Op-amp Characteristics

Few electronic devices approach their ideal behavior to the extent that op-amps do, as designers familiar with their application will thankfully acknowledge. Nothing is perfect, however, and these imperfections, according to Murphy's law,† are likely to lead to problems if ignored.

Measurement techniques for these characteristics are described here primarily to clarify their meaning and suggest their minimization in circuit applications. In practice the manufacturer's specification sheet is consulted for typical, maximum, or minimum values. If individual compensation is required, an in-circuit adjustment procedure must usually be devised. Examples are given here for feedback circuits, which will be discussed in more detail.

Output Saturation

An excessive input voltage will drive the output to positive or negative saturation, a limiting voltage which is somewhat less in magnitude than the power-supply voltage. For the 741 op-amp with a ±15-V supply, the output will saturate at about +14 or −14 V. If the output is shorted or loaded with an excessively low resistance, the output current may saturate instead. Of course, the maximum output voltage under current saturation is less, possibly much less, than when properly loaded. For the 741, current saturation occurs at 10 to 20 mA.

Offset Voltage

The output of a nonideal op-amp will not be zero if the input voltage difference is zero (both inputs shorted to ground) but can be made arbitrarily close to zero by applying a small dc voltage or offset voltage of the proper magnitude and sign to either input. Offset voltage can be a problem when amplifying small dc signals and can usually be neglected at high signal levels. An equivalent circuit which incorporates the effect of

---

†Murphy's law: If anything can go wrong, it will.

the voltage offset is shown in Fig. 3-3. It utilizes the concept of an ideal op-amp as a subunit of the nonideal amplifier equivalent circuit. Offset voltage is equivalent to a dc voltage generator or battery (typically 1 to 10 mV) in series with the input. By convention the generator is placed in series with the noninverting input (a generator of opposite sign in series

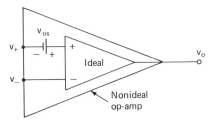

Fig. 3-3   Offset-voltage equivalent circuit.

with the inverting input is physically indistinguishable). Care must be taken to distinguish the effect of offset voltage from the similar effects of offset and bias current, discussed in the next section. Measurement techniques are described below.

When voltage-offset compensation is needed, an individual in-circuit null adjustment is usually made without bothering to measure the offset. The examples of external in-circuit compensation methods shown in Fig. 3-4a are a variation of the measurement circuits discussed in Sec. 3-5.

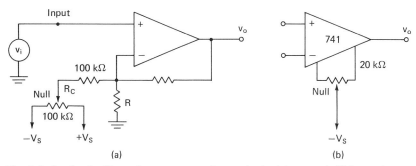

Fig. 3-4   In-circuit offset-voltage compensation methods: (a) external; (b) internal.

Here current flowing through R_C produces a voltage across R which bucks the offset voltage. Either sign of offset voltage can be compensated by adjustment of the null potentiometer. Many op-amps have internal compensation circuits which require the addition of a single potentiometer, following manufacturer's specifications. In either case, the null potentiometer is adjusted so that the output voltage v_o is essentially zero.

Bias and Offset Current

As pointed out above, a small bias current (dc) flows out of both inputs of an op-amp to ground. For the usual BJT transistor input the bias current is identical with the base current and is in the range of 0.01 μA. For FET input operational amplifiers, the bias current is much smaller, often below 0.01 pA. Bias current is sometimes a problem with high input impedance and/or low-level signal amplification, but it can be ignored in most applications.

An equivalent circuit which expresses the effect of bias current in terms of dc generators is shown in Fig. 3-5. Offset current is the difference in

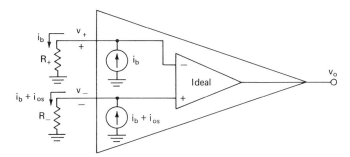

Fig. 3-5   Bias- and offset-current equivalent circuit.

bias currents of the two inputs and is often an order of magnitude smaller than the bias current. For simplicity, in this equivalent circuit, the input current of the noninverting input is defined as the bias current; it could have been defined as the average of the two inputs, but the distinction is unimportant. Bias current produces an effect similar to that of offset voltage because of the voltage drop across the resistors R_+ and R_- attached to the inputs of the op-amp of Fig. 3-5. It will not be zero even if the offset voltage is zero. The voltages are

$$v_+ = i_b R_+ \qquad v_- = (i_b + i_{os})R_- \qquad (3\text{-}2)$$
$$\Delta v = v_+ - v_- = i_b(R_+ - R_-) + Ri_{os} \qquad (3\text{-}3)$$

It can be seen that Δv acts as an offset voltage except for the dependence on resistance. Bias current can be neglected if the dc resistance to ground is low. Note also that the i_b term is zero and Δv small if $R_+ = R_-$.

Compensation for offset and bias current in many cases can be made by the same circuits which compensate for offset voltage (Fig. 3-4). Where high input resistance is desired with a BJT op-amp, the current-compensation circuit of Fig. 3-6a can be used, but because the bias current increases rapidly with temperatures, accurate compensation with a varying ambient temperature cannot be achieved. In this case, FET input amplifiers should be used.

A simple, effective, and common method of bias-current compensation is to design the circuit so that the dc resistance to ground is the same for both inputs [$R_- = R_+$ in Eq. (3-3), neglecting i_{os}]. This can always be

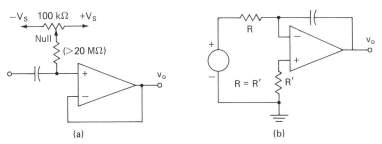

(a)   (b)

Fig. 3-6 Current offset and bias compensation: (*a*) external null; (*b*) balanced input resistor.

done, if necessary, by adding a resistor in series with one input (Fig. 3-6*b*). An exact null (at one temperature) can be achieved by making one resistor variable.

A dc resistance or current path to ground is always required, possibly through the signal source. The op-amp will go to saturation without a ground return.

Open-Loop Gain and Frequency Response

Open-loop gain A_o, as expressed in Eq. (3-1), is a function of frequency, as Fig. 3-7 indicates. Only infrequently must the open-loop gain be known to be better than an order of magnitude, and for many calculations the gain is assumed to be infinite. All op-amps respond to dc gain or zero frequency.

While the high-frequency breakpoints for most operational amplifiers occurs at a surprisingly low frequency (5 to 100 Hz), useful gain can be obtained at much higher frequencies. With the unity-gain (feedback) configuration, for example, the high-frequency response extends four or five orders of magnitude higher. The unity-gain (small-signal) frequency response F_U or gain-bandwidth product is a useful parameter which can be thought of as the highest usable frequency of the amplifier.

Common-Mode Rejection Ratio

The output of an ideal differential-input op-amp will depend only on the difference in input voltages, $v_+ - v_-$, that is, the difference mode, and is independent of the average input voltage ($v_+ + v_-$)/2, or common mode.

A real amplifier will depend slightly on the common mode according to

$$v_o = A_o(v_+ - v_-) + \frac{A_o}{m}\frac{(v_+ + v_-)}{2} \tag{3-4}$$

where m is the common-mode rejection ratio (CMRR). Typically m is of the order of 1000 (60 dB).

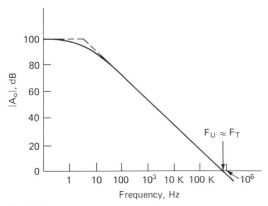

Fig. 3-7   Frequency response of an op-amp.

Problems with low common-mode rejection ratio may occur in applications demanding high linearity, but ordinarily they can be ignored. Measurement and adjustment are discussed in connection with the instrumentation amplifier (Chap. 4).

Input (Common-Mode) Voltage Range

Op-amps work properly only if input voltages are within a specified range, termed the *common-mode operating range*. If the input voltage exceeds this range slightly, the output will swing to voltage saturation. If the voltage is exceeded by a large amount, i.e., if the maximum allowable input voltage is exceeded, the device will be destroyed. Newer model ICs have a permissible input-voltage range roughly equal to the output-saturation range, for example, ± 14 V for the 741 with $V_S = \pm 15$ V, so that if the output of one stage is in the linear range, although close to saturation, the next stage has a wide enough common-mode range to follow it.

A few op-amps, especially older models, stay in saturation indefinitely even when the input voltage returns to the normal range, an undesirable condition known as *latch-up*. To unlatch, the power must be turned off. These devices require some sort of input-voltage limiting for reliable operation.

Full-Output Frequency Response and Slew Rate

Many op-amps cannot deliver anything near full voltage output at higher frequencies. In this respect, IC amplifiers are inferior to discrete transistors. One way of expressing this limitation is to specify a maximum peak-to-peak output-voltage swing at higher frequencies (Fig. 3-8a). If the

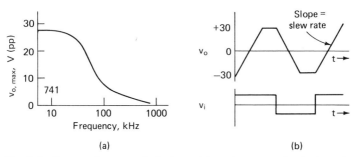

(a)   (b)

Fig. 3-8 Op-amp frequency limitations: (*a*) maximum-amplitude sinusoidal response; (*b*) square-wave response (slew-rate-limited).

output is driven beyond this point, the signal becomes distorted and may clip. Often the onset of distortion is gradual, depending on load conditions, and otherwise difficult to characterize.

A closely related parameter is the slew rate, defined as the maximum output-voltage change per unit time. If a square wave is applied to the input, the output initially will be a ramp (linear rise) at the slew rate even though a more rapid (exponential) rise time would be implied by the frequency response (a small-signal parameter). This is pictured in Fig. 3-8b.

Input and Output Impedance

An ideal amplifier draws no current from the signal source, i.e., acts as an infinite input impedance. Of course, this is not strictly true, as indicated by Fig. 3-9. Resistances, both to ground R_{g+}, R_{g-} and between inputs R_d, are present. At higher frequencies the input capacitances C_{i-} and C_{i+} may become important. In most practical applications these parameters can be neglected. Although R_d itself is not large compared with nominal circuit values, the voltage difference between inputs is normally held to very small values by negative feedback, as shown below, so that the effective value of R_d as referred to the input is normally very high and thus negligible. Offset voltage and current are more common sources of concern. An ideal op-amp has a zero output impedance; for the nonideal op-amp it is R_o in the equivalent circuit of Fig. 3-9.

The effect of feedback on stage input impedance R_i and output impedance R_{os} is calculated here for the noninverting configuration discussed in the next chapter. An amplifier input resistance (between inputs) of R_d and an open-loop gain of A_o are assumed. From Fig. 3-10 it can be seen that v_- is equal to βv_o, so that $v_o = A_o(v_i - v_-)$. Now $i_i R_d = v_+ - v_-$, so that $v_o = A_o R_d i_i = v_i/\beta$. But by definition, $R_i = v_i/i_i$ or

$$R_i = A_o R_d \beta \tag{3-5}$$

Thus the stage input impedance is usually several orders of magnitude higher than the amplifier difference resistance.

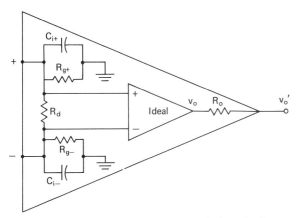

Fig. 3-9   Input and output impedance equivalent circuit.

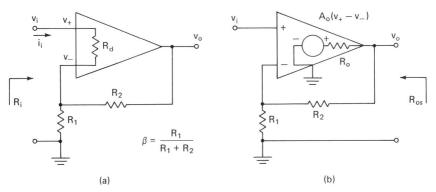

(a)   (b)

Fig. 3-10   Stage input and output impedance calculations.

Stage output impedance R_{os} is found by a similar analysis to be

$$R_{os} = \frac{R_o}{A_o\beta} \qquad (3\text{-}6)$$

As the frequency rises, R_{os} will rise (and R_i will fall) since A_o decreases with frequency. Because A_o is usually high, the output impedance of any op-amp configuration with negative feedback is typically well below 1 Ω, so low it is difficult to detect a drop in output voltage with loading unless excessive current is drawn, causing the amplifier to go into current limit.

3-4   Types of Op-amps

Standard op-amps produce an output voltage proportional to the difference in input voltages. Most are built from BJTs and can be termed voltage-controlled BJT amplifiers. The popular 741 and its dual versions (1458, etc.) and quad versions (4136, etc.) fall into this category. These devices and the FET op-amp were described above. An amplifier combining the advantages of both the BJT and FET is the BIFET amplifier. The input has an FET, but the remaining transistors are BJT type. It can be used for circuits in this text wherever FET-type op-amps are specified because it has the necessary high input impedance and near zero input (bias) current.

Op-amps can also be divided into internally compensated and externally compensated (or uncompensated) types. Without compensation oscillation at a high frequency is likely (as discussed in Sec. 3-7). Most newer types of op-amps are internally compensated, except for those intended for high-frequency operation, and therefore the user need not be concerned with compensation.

Significant differences in characteristics of the various models of op-amps are (1) input impedance and/or current (especially between the BJT and FET/BIFET types), (2) frequency response, (3) compensation requirements, and (4) voltage offset, including drift with time or temperature. The large and sometimes confusing variety of op-amps available seems to be a reflection of the competitive nature of manufacturers rather than a genuine need for a high degree of diversity. The reader is reminded that since most op-amp configurations will work with almost any op-amp, cost and availability may be the major factor in choosing a particular model rather than some subtle difference in characteristics.

Two types of op-amps differ functionally from the amplifiers discussed above. The first is the operational transconductance amplifier (OTA), typified by the CA3080, which differs in two respects. Instead of an output being a voltage source, it is a current source. As a consequence, the gain relation corresponding to Eq. (3-1) becomes

$$i_o = G(v_+ - v_-) \qquad (3\text{-}7)$$

where G is a transconductance (gain) with the units of siemens or millisiemens. If a load resistor R_L is added (Fig. 3-11), the circuit acts like a

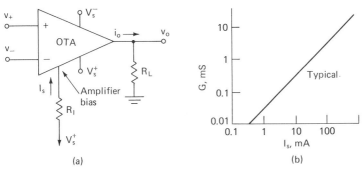

(a)   (b)

Fig. 3-11 Operational transconductance amplifier (OTA); (*a*) amplifier shown with load resistor and fixed set current I_S; (*b*) transfer characteristics.

standard op-amp with an output voltage $v_o = R_L i_o$, so that now the performance described by Eq. (3-1) is obtained with $A = GR_L$. While the open-loop output impedance R_L is higher than in standard amplifiers, it is reduced considerably by the negative feedback employed by most circuit configurations and is not likely to be a problem.

The second function difference of the OTA is that its gain G is variable. Specifically it is proportional to a set current I_s injected into an input lead

$$G = \gamma I_s \qquad (3\text{-}8)$$

where γ is a gain-conversion constant for the particular device. As Fig. 3-11 indicates, this relation is quite linear and holds over a number of decades. The set current should always be positive (inward) and not exceed 1 mA. The device is excellent as an amplifier for circuits requiring precise voltage control of gain, as a linear modulation device, or as an analog multiplier. Problems with the CA3080 are that the gain-conversion constant γ varies with the device and the device is undercompensated, which under some conditions results in oscillation.

Another significantly different op-amp is the operational current-controlled amplifier (OCA), also termed a current-difference or Norton amplifier, which has an output voltage proportional to the difference in input currents (i_+ and i_-), that is,

$$v_o = B(i_+ - i_-) \qquad (3\text{-}9)$$

where B is a gain ($B \approx 3 \text{ G}\Omega$ for the LM3900). Currents must always flow inward. Either dual- or single-ended supply (one side grounded) is permissible, and the most common applications utilize a grounded supply. The inputs are, in fact, transistor bases with emitters connected to the ground (negative-supply) lead. Thus the input voltages are 0.4 to 0.6 V more positive than ground. The amplifier symbol is similar to the op-amp symbol with the addition of an arrow on the input (Fig. 3-12) to signify the

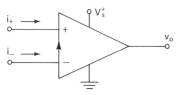

Fig. 3-12 Operational current-controlled amplifier (OCA).

current-difference aspect. Maximum output voltage swing is between 0.2 V above ground (most negative) and $(V_s^+ - 1)$ volts (most positive).

Circuit configurations for the current-controlled amplifiers are similar or analogous to those for voltage-controlled amplifiers but not identical. Examples of circuit adaptation are given in Chap. 4. It should be pointed out that currently available OCA ICs have less satisfactory differential (common-mode) characteristics and are otherwise less ideal than standard op-amps. Main advantages are single-ended (grounded) supply operation and low cost.

3-5 Measurement of Characteristics

In practice there is seldom any need for the average user to measure the open-loop characteristics of op-amps since they are given by the manufacturer. This is particularly true because the performance of an amplifier configuration (with feedback) depends largely (and ideally entirely) on the values of the external components (resistors, capacitors) rather than on the individual amplifier characteristics. Sometimes measurements are made, however, to test a critical parameter or to check for proper amplifier function. Because the open-loop amplifier is unstable in the sense that the output drifts rapidly into saturation unless the input offset is carefully adjusted, the test circuits actually are closed-loop. From the test data the open-loop characteristics are inferred.

Offset Voltage and Bias Current

Measurement of offset voltage can be made by the circuit of Fig. 3-13. It is basically a high-gain (1000) noninverting amplifier. With the switches closed, the offset voltage v_{os} is amplified by a factor of 1000 at the output

v_o. By simply measuring the dc voltage at the output, the offset is immediately known.

To determine the bias currents i_{b1} and i_{b2}, the switches are opened, thus allowing the current to flow through the resistors. The resulting voltage drop across the resistors ($i_{b1}R_1$ and $i_{b2}R_1$) could be measured directly with

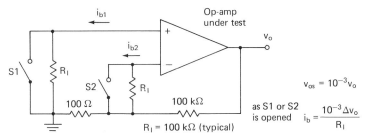

Fig. 3-13 Test setup for op-amp offset-voltage and bias-current measurement.

a sensitive voltmeter but is more conveniently determined at the amplifier output v_o since the voltage is 1000 times higher and the impedance is low. The change in voltage Δv_o as switch 1 is opened is due to i_{b1} (inverting input) and the change in voltage $-\Delta v_o$ as switch 2 is opened is due to i_{b2} (noninverting input). The change Δv_o as both switches are opened simultaneously is proportional to the offset current i_{os}.

Open-Loop Gain

Accurate measurements of dc open-loop gain are rather difficult because of amplifier drift; ac measurements are easier and generally approximate the dc gain if extended to low frequencies (1 to 10 Hz) since the frequency response of op-amps is generally flat below 10 to 100 Hz. The objective of the open-loop-gain measurement is usually to measure the gain magnitude and phase from low frequencies (extrapolated to dc) up to unity-gain frequency (open-loop gain less than 1). For this purpose, the circuit of Fig. 3-14 will suffice. Note that the dc feedback is provided through R_b (inverting configuration) but that the feedback is ineffective for alternating current. The gain measurement is therefore effectively open-loop (above $f \approx 5$ Hz and $A_o < 10^5$). Measurements of the ratio of output- to input-voltage magnitude $|A_o|$ and phase of the output with respect to the input θ_0 are made on an oscilloscope and presented on Bode plots ($|A|$ in decibels and θ_0 in degrees vs. frequency on a log scale, such as Fig. 3-17). Any dc offset at the amplifier output is ignored.

Slew Rate and Full-Output Response

Slew rate is determined largely by internal transistor saturation and other limitations of the op-amp and is not much affected by feedback. The peak amplitude at the output v_o at which the signal begins to clip or distort is also only moderately dependent on the extent of negative feedback although the waveshape is quite dependent on frequency and feedback. For these reasons the closed-loop test circuits of Fig. 3-15 can be used. With a square-wave input the slope of the linear rise and fall of the

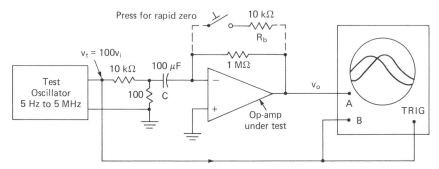

Fig. 3-14   Test setup for open-loop op-amp gain and phase measurements (C acts as a short circuit).

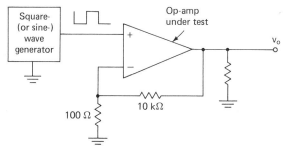

Fig. 3-15   Test setup for slew rate and full-voltage output response of op-amps.

resulting output signal is measured. The slope, usually expressed in volts per microsecond (see Fig. 3-8), is the desired slew rate S_L. With a sine-wave-generator input, the output will be a sine wave up to a certain amplitude at a given frequency, at which point a distortion of the peaks will be noted. This maximum linear output is plotted as a function of frequency.

3-6   Feedback Fundamentals

Virtually all linear operational circuits utilize negative feedback. A high gain A_o is very desirable for feedback configurations, and manufacturers design the amplifiers with this requirement foremost. Without feedback, an op-amp output will rapidly drift into voltage saturation as a consequence of its high gain. Additionally, frequency response and linearity are greatly improved by feedback.

A general block diagram of a voltage feedback circuit is shown in Fig. 3-16. Here the amplifier, feedback network, and subtraction circuits have been separated for the sake of clarity. The combination of the three blocks

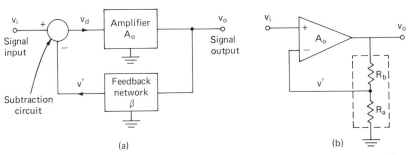

Fig. 3-16   Negative-feedback circuits: (a) general block diagram; (b) op-amp example.

is one stage of amplification for a closed-loop amplifier. It is assumed that the amplifier has an open-loop gain of A_o, which normally has a positive value, or

$$v_o = A_o v_d \qquad (3\text{-}10)$$

where v_o is the output voltage and v_d is the input voltage. The subtraction circuit has an output v_d equal to the difference in the two inputs v_i and v', or

$$v_d = v_i - v' \qquad (3\text{-}11)$$

In the standard differential-input op-amp the amplifier and subtraction circuit are combined [compare Eqs. (3-10) and (3-11) with Eq. (3-1), $v_+ = v_i$ and $v_- = v'$]. The feedback circuit allows some fraction β of the output voltage v_o to be fed back so that v' is

$$v' = \beta v_o \qquad (3\text{-}12)$$

For the purpose of this discussion it is helpful to think of β as a frequency-independent constant less than unity, but these restrictions are not necessary and in fact do not hold for many feedback configurations. By combining Eqs. (3-5), (3-11), and (3-12) to eliminate v' and v_d one obtains

$$v_o = \frac{A_o v_i}{1 + \beta A_o} \qquad (3\text{-}13)$$

Thus the output voltage v_o for a given input v_i is reduced by the factor $1 + \beta A_o$. Defining the closed-loop gain A as v_o/v_i and dividing by A_o, we get the standard form of the gain of a voltage feedback amplifier

$$A = \frac{1}{(1/A_o) + \beta} \tag{3-14}$$

When the open-loop gain A_o is high, as is normally the case,

$$A \approx \frac{1}{\beta} \quad \text{for } A_o\beta \gg 1 \tag{3-15}$$

This relation holds even if $\beta = \beta(j\omega)$ or $\beta = \beta(s)$. Note that this relation will not hold to arbitrarily high frequency since A_o of all amplifiers decreases as the frequency increases.

The main reason for using a feedback amplifier is that the stage gain A or transfer characteristic depends mainly on the feedback network, which is usually passive and therefore generally linear and constant. Often β is determined by the ratio of two resistors. If the ratio is accurate to 1 percent, the gain A is accurate to 1 percent even if A_o varies greatly, provided, however, $A_o\beta \gg 1$. Feedback also greatly increases amplifier input impedance, decreases output impedance, reduces sensitivity to power-supply variations, and generally makes the amplifier act more ideally.

Sometimes feedback is easier to understand if we follow the effect of a voltage change on the input around the loop. Suppose the input signal and thus all voltages are zero, A_o is 1000, and β is 0.1. Now suppose that the input v_i suddenly becomes $+0.01$ V. If the feedback network were disconnected (loop broken), v_o would be 10 V (and $v_o = 0.01$ V). At the instant the loop was completed, v' would be 1 V, and therefore v_d would change to -0.99 V, assuming that the amplifier response time is finite and v_o cannot change instantaneously. Shortly, however, v_o moves to a negative value, but as it approaches 0.1 V (or $v' = 0.01$ V) the difference voltage v_d becomes smaller, thus slowing the rate of change of v_o. At equilibrium $v = 0.101$ V, $v' = 0.0101$ V, and $v_d = 1.01 \times 10^{-4}$ V for $v_i = 0.0100$ V. Should v_o overshoot below (more negative) than its equilibrium value, the difference voltage will rise and correct the error.

Another way of looking at the feedback circuit is to think of the function of the amplifier as minimizing the difference voltage v_d by increasing the correction-voltage feedback. As the gain A_o becomes high, the correction increases rapidly if $v' \neq v_i$. In the limit of infinite gain, $v' = v_i$, or the condition where $\beta v_o = v_i$ is forced by the amplifier.

An example of negative feedback is given in Fig. 3-16b, where $\beta = R_a/(R_a + R_b)$. The differential-input op-amp includes both the subtraction circuit and amplifier. Alternatively the amplifier gain may be considered to have a negative sign and the subtraction circuit can be replaced by a summation circuit. This and many other examples are discussed in Chap. 4.

It has been implicitly assumed that the A_o has a positive value and/or that the feedback applied to the subtraction circuit is negative. If the feedback were positive, the circuit would either oscillate or force the amplifier to voltage saturation. The latter condition generally occurs if the feedback is positive at dc while the former condition occurs if the feedback is negative at dc but changes to positive at the oscillation frequency due to a frequency-dependent phase shift of A_o and/or β. Circuits with positive dc feedback may be bistable and very useful in digital or switching applications, as discussed in subsequent chapters. Circuits with frequency-dependent positive feedback are oscillators. Oscillator design is discussed in Chap. 12. When the oscillations are unwanted, the circuit is considered unstable. A discussion of instability and means of stopping the oscillation through frequency compensation is discussed in the next section.

3-7 Stability and Compensation

Any high-gain amplifier is likely to become unstable when feedback is employed unless proper compensation is used. Op-amps are no exception, but fortunately the user generally need not be concerned with compensation-network design because nearly all op-amps in use, including the popular 741, are internally compensated. A price is paid for the convenience of internal compensation, specifically poorer frequency response than obtainable by an optimally compensated amplifier. For applications where best frequency response is required, the user must tailor the compensation to the specific op-amp configuration.

Oscillation and instability of feedback circuits occur because the feedback, which is negative over most of the operating range, becomes positive at high frequencies due to amplifier phase shift. Because both the amplifier gain A_o and feedback network β can produce phase shifts in general, it is the phase shift of the total loop transfer function $A_L = A_o\beta$ which is important.

Consider first the open-loop transfer characteristics of the amplifier. Practically every amplifier shows an increasingly negative phase shift as the frequency increases (Fig. 3-17). Most show two or three break frequencies (ω_1, ω_2, ω_3), which can be traced to base-emitter and/or collector capacitors and resistances. Each break frequency is associated with a low-pass filter (Chap. 11). Also associated with each filter is a phase shift, which equals $-45°$ at the break frequency (Fig. 3-17b) and 90° asymptotically. When the phase shift reaches $-180°$, oscillation will occur. Thus the maximum shift possible with a two-section low-pass filter (two break frequencies) is $-180°$ and with a three-section filter (three break frequencies) is $-270°$. Phase shifts above the open-loop unity-gain frequency f_u are of little practical interest. From these plots, A_o in phasor form is obtained

$$A_o(\omega) = |A_o| e^{j\theta_0} \tag{3-15}$$

Another way of looking at system stability is through Eq. (3-13). At low

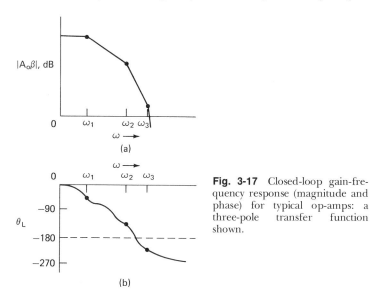

Fig. 3-17 Closed-loop gain-frequency response (magnitude and phase) for typical op-amps: a three-pole transfer function shown.

frequencies the loop gain $A_L = A_o\beta$ is assumed to be positive (negative feedback), and therefore the denominator, $1 + A_o\beta$, is always greater than unity. If $A_o\beta$ has a phase shift so that at some (usually high) frequency the loop gain becomes negative ($A_o\beta = 1$), the closed-loop amplifier gain becomes infinite and the amplifier is unstable.

Techniques for improving the stability of an amplifier (at any value of $A_o\beta$) are referred to as compensation. In most cases the response of the IC amplifier is altered by capacitors connected to leads brought out of the IC package. The manufacturer's recommendations should be followed in selection of specific values. Alternatively the feedback network β can be altered by the addition of capacitors. Common methods of compensation are phase lag, phase lead, and dominant pole.†

---

†A detailed analysis of amplifier stability and compensation is given in R. Roberge, "Operational Amplifiers," Wiley, New York, 1975.

Chapter Four

Operational-Amplifier Configurations

This chapter discusses the basic op-amp configurations, or building blocks, out of which numerous more complex circuits are composed. Most involve feedback to carry out their particular function. The characteristics of these feedback circuits depend primarily on the configuration, i.e., how the circuit is connected, and on the resistors and other passive components. Ideally, and to a great extent in practice, these circuits are independent of the particular op-amp characteristics except for the general requirement of high open-loop gain. Some previous acquaintance with feedback is helpful (see Chap. 3), but a working knowledge can be acquired by studying the first few configurations in detail.

4-1 Noninverting Amplifier

Perhaps the most useful op-amp configuration is the noninverting amplifier (Fig. 4-1). Its purpose is to amplify an input voltage v_i by a factor A to give an output voltage of v_o. A fraction β of the output voltage is fed back to the inverting input of the op-amp. Because the sign of the feedback is such as to decrease the magnitude of the output, it is an example of negative feedback.

Calculation of the closed-loop gain ($A = v_o/v_i$) can be made with Eq. (3-1) ($v_+ = v_i$), repeated here

$$v_o = A_o(v_i - v_-) \qquad (4\text{-}1)$$

The voltage of the inverting input is a fraction β of the output

$$v_- = \beta v_o = \frac{R_a v_o}{R_a + R_b} \qquad (4\text{-}2)$$

where the very small current flow into the input has been neglected. Inserting v_- from Eq. (4-2) into Eq. (4-1) gives an expression for gain

$$A = \frac{v_o}{v_i} = \frac{1}{\beta + 1/A_o} \qquad (4\text{-}3)$$

For a properly designed amplifier β is chosen to be at least 100 times greater than $1/A_o$. This usually presents no difficulty because A_o is typi-

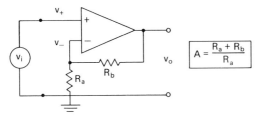

Fig. 4-1   Noninverting amplifier.

cally very high (10^4 to 10^6). To a very good approximation, the gain [Eq. (4-3)] becomes

$$A \approx \frac{1}{\beta} = \frac{R_a + R_b}{R_a} \qquad (4\text{-}4)$$

It should be noted that this expression is independent of A_o, but it is true only if A_o is sufficiently large. At sufficiently high frequencies this relation will fail because A_o decreases with frequency, but at lower frequencies Eq. (4-4) is quite accurate in practice. Where accurate gains are desired, close-tolerance resistors (1 percent) are used, and closed-loop gains A are limited to 100 or less.

It is instructive to estimate the voltage difference between v_+ and v_- when feedback is present and the op-amp is in the normal linear range (not in saturation). If the maximum linear output is of the order of 10 V and A_o is 10^5, the input voltage difference ($\Delta v = v_+ - v_-$) by Eq. (3-1) must be 10^{-4} V or less, i.e., very small under all conditions. If A_o increases with negative feedback, Δv decreases and v_o remains almost constant (if v_i

is constant). In the limit of infinite gain a very good approximate in most cases is to set $\Delta v = 0$ or $v_+ = v_-$. The infinite-gain approximation can be used for other configurations as well and is always equivalent to setting $v_+ = v_-$. This assumption, together with zero current flow into the inputs, generally simplifies the gain derivation considerably.

An advantage of the noninverting amplifier is its high input impedance. The minimum gain is unity ($R_a = \infty$, $R_b = 0$), and the maximum useful gain about 10^2 to 10^3. The best range of R_b is 2 to 100 kΩ.

4-2   Inverting Amplifier

A very useful amplifier is the inverting amplifier, shown in Fig. 4-2. The gain can easily be calculated, utilizing the infinite-gain approximation ($v_+ = v_-$). The current i, which flows from the input to the output, is

$$i = \frac{v_i}{R_a} = -\frac{v_o}{R_b} \tag{4-5}$$

Therefore, the closed-loop gain A is

$$A = \frac{v_o}{v_i} = -\frac{R_b}{R_a} \tag{4-6}$$

A disadvantage of the inverting amplifier is its relatively low input impedance, which is equal to R_a because the inverting input is at virtual ground

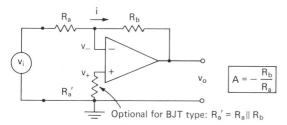

Optional for BJT type: $R_a{}' = R_a \| R_b$

Fig. 4-2   Inverting amplifier.

potential. Input impedance, however, is ordinarily much higher than the op-amp output impedance and therefore rarely presents a problem when driven by another op-amp. The input resistance R_a must not be too high (over 100 kΩ) with BJT-type op-amps or the effect of the bias and offset current may become too high, especially if R_a' is omitted. Closed-loop gains A of 0.01 to 100 or 1000 are practical. The unity-gain inverter ($R_a = R_b$) is popular.

4-3 Differential Amplifier

The differential-amplifier stage has the same properties as the differential-input op-amp except that it employs negative feedback to stabilize the gain A. Following Eq. (3-1), with A replacing A_o, we define the gain as

$$v_o = A(v_1 - v_2) \qquad (4\text{-}7)$$

The simplest of several circuits is shown in Fig. 4-3. It may be thought of

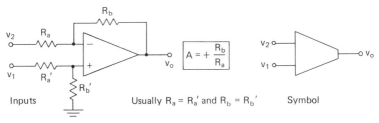

Fig. 4-3 Differential-amplifier stage.

as a combination of the inverting amplifier and noninverting amplifier previously discussed except that the input signal on the noninverting side is slightly reduced by the factor $\alpha = R_b/(R_b + R_a)$; that is, $v_+ = \alpha v_1$. The attenuation is required because the gain of the noninverting side is higher. It can be seen that if Eq. (4-4) is multiplied by α, it becomes identical to Eq. (4-6).

For many applications, the CMRR m must be made as small as possible, and in this case R_a' or R_b' may be made variable over a narrow range (± 5 percent with 1 percent tolerance resistors). To maximize m, the test circuit of Fig. 4-13 can be used. As with the inverting amplifier, it is good practice to keep R_b in the range of 2 to 100 kΩ to reduce the offset-bias-current effects.

If the input impedance of this circuit is too low for the desired application, noninverting amplifiers, perhaps unity-gain, are added preceding each input. An improved version of the differential-input amplifier, termed an *instrumentation amplifier*, is discussed in Sec. 4-8.

4-4 Summing Amplifier

Occasionally the need arises to sum two or more signals. A variation of the inverting amplifier (Fig. 4-4) is well suited for this purpose. Because of the feedback, the summing point v_- is held at zero. As a consequence the current in any input (i_1, i_2, etc.) is independent of the current in any other, while the output current i is the sum of the inputs. If the resistors are unequal, an extension of Eqs. (4-5) and (4-6) yields

$$v_o = - R_b \left(\frac{v_1}{R_1} + \frac{v_2}{R_2} + \cdots \right) \tag{4-8}$$

where any number of inputs may be attached.

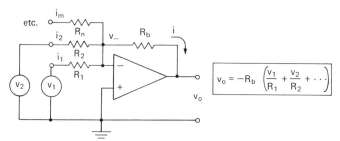

Fig. 4-4 Inverting summing amplifier.

As with the standard inverting amplifier, signal voltage sources must have a low impedance, such as provided by the output of a previous feedback amplifier. To correct for the signal inversion, a unit gain inverting amplifier can be used following the summing amplifier. If a signal is to be subtracted instead of added, a unit-gain inverter can be used before the summing amplifier in line with that signal input.

It should be noted that a two-signal input amplifier, with one signal added and the other subtracted, is equivalent to the differential-input stage discussed above, but that amplifier configuration cannot be extended to more inputs because the summing points (v_+ and/or v_-) are not held at ground potential.

4-5 Integrator

The output of an integrator is proportional to the integral over time of its input signal. If the input voltage v_i is constant, the output will be a ramp. In Fig. 4-5 an integrator circuit related to the inverting amplifier is given. Calculation of the output voltage as a function of time $v_o(t)$ or, alternatively, its Laplace transform $V_o(s)$ can be made by noting that the input and output currents are equal and that v_- is at virtual ground ($A_o \rightarrow \infty$ approximation)

$$i(t) = - \frac{C \, dv_o}{dt} = \frac{v_i}{R} \tag{4-9a}$$

$$I(s) = - CsV_o(s) = \frac{V_i(s)}{R} \tag{4-9b}$$

Rearranging, we obtain

$$v_o(t) = \frac{-1}{RC} \int_0^t v_i(t')\,dt' + v_{oo} \qquad (4\text{-}10a)$$

$$V_o(s) = \frac{-V_i(s)}{RCs} \qquad (4\text{-}10b)$$

where v_{oo} is the output at $t = 0$. Frequently v_{oo} is set equal to zero at $t \leqslant 0$ by keeping the capacitor shorted before $t = 0$. The factor $1/RC$ is the

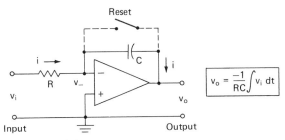

Fig. 4-5 Basic integrator.

equivalent of a gain or sensitivity. Note that the output is inverted and either polarity input is permissible.

Output saturation will eventually occur unless the integrator is reset or the input is of alternating polarity with a strictly zero time average. Bias current can be a serious problem at longer integration times and may cause saturation even with a grounded input unless current compensation (Fig. 3-5) is used. A high-value resistor in parallel with the capacitor will greatly reduce the drift to saturation, but the resulting circuit is, strictly speaking, a high-pass filter rather than an integrator (see Chap. 11). FET input op-amps are much superior if the integration time is long.

A practical integrator is shown in Fig. 4-6. It has an FET which acts as a reset switch, connected so as to short the capacitor and thus force the output to virtual ground when the FET is turned on (positive gate for an n-channel FET). In some applications a simple momentary-contact push-button switch will suffice. If a JFET is used, a limiting resistor (typically 100 kΩ to 10 MΩ) must be inserted in series with the gate to prevent excessive drain current when the gate becomes forward-biased. Care must be taken to bias the FET to cutoff. Because $R_{on} \neq 0$, the capacitor will not discharge or reset instantaneously but will require several time constants CR_{on} (typically $R_{on} = 10$ to $1000\ \Omega$). Further, it is desirable to choose R at least two orders of magnitude higher than R_{on} for accurate reset unless the input voltage is known to be close to zero when reset. Other versions of the integrator are discussed in Chap. 11. FET biasing circuits are discussed in Sec. 14-2.

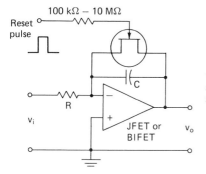

Fig. 4-6 Practical integrator. R is typically 10 to 100 kΩ (BJT op-amp) or up to 100 MΩ (FET op-amps).

4-6   Current-to-Voltage Converter

A current-to-voltage converter has an output voltage v_o which is proportional to an input current i_i. Signals produced by a very high impedance source, e.g., the photodiode, act as current generators. The circuit is very simple (Fig. 4-7) and again based on the inverting amplifier. An expression for the output voltage in terms of the input current has already been derived [Eq. (4-5)].

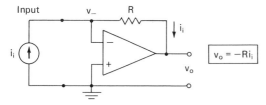

Fig. 4-7 Current-to-voltage converter.

A simple resister also acts as current-to-voltage converter (Ohm's law), but the op-amp circuit reduces the voltage drop at the input v_- to nearly zero while producing an easily measurable output voltage v_o. Thus it approaches an ideal milliammeter characteristic of zero voltage drop.

Strictly speaking, the current which flows through the resistor is a sum of i_i and the op-amp bias current, so that the output voltage will have a zero error unless the bias current is compensated or, better, an FET input op-amp is selected.

A voltage-to-current converter has an output current i_o which is proportional to an input voltage v_i and independent of the output load resistance R_L or voltage v_o. In other words, the circuit acts as an ideal current generator. If the load resistance R_L is floating (neither side grounded), the circuit of Fig. 4-8 will work well. Actually, it is just the

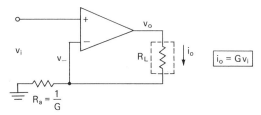

Fig. 4-8 Voltage-to-current converter (floating load).

noninverting amplifier reinterpreted in terms of output current. Examination of the circuit will reveal that v_-, which is equal to v_i by the infinite-gain approximation, is proportional to i_o for any value of R_L and is even independent of v_o provided v_o is not high enough to cause saturation. This remains true even if R_L is variable or has a quite nonlinear current-voltage relation.

The load resistance R_L can be replaced by reactive load or impedance Z_L, although with some op-amps the circuit will become unstable and oscillate at a high frequency. Should this occur, the oscillation can often be eliminated by altering the external frequency-compensation network, if present, or by adding a small capacitor across the load impedance.

Grounded loads can be driven by the circuit of Fig. 4-9. With the proper choice of resistors, the current into the load i_o becomes proportional to v_i, as seen by the following analysis (again assuming the infinite-gain approximation, $v_- = v_+ = v_o$):

$$i' = \frac{v_i - v_o}{R_a} = \frac{v_o - v'}{\alpha R_a} \tag{4-11}$$

$$i_o = \frac{v' - v_o}{\alpha R_b} - \frac{v_o}{R_b} \tag{4-12}$$

Note that the resistance ratio α is arbitrary (at least at small-signal levels). Solving for v' from Eq. (4-11) and substituting into Eq. (4-12) gives the gain relation

$$i_o = \frac{-v_i}{R_b} \tag{4-13}$$

Output currents of either polarity may be obtained. If the resistors are not accurately balanced, a small output current will be present when $v_i = 0$. Usually α is chosen between 0.2 and 1.0 and R_b is chosen such that v_i is about half the output voltage at saturation when the output current is at maximum.

Another type of voltage-to-current converter utilizes an OTA, an example of which is shown below in connection with the voltage-controlled gain amplifier (Fig. 4-16).

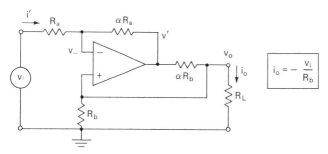

Fig. 4-9 Voltage-to-current converter (grounded load).

4-7 Bridge Amplifier

A convenient op-amp configuration which produces an output voltage proportional to the resistance change ΔR of a resistance sensor is shown in Fig. 4-10. It can be thought of as a combination of an inverting amplifier

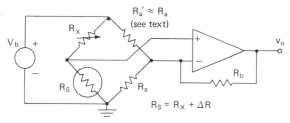

Fig. 4-10 Bridge amplifier.

(Sec. 4-2) and a Wheatstone bridge. The two resistors R_a and R_a' not only are two elements of the bridge but also are a part of the feedback resistance. Either an ac or dc bridge voltage supply V_b is common. Usually the dc supply is obtained from a pair of resistors connected to the primary regulated power supply (+15 V).

Analysis is simplified by the equivalent circuit of Fig. 4-11, intended to apply when the bridge is near balance ($R_X = R_S$). Recall that any two-terminal network can be replaced by its Thevenin equivalent; the resistance-sensor arm (Fig. 4-11b) is replaced by a voltage source v_1 and a series resistance, which is equal to $R_S/2$ at balance ($\Delta R = 0$ or $R_S = R_X$). Note that *at* balance, $v_+ = V_b/2$ and that *near* balance

$$v_1 = \frac{(R_X + \Delta R)V_b}{R_X + (R_X + \Delta R)} = \frac{V_b R_X(1 + \Delta R/R_X)}{2R_X(1 + \Delta R/2R_X)} \approx \frac{V_b}{2}\left(1 + \frac{\Delta R}{2R_X}\right) \quad (4\text{-}14)$$

where the binomial expansion for $(1 + x)^{-1}$ with $x = \Delta R/2R_X$ has been used to obtain the last term. Terms of the order x^2 are neglected. The

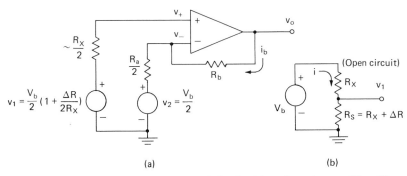

Fig. 4-11   Bridge-amplifier equivalent circuit for the (a) noninverting amplifier; (b) sensor-arm subcircuit.

Thevenin equivalent of the other arm (R_a, R_a') is found similarly. Because of the high amplifier input impedance, the series impedance of the sensor arm $R_S/2$ is unimportant so that and $v_1 = v_+$ (unless R_S is very high, in which case the op-amp bias current may have to be taken into account). Only if the feedback current i_b is zero will $v_2 = v_-$; this is the strict balance condition for which $v_o = V_b/2$. Calculation of the output voltage v_o from the equivalent circuit shows

$$v_o \approx \frac{AV_b}{4}\frac{\Delta R}{R_X} \quad \text{where } A = \frac{R_a + 2R_b}{R_a} \quad (4\text{-}15)$$

Although $v_o = V_b/2$ at strict balance, in practice R_X is usually adjusted so that $v_o = 0$ rather than $R_X = R_S$ as is assumed for Eq. (4-15). Alternatively R_a' can be adjusted so that $v_o = 0$ when $R_X = R_S$ (rather than $R_a = R_a'$).

The output voltage can be expressed in terms of a sensitivity factor S for a change in the variable Δy, which the sensor is measuring if the sensor-resistance change due to the variable $\gamma = (\Delta R/R_S)\,\Delta y$ is known

$$v_o = S\,\Delta y \quad \text{where } S = \frac{AV_b\gamma}{4} \quad (4\text{-}16)$$

4-8   Instrumentation Amplifier

An instrumentation amplifier is a fixed-gain differential-input amplifier consisting of three op-amps, as indicated in Fig. 4-12. The gain expression is formally the same as that for an op-amp, i.e.,

$$v_o = A(v_a - v_b) \qquad (4\text{-}17)$$

except that the open-loop gain is replaced by the gain with feedback A. Specifically,

$$A = \frac{(2R_b + R_a)\,R_d}{R_a\,R_c} \qquad (4\text{-}18)$$

It is basically an improved version of the differential amplifier of Fig. 4-3.

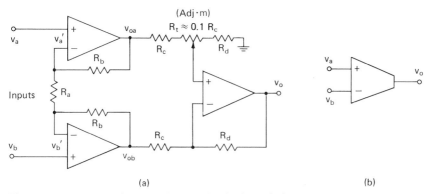

Fig. 4-12   Instrumentation amplifier: (a) circuit, (b) symbol.

Features are (1) high input impedance, especially with FET op-amps on the input, (2) high CMRR, or m, and (3) precision high gain. High input impedance is achieved by using the noninverting amplifier configuration on the inputs. Precision high gain is achieved by two stages of feedback amplifiers. High common-mode rejection is achieved by the dual noninverting-configuration circuit, which utilizes a common feedback resistor R_a.

The dual-input stage can easily be understood by looking at the response of the individual sections with a signal applied to one input with the other grounded. If the input v_b is grounded, using the infinite-gain approximation shows that the corresponding noninverting input v_b' will be practically at ground potential. In other words, the lower part of R_a will then be at virtual ground, and the upper input amplifier will be identical to the noninverting configuration Fig. 4-1. The first-stage output voltage v_{oa} will be $v_a(R_a + R_b)/R_a$. Since the lower IC acts as an inverting amplifier, its output will be $v_{ob} = -v_aR_b/R_a$. Applying a signal to the v_b input instead (v_a grounded) results in output voltages $v_{ob} = v_b(R_a + R_b)/R_a$ and $v_{oa} = -v_bR_b/R_a$. Because the amplifier is linear, the difference in voltage $v_{oa} - v_{ob}$ will be the same if neither input is grounded and is equal to the first-stage gain multiplied by $v_a - v_b$. The output stage is a differential amplifier (Fig. 4-3) with a gain of R_d/R_c, so that the total gain is that given by Eq. (4-18).

Consider now the CMRR, m. Suppose the inputs are tied together ($v_a = v_b$) and a common-mode signal v_a applied (with respect to ground). In this case, again using the infinite-gain approximation, we have $v'_a = v_a = v'_b$ and therefore $v_{oa} = v_{ob} = v_a$. In other words, the common-mode gain is unity while the differential gain A_d is $(R_a + 2R_b)/R_a$. This means that m is improved by a factor of A_d by the first stage. The second stage m is determined primarily by the accuracy of the resistor ratio (which in turn determines the gains of the inverting and noninverting sides) and, to a lesser extent, by op-amp nonlinearity. A resistor trim R_t allows maximization of m. As the circuit of Fig. 4-13 suggests, m can easily be deter-

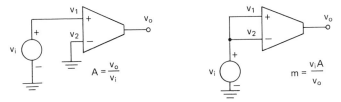

Fig. 4-13   CMRR m test circuit: (a) measurement of voltage gain A; (b) measurement of common-mode response.

mined by first measuring the gain A (one input grounded) and then the common-mode response $v_o/v_i = A/m$ with both inputs tied together.

The complete instrumentation amplifier is also available in one IC package, a modest convenience. Care must be taken to examine the circuit of any device called an "instrumentation amplifier" since sometimes the term is applied to any quality amplifier.

4-9   Logarithmic Amplifier

A logarithmic amplifier or the inverse, the antilog amplifier, is useful for compensating transducer nonlinearities or for analog computations. The general form of the circuit can be applied to other nonlinear functions if a component with the appropriate nonlinear functional dependence (or its inverse) is available. Logarithmic amplifiers deliver an output voltage v_o proportional to the logarithm of the input voltage v_i; that is,

$$v_o = K \log v_i + v_{oo} \qquad (4\text{-}19)$$

where K is a gain factor and v_{oo} is a constant (voltage offset) which is normally set equal to zero at the minimum input voltage. An antilog amplifier does the inverse, or

$$v_o = B e^{\alpha v_i} + v_{oo} \qquad (4\text{-}20)$$

where α and B are gain parameters. Choice of number base (10, e, etc.) is made through adjustment of the gain parameters.

A circuit configuration with a logarithmic transfer function utilizes the exponential increase in diode current with the voltage across the diode. As Fig. 4-14a indicates, the basic circuit (which does not have temperature

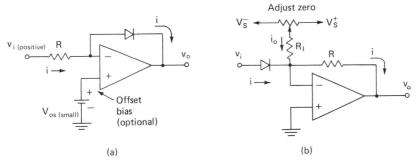

Fig. 4-14   Basic (a) logarithmic and (b) antilogarithmic amplifiers.

compensation) is a variation of the inverting-amplifier configuration. Using the infinite-gain approximation, we find the current i to be

$$i = \frac{v_i}{R} = \beta_o e^{-\alpha(v_o + v_{os})} \qquad (4\text{-}21)$$

where β_o and α are constants for the forward-based diode ($\alpha = 39$ V for silicon). Note that the diode is forward-biased only for positive input voltage and the output is then negative. Taking the logarithm of both sides of Eq. (4-21), we obtain the general expression

$$v_o = \frac{-\ln v_i}{\alpha} + \frac{\ln R\beta_o}{\alpha} - v_{os} \qquad (4\text{-}22a)$$

which is the same form as Eq. (4-19).

If we adjust the offset bias voltage v_{os} so that the last two terms cancel and insert the numerical value for α, we obtain the simpler expression valid for a silicon diode amplifier

$$v_o = -0.059 \log v_i \qquad (4\text{-}22b)$$

If v_i is expressed in millivolts, it is convenient to adjust v_{os} so that $v_o = 0$ when $v_i = 1.00$ mV. Offset-voltage adjustment is critical since a small voltage change at the input can mean the difference between the desired forward bias and reverse bias (amplifier swings to saturation). Both polarity inputs, however, can be accommodated by adding a second diode in parallel (reverse polarity).

The antilogarithmic (or exponential) amplifier (Fig. 4-14b) is similarly analyzed:

$$i = \beta_0 e^{\alpha v_i} + i_o = -Rv_o \qquad (4\text{-}23)$$

except that an adjustable offset bias current i_o rather than an offset voltage is applied (note that the value of R_I is unimportant if it is reasonably high). Equation (4-23) can be put into the form of Eq. (4-20) with $B = -\beta_0/R$ and $v_{o0} = -i_o/R$.

A more practical logarithmic amplifier on which temperature compensation is provided is shown on Fig. 4-15. In place of the diode is a diode-

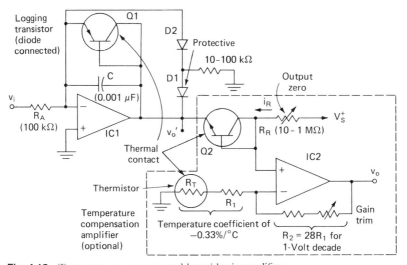

Fig. 4-15   Temperature-compensated logarithmic amplifier.

connected logging transistor Q1 which exhibits a logarithmic conformity over a wide range (6 decades). Also included are protective diodes to limit current should reverse polarity be applied. Capacitor C prevents oscillation by providing frequency compensation. Compensation of the rather strongly temperature-dependent diode parameters α and β_o is provided.

To compensate for the β_o variation, a second logging transistor Q2 is connected so that its voltage drop exactly cancels that of Q1 at a particular temperature, for example, 25°C, and input voltage. A bias current i_R is passed through Q2 to establish the compensating voltage, thus providing an offset voltage (zeroing of v_o' or v_o) with the same circuit that compensates for temperature changes. The choice of logging transistor is not critical.

Compensating for the temperature coefficient of the parameter α, which varies inversely as the absolute temperature, is done by the thermistor and series resistor, which together constitute R_1. The composite temperature coefficient should be the same as that of α, that is, 0.33 percent per Celsius degree. As with any noninverting amplifier, R_2 is chosen to provide the desired gain, in this example such that the net gain factor K of Eq. (4-19) is unity. For many applications, the temperature coefficient of α is negligible, and in this case R_1 can be replaced by a fixed resistor.

Somewhat more complex circuits are available in which the transistor is not diode-connected. These offer a wider dynamic range (8 decades).†

4-10 OTA Multiplier or Gain Control

The multiplication circuits discussed here utilize the current-controlled gain feature of the operational transconductance amplifier (OTA). Analog multiplication implies that output voltage v_o is the product of two input voltages v_x and v_y, or

$$v_o = B v_x v_y \qquad (4\text{-}24)$$

where B is a gain factor with units of V^{-1}. Alternatively the amplifier voltage gain $A = B v_y$ may be thought of as variable, with v_y the control voltage and v_x the signal input voltage. The circuit shown (Fig. 4-16) functions only if v_y is positive although v_x may have either sign (two-quadrant multiplication).

Gain control current for IC1 is derived from IC2 which is utilized as a voltage-to-current converter. The gain of IC2, and thus the total circuit gain, is set by R_B The output of IC1 is a current i_o proportional to i_y and v_x. This current is converted into an output voltage by IC3, and Eq. (4-24) thus follows. Linearity is within 1 percent, but individual gain adjustment is required. If v_x and v_y are tied together as the (positive) input, the output is proportional to the square of the input.

---

†Logarithmic circuits are discussed in detail by Y. J. Wong and W. E. Ott, "Function Circuits: Design and Applications," McGraw-Hill, New York, 1976.

Division is accomplished by incorporating the multiplying circuit into a feedback loop, as indicated in Fig. 4-17. This technique of obtaining the inverse function by incorporating the function into a feedback network is similar to that employed by the logarithmic amplifiers (Sec. 4-9). If the v_x and v_y inputs are tied together, the unit acts as a square-rooting circuit.

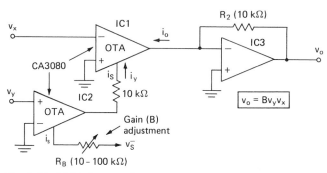

Fig. 4-16 Multiplier, or voltage-controlled gain amplifier.

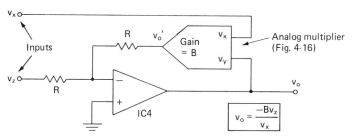

Fig. 4-17 Analog division utilizing a multiplying amplifier.

4-11 Current-Differencing-Amplifier (CDA) Configurations

Circuit configurations based on the current-differencing or current-controlled type of op-amp are roughly analogous to the ordinary voltage-controlled op-amp described above provided that (1) input voltages are converted into input currents by resistors and (2) the bias (input) currents are chosen so that the amplifier output does not go to saturation. The circuits here use a single-ended (grounded) supply. If dual-supply operation is desired instead, the V_S terminal replaces ground. For operation in the linear region, the output usually is biased so that its dc (no signal or quiescent) value is somewhat less than half the supply voltage V_S.

The input-resistor biasing method can be understood by examining the comparator circuit (Fig. 4-18a). Resistor R_B establishes the current threshold by injecting a fixed current $i_- = V_S/R_B$ into the inverting input of the amplifier. When the current into the noninverting input, $i_+ = v_i/R_A$, exceeds i_-, the output will swing to positive saturation (**1**),† while if it is less, the output will be close to ground potential (**0**). Thus the voltage threshold V_T at the input v_i is (neglecting the input voltage drop to ground)

$$V_T = V_S \frac{R_A}{R_B} \qquad (4\text{-}25)$$

Hysteresis may be added by the positive-feedback resistor R_H. With a +5-V supply, the output is TTL-compatible (see Chap. 5) provided the current difference $i_+ - i_-$ exceeds 10 μA.

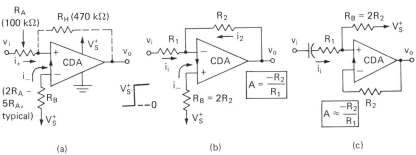

(a)   (b)   (c)

Fig. 4-18 Current-differencing-amplifier configurations: (*a*) comparator; (*b*) inverting amplifier (dc-coupled); (*c*) noninverting amplifier (ac-coupled).

It is convenient to analyze CDA circuits by the infinite-gain approximation, which implies that $i_+ = i_-$, provided of course that the amplifier is in the linear operating region. A linear amplifier requires that the output be biased to about half the supply voltage, as indicated above. For the inverting configuration of Fig. 4-18*b*, this is done by choosing R_2 to be half the value of that bias resistor R_B with $i_i = 0$, or

$$\frac{V_S}{R_B} = i_+ = i_- = \frac{v_o}{R_2} \qquad \text{for } i_i = 0 \qquad (4\text{-}26)$$

which for $R_B = 2R_2$, implies that $v_o(\text{dc}) = V_S/2$. It is assumed that the input v_i has no dc component. Likewise the output voltage is $v_o = v_o' + V_S/2$, where v_o' refers to a small-signal component such that $i_i = 0$ when $v_o' = 0$.

If a signal $i_i = v_i'/R_1$ is now applied, the output-voltage change v_o is found by the setting $i_i = i_2$ (infinite-gain approximation), and the voltage gain is

$$A = \frac{v_o'}{v_i} = -\frac{R_2}{R_1} \qquad (4\text{-}27a)$$

†The binary digits **0** and **1** for logic 0 and logic 1 are set in boldface to distinguish them from scalars (except in truth tables).

as expected for the inverting amplifier. This circuit can be converted into an ac amplifier by the addition of an input capacitor. It may be observed that establishment of a bias point is much less precise and more difficult for the CDAs than for the standard op-amps which operate around ground potential.

The noninverting amplifier, shown ac-coupled in Fig. 4-18c, is biased similarly to the inverting amplifier. Signal current i_i is injected (through R_1) into the noninverting input. Here v_i' and v_o' refer to the ac components of the input and output signals. Again assuming the infinite-gain approximation ($v_i/R_1 = \Delta i_+ = v_o'/R_2$), we obtain the stage ac voltage gain

$$A \approx \frac{v_o'}{v_i} = - \frac{R_2}{R_1} \qquad (4\text{-}27b)$$

Numerous circuit configurations can be implemented with the CDA, but they generally do not perform as well as the standard op-amp configurations. Where performance is not exacting, these low-cost configurations can be attractive, especially when single-ended supply operation is required.

4-12 Voltage-Offset–DC-Bias-Adjustment Methods

Adjustment of the dc level of an op-amp output to zero or a specified bias value can be accomplished by the circuits of Fig. 4-19. An adjustable

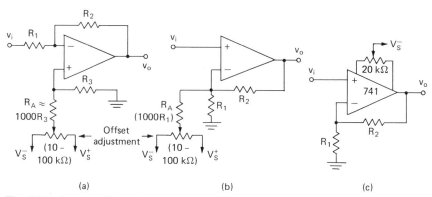

(a)   (b)   (c)

Fig. 4-19 Op-amp offset and dc bias circuits: (a) external adjustment for inverting amplifier; (b) external adjustment for noninverting amplifier, and (c) internal-offset method.

current flows through a resistor connected from either input to ground in Fig. 4-19a and b. The maximum current (and thus offset-voltage range) is limited by R_A, which in combination with the input resistor acts as a voltage divider. The adjustment range of about ± 15 mV for the value of

R_A suggested is adequate to compensate for nearly any op-amp offset. If a larger dc bias is required, R_A can be made smaller.

Some op-amps, e.g., the 741 (Fig. 4-19c), have pins to which an adjustment potentiometer can be connected. The adjustment range of ± 10 mV is adequate to compensate for the input offset but normally is not large enough to bias the output to several volts, as the other circuits shown will do.

If two dc amplifiers are connected in cascade (one driving the next), it is usually necessary to provide only voltage offset to the first, or input, stage since the effect of the offset voltage (and offset compensation) is amplified when applied to the later stage. Offset is usually not needed for ac amplifiers because the undesirable dc component can be taken out by a dc blocking capacitor or high-pass filter.

4-13 Comparator

The response of an op-amp comparator is full positive output (saturation) when the input difference $v_+ - v_-$ is positive. When the input difference is negative, the output swings to negative saturation. It is similar but not identical to the digital-comparator IC (Chap. 5).

Three versions of the comparator are shown in Fig. 4-20. In Fig. 4-20a the threshold voltage V_T is adjustable. Because of the high open-loop gain of the op-amp, an input-voltage rise to only slightly more positive than V_T will cause the output to swing from negative to positive saturation. The output polarity can be reversed by interchanging the inputs v_+ and v_-.

Positive feedback is utilized in Fig. 4-20b to speed the transition and to provide hysteresis. It can be thought of as a noninverting amplifier with the inputs interchanged. With hysteresis the input voltage required to produce a positive output transition is slightly higher than that required to produce a negative transition (Fig. 4-20d). False transitions due to noisy signals are reduced by hysteresis. Hysteresis is increased by decreasing the ratio of R_b to R_a. Typically R_a/R_b is 10 to 1000.

A small capacitor across the feedback resistor will speed the response somewhat, but the slew-rate limitation cannot be overcome and rather sluggish rise times of 2 to 20 μs are typical. If faster response is required, a fast-switching transistor can be connected to the amplifier output or, better, a high-speed digital comparator can be substituted for the op-amp.

4-14 Multivibrator

In this section several op-amp versions of the astable, monostable, and bistable multivibrators are described, which are convenient for many control, switching, and timing functions. Digital ICs (Chap. 5) and IC timer circuits (Chap. 17) can perform the same function at higher speed and are ordinarily far superior but in some respects not as versatile. Also the power-supply voltages and signal levels differ from those of most op-amps, an annoyance which can sometimes be avoided by not mixing the two types of ICs.

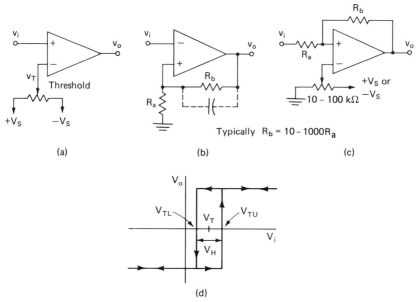

Fig. 4-20   Op-amp comparators: (*a*) simple comparator with variable threshold; (*b*) inverting comparator with hysteresis; (*c*) noninverting comparator with hysteresis and variable threshold; (*d*) response showing hysteresis (V_h).

The monovibrator, or one-shot multivibrator of Fig. 4-21, is one of many versions. It produces a pulse of width T_p following the application of a fast trigger pulse. Before the input is applied, v_- is zero, v_+ is $-\beta V_S$, where β is the voltage-divider ratio of R_a and R_b, and v_o is at negative saturation. If a negative-going input pulse of magnitude greater than the threshold $v_+ = -\beta V_S$ is applied, the output will swing to positive saturation, strongly reinforced by positive feedback through the capacitor C. Not until C discharges to zero, after a delay of T_p, will the output return to negative saturation. If R_b is returned to a positive supply $+V_S$ instead of $-V_S$, the output will be inverted and will be triggered by a positive-going pulse. The input capacitor C and resistor R form a differentiating circuit whose values are chosen along with the threshold voltage for reliable triggering.

An astable, or free-running, multivibrator is shown in Fig. 4-22. The output alternately swings from negative to positive saturation with a period determined by the time constant RC and the positive-feedback resistance ratio. If the output v_o is at positive saturation, $v_+ = 0.39v_o$, and until the capacitor charges to a value more positive than this, the output remains at positive saturation. When v_- exceeds v_+, the output swings to

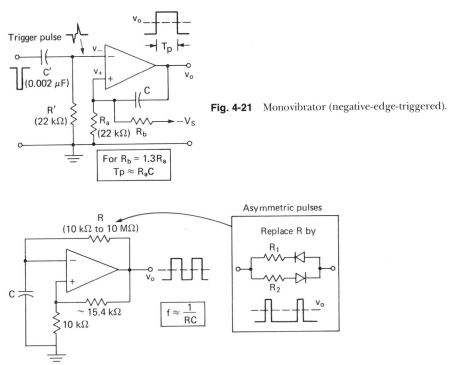

Fig. 4-21 Monovibrator (negative-edge-triggered).

Fig. 4-22 Astable multivibrator. For asymmetric pulses, add diode circuit shown.

negative saturation, the capacitor charges in a negative direction until it reaches $-0.39v_o$, and the cycle repeats. The resistor ratio is chosen so that the charging time (half-cycle) is exactly $RC/2$, and thus the full period is RC. Variable frequency can be achieved by making R variable. If an asymmetric wave is desired, the diode-resistor network, which causes the negative and positive charging rates to differ, replaces R.

A bistable multivibrator, or flip-flop, is shown in Fig. 4-23. The set-reset is similar to the monostable multivibrator and the comparators with hysteresis, discussed previously. Following a positive pulse applied to input S, the output goes to, and remains at, positive saturation until a positive pulse is applied to input R. Pulses should not be applied to R and S simultaneously.

Bistable toggle-type flip-flops based on op-amps are rather impractical compared with the simplicity of a standard digital circuit.

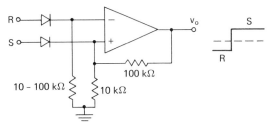

Fig. 4-23   *SR* flip-flop.

4-15   Peak Reader

Measurement of a signal peak voltage is often simplified if the peak value is held for some time following the peak. The circuit of Fig. 4-24 is a

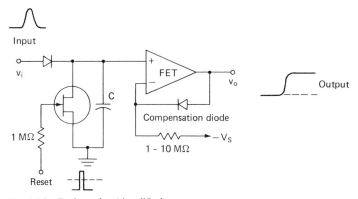

Fig. 4-24   Peak reader (simplified).

positive-peak reader. The capacitor will charge through the input diode following the output voltage while the voltage is rising but cannot discharge through the diode and thus remains at the peak voltage. A unity-gain amplifier or follower prevents discharge of the capacitor by the output load. Diode voltage drop (~0.5 V) is compensated by a second diode in the feedback path. The capacitor is discharged or reset by the FET, normally biased to cutoff and switched on by the reset pulse. Sometimes it can be replaced by a simple momentary-contact switch. Unless an FET input op-amp is used, compensation is normally required to minimize the slow charge or discharge of the capacitor, seen as a drift of the output peak reading with time. It can be converted into a negative-peak reader by reversing the diodes (with the compensated diode returned to $+V_S$). This circuit works best if the signal level is over 1 to 2 V.

An improved peak reader shown in Fig. 4-25 utilizes two op-amps. Peak reading is done by diode D1 as for the simple peak reader described above. A unity-gain amplifier IC2 follows the capacitor C voltage. Feed-

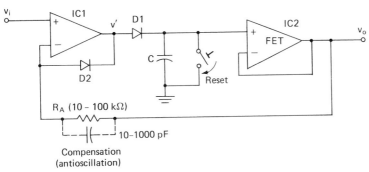

Fig. 4-25   Precision positive-peak reader.

back from v_o to the noninverting input of IC1 forces v_o to equal v_i (infinite-gain approximation). Diode D2 keeps the IC from saturating during negative-going signals. The feedback resistor R_A limits the current from IC2 during transients. Conversion from a positive- to a negative-peak reader requires only a reversal of diode polarity.

4-16   Output Limiting and Clipping

Sometimes it is desirable to limit an op-amp circuit output voltage to a point below the saturation value. One reason might be to clip the signal at a precise voltage. The saturation voltage might be used, but it is somewhat variable and dependent on load. Further, limiting circuits keep the op-amp in a linear region, an advantage at higher frequency because the time delays associated with bringing the op-amp out of saturation are avoided and the slew-rate limitation is less severe. Two Zener-diode bounding or clipping circuits for the inverting-amplifier configuration are shown in Fig. 4-26. Above the Zener breakdown voltage, the effective feedback resistance R_b, and thus the closed-loop gain, approaches zero. A zero gain at a particular voltage level implies that the output voltage will not rise farther as the input rises; i.e., the output signal is clipped. In the circuit in Fig. 4-26a if we select Zeners of different value, the negative and positive clipping voltage can be different. A disadvantage of this circuit at higher frequencies is the rather poor frequency response of Zener diodes.

The poor frequency response is overcome by the circuit in Fig. 4-26*b* because the Zener diode always has the same polarity and only the faster bridge diodes reverse polarity with an ac signal. The Zener-diode bias current from the power supply shown in Fig. 4-26*b* (an optional feature) maintains the Zener at a constant voltage.

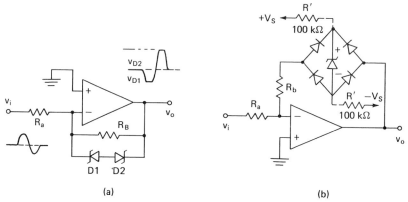

(a)   (b)

Fig. 4-26 Clipping circuits, Zener-diode versions.

Although it is not immediately obvious, the Zener resistors R' do not affect the feedback or gain calculation. This is because when the diodes connected to the output conduct, the resistors in effect appear between the output and power supply as load resistors and not across the feedback resistor R_b. Of course when the output voltage is too low for diode conduction, both resistors R' are effectively out of the circuit.

Chapter Five

Basic Digital Devices:
TTL and CMOS

The dominant application of digital ICs is, of course, in computers, but their role in general instrument design is also very important. The most extensive digital series are TTL and CMOS. Several other less common IC series are also useful in instrumentation applications and will be discussed briefly here.

Digital IC devices have two stable states, a higher positive-voltage state termed *high* or **1**† and a lower voltage state termed *low* or **0**. This definition applies to conventional binary positive logic. Negative logic, an alternate notation, is discussed in Sec. 5-20. Within a series the voltage range of all devices has the same value and can be interconnected in any order; i.e., the devices are fully compatible. Specification of the logic outputs for various combinations of the inputs is done by means of a *truth table* and is frequently the only information about a device needed (aside from pin numbers and the voltage levels of the series).

---

†The binary digits **0** and **1** for logic 0 and logic 1 are set in boldface to distinguish them from scalars (except in truth tables).

5-1   Inverter/Drivers

The simplest digital device, the digital equivalent of a unity-gain inverter, is the inverter. Its output is at **0** when the input is at **1** and at **1** when the input is at **0**. Its symbol and truth table are shown in Fig. 5-1.

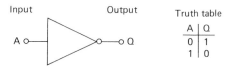

Fig. 5-1 Inverter.

A driver (Fig. 5-2) is the digital equivalent of a unity-gain amplifier. The output simply repeats the logic state of the input. However, the output presumably has a greater current (and/or voltage) capacity than the input source or there would be no point in using a driver (except perhaps for

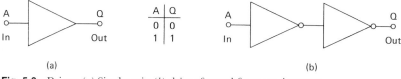

Fig. 5-2 Driver: (*a*) Single unit; (*b*) driver formed from two inverters.

waveshaping). Increased output-voltage capacity is needed only for interfacing between different series of devices. Increased current capacity allows multiple loads to be connected. Two inverters in series are equivalent to a driver. Inverters (and drivers) are usually manufactured in IC form with six units per case or package, referred to as *hex inverters.* Note that all 14 leads of a standard 14-pin package are used (including 2 for power supply).

5-2   OR, NOR, and XOR Gates

An OR gate has two or more inputs and one output. Its output will go to **1** if any input is at **1**, as indicated by the truth table in Fig. 5-3.

A NOR gate is the equivalent of an OR gate followed by an inverter. Note that the small circle on the output side (Fig. 5-4) as elsewhere indicates

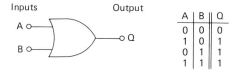

Fig. 5-3 Two-input OR gate.

Inputs   Output

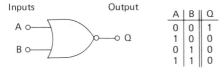

A	B	Q
0	0	1
1	0	0
0	1	0
1	1	0

Fig. 5-4 Two-NOR-gate OR.

Inputs   Output

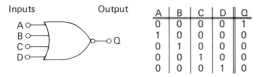

A	B	C	D	Q
0	0	0	0	1
1	0	0	0	0
0	1	0	0	0
0	0	1	0	0
0	0	0	1	0

Fig. 5-5 A quad-input NOR.

Inputs   Output

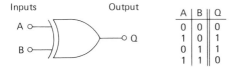

A	B	Q
0	0	0
1	0	1
0	1	1
1	1	0

Fig. 5-6 Two-input EXCLUSIVE-OR (XOR).

inversion. NOR gates with more than two inputs are available. A four-input (quad) NOR is shown in Fig. 5-5. The output is **1** only if all inputs are **0.**

An exclusive-OR gate (XOR) has two or more inputs and one output (Fig. 5-6). As with the regular OR gate, the output will go to **1** if any input is at **1,** except that the output will go to **0** if all inputs go to **1.**

5-3 AND and NAND Gates

An AND gate has two or more inputs and one output. Its output will go to **1** only if all inputs are at **1,** as indicated by the truth table in Fig. 5-7. A NAND (Fig. 5-8) gate is the equivalent of an AND gate followed by an inverter.

Inputs   Output

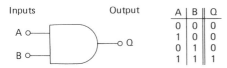

A	B	Q
0	0	0
1	0	0
0	1	0
1	1	1

Fig. 5-7 Two-input AND gate.

Inputs   Output

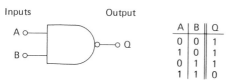

A	B	Q
0	0	1
1	0	1
0	1	1
1	1	0

Fig. 5-8 Two-input NAND gate.

5-4   Transmission Gates

A transmission gate (Fig. 5-9) allows bidirectional flow of pulses when on and no flow when off. It is equivalent to a voltage-controlled switch. Although BJT gates are sometimes used, most transmission gates are FET type, commonly CMOS. When the FET is on, the resistance R_{on} is typically 400 Ω, but devices with resistances below 25 Ω are available. These devices are symmetric, and either terminal can be connected to the input. Analog as well as digital signals can be switched, but the input and output voltages must fall within a certain range, generally between the most negative terminal V_{SS} and most positive V_{DD}. Voltages outside the range may cause the control gate to lose control (turn off or on improp-

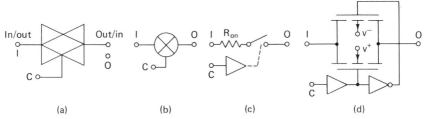

Fig. 5-9   Transmission gates: (*a*) symbol, (*b*) alternate symbol, (*c*) electrical equivalent, and (*d*) CMOS analog digital gate.

erly) or damage the device. This restriction must be observed if the FET substrates, which must not be reverse-biased, are connected to V_{SS} or V_{DD} (see Chap. 2). If V_{SS} is grounded, the input or output voltage cannot be negative, a disadvantage for some analog applications.

Transmission gates are often manufactured with four gates per package. A wide variety of devices is available, some with built-in decoders or features important to specific applications. Selection should consider the permissible voltage range of the switched signal when the digital control supply is connected as required, for example, V_{SS} = GND. When a transmission gate is connected in series with the output of a digital device, the three states are **0, 1,** and off (or high impedance). Applications are discussed in Chap. 18.

5-5   Set-Reset (*SR*) Flip-Flops

An *SR* flip-flop has two inputs and two outputs, as indicated in Fig. 5-10. Application of a pulse to input *S*, that is, momentarily bringing *S* to **1,** will cause the output *Q* to go to **1** until reset by bringing input *R* to **1.** The second output $\bar{Q}$ is the inverse of *Q*. Unlike the gates discussed above, the flip-flop outputs remain in a state indefinitely unless altered by an input pulse, however brief. In other words, if both inputs are at **0** (no pulse), the outputs remain unchanged, as indicated by the truth table of Fig. 5-10. If both inputs are brought to **1,** the outputs may go to either state, and in practice the first input to return to **0** loses control. In a properly designed system both inputs will not be brought to **1** simultaneously.

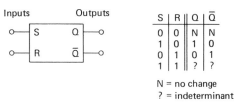

S	R	Q	Q̄
0	0	N	N
1	0	1	0
0	1	0	1
1	1	?	?

N = no change
? = indeterminant

Fig. 5-10 *SR* flip-flop.

Although a few *SR* flip-flop integrated circuits are available, in most cases a pair of two-input NOR or NAND gates are connected to act as a flip-flop. A verification of the truth table for the two-NOR-gate circuit of Fig. 5-11 demonstrates that it has the properties of an *SR* flip-flop.

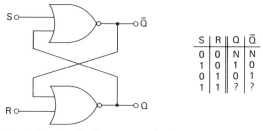

S	R	Q	Q̄
0	0	N	N
1	0	1	0
0	1	0	1
1	1	?	?

Fig. 5-11 *SR* flip-flop composed of two NOR gates.

A NAND-type *SR* flip-flop (Fig. 5-12) is made by substituting NAND gates for the NOR gates. However, the input logic sense is inverted. Logic **1** on both inputs produces no change while a brief **0** on the *A* (or *B*) input results in a **1** on the *Q* (or *Q̄*) outputs.

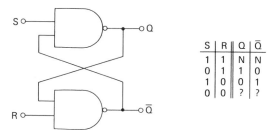

S	R	Q	Q̄
1	1	N	N
0	1	1	0
1	0	0	1
0	0	?	?

Fig. 5-12 *SR* flip-flop composed of two NAND gates.

5-6 *JK* Flip-Flops

A *JK*, or master-slave, flip-flop has two logic inputs *J* and *K*, two outputs (Q and $\bar{Q}$), at least two control inputs, a clock CK, and a clear (Fig. 5-13). The *JK* inputs are similar to the *SR* inputs of the *SR* flip-flop, but the outputs do not change until a pulse is applied to the clock input. In digital computers the clock synchronizes the various logic operations. It is convenient to think of the *JK* flip-flop as two *RS* flip-flops, as illustrated in Fig. 5-14. When the clock is low (**0**), the input AND gates pass the *JK* input signals to the master flip-flop, which follows the inputs, changing state as the set or reset pulses are applied. However, the slave does not follow the master because the transfer AND gates do not pass the signal until the clock goes high (**1**). The *JK* inputs are first disconnected to prevent the master from changing during the clock pulse. Next the transfer AND gates

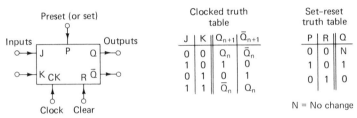

Fig. 5-13 *JK* flip-flop.

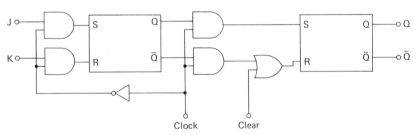

Fig. 5-14 *JK* flip-flop logic diagram (simplified).

connect the slave to the master and remain connected while the clock is high. If the clock returns to **0**, the slave does not change state although the master will again follow the input. The state of the master flip-flop cannot be detected externally except through its effect on the slave because its leads are not brought out from the IC. At any time the slave can be reset (Q to **0**) by applying a pulse (**1**) to the clear input. Ordinarily a flip-flop is set to **0** or cleared before beginning an operation.

In a number of details the logic diagram of Fig. 5-14 has been simplified for the discussion. Actually the input ANDs always isolate the input slightly before the transfer ANDs operate (the simplified diagram is ambiguous in this respect), and the exact point where the transfer occurs, e.g., on the rising edge or falling edge of the clock pulse, is not indicated. Further, no indication of the removal of the indeterminate state ($J = 1$, $K = 1$), a key feature of the JK flip-flop, is suggested. These flip-flops are actually designed so that they reverse their output state upon application of a clock pulse if the inputs are both at **1**. In terms of the truth table, the output state Q_n at clock pulse n is changed to $Q_{n+1} = \bar{Q}$ at clock pulse $n + 1$, where the bar indicates the inverse state ($1 \rightarrow 0$ and $0 \rightarrow 1$).

As Fig. 5-13 suggests, most flip-flops have both preset and clear inputs, which allow the output Q to be set to **1** in a manner similar to the SR flip-flop. Various types of JK flip-flops differ in a number of details such as clocking on the rising instead of falling edge or requiring inverted logic to clear ($\bar{R} = 0$ to clear, $\bar{R} = 1$ to count). Especially with flip-flops, the particular specification sheets for the device should be consulted for truth tables and timing diagrams.

Actually, JK flip-flops are not used much as individual units and should be thought of primarily as a building block for more complex counters. The type-D flip-flop (discussed below) is more popular where a single flip-flop is required.

5-7 Binary Counters

The JK flip-flop can be operated as a binary counter because it reverses states upon application of clock pulse when both JK inputs are at **1**. For this application as Fig. 5-15 indicates the clock becomes the input and the JK terminals are permanently wired to a **1** voltage supply. Before counting, a pulse is applied to all the clear inputs, so that all the outputs (Q_A, etc.) are at **0**. If a series of pulses is applied to the A input (clock), its output will reverse each time the clock pulse goes through one complete cycle, which for many types of flip-flops occurs on the falling edge of the clock pulse. Note that two complete input cycles are required to produce one complete output cycle. There is no requirement that the input pulse be periodic or symmetric.

If indicator lamps are connected to monitor the outputs, the output can be interpreted as a binary counter. A package of four flip-flops with the JK inputs internally connected to **1** is referred to as a *4-bit binary* or *ripple counter* (Fig. 5-16) with the clock input renamed the *toggle* or *count input T*. Decimal equivalents for several binary states are shown in the table. Note that a 4-bit binary counter ranges from 0 to 15 decimal and has an output or carry pulse which can be connected to the toggle input of additional stages.

Binary counters are available in many forms, two of which are indicated in Fig. 5-17. All have clock and clear inputs. Some also have an enable input which can inhibit the count. Outputs of the various flip-flops are usually labeled, in order starting with the least significant bit (LSB), by any of at least four sets of notation: Q_A, Q_B, Q_C, ..., $A, B, C, ..., Q_1, Q_2$,

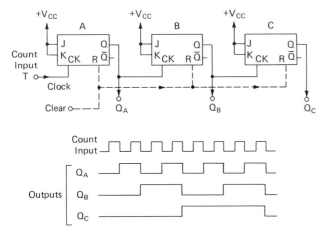

Fig. 5-15 Basic three-stage binary counter.

Q_3, ... or Q_0, Q_1, Q_2, ... In nearly all cases the output state changes on falling edge of the clock pulse although for few devices change on the rising edge is an option.

5-8 Decimal (BCD) Counters

A binary-coded decimal (BCD) counter is identical to the 4-bit binary counter except for an internal circuit which resets all the flip-flops to **0** on the tenth count (binary **1010**) rather than the sixteenth. This is done by an AND gate with its output connected to clear and its inputs connected to the Q_B and Q_D outputs. Each 4-bit counter is equivalent to one decimal digit and is separately decoded from the binary as a digit from 0 to 9. No 4-bit binary numbers higher than **1001** exist. A block diagram is shown in Fig. 5-18.

The decimal digit can be read by a seven-segment display after decoding. Each segment of the display can be independently illuminated and the decoder/driver selects the proper combinations of segments to form the desired digit, as determined by the BCD input. Illumination is provided by light-emitting diodes, neon lamps, or incandescent bulbs. Detailed decoder display circuits are given in Chaps. 7 and 17.

Counters with a modulus or base other than 10 can be constructed in a

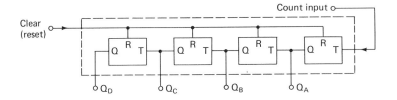

Q_D	Q_C	Q_B	Q_A	Decimal equivalent
0	0	0	0	0
0	0	0	1	1
0	0	1	0	2
0	0	1	1	3
1	0	0	1	9
1	0	1	0	10
1	1	1	1	15

Fig. 5-16 Basic 4-bit binary or ripple counter.

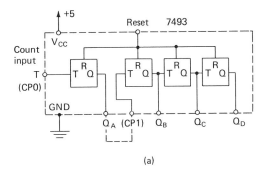

(a)

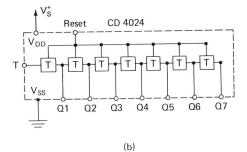

(b)

Fig. 5-17 Binary counters: (*a*) 4-bit TTL and (*b*) 7-bit CMOS.

similar fashion. For example, if the AND-gate inputs are connected to the Q_C and Q_B counter outputs and the gate output to reset, the counter acts as a base 6 or divide-by-6 counter (the Q_D output is not used unless a divide-by-12 counter is desired). Some counters, for example, 7490, have an internal AND gate on the reset input for convenience.

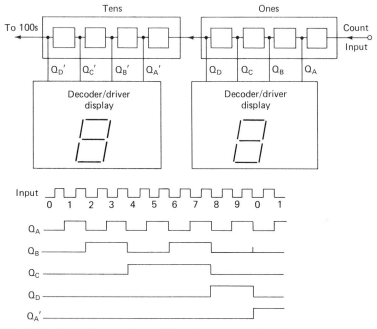

Fig. 5-18   Block diagram of two-digit BCD counter and readout.

Two of many available decade counters are shown in Fig. 5-19. The LS-TTL counter (74LS90) is pin-compatible and functionally identical to the popular TTL counter 7490. It features a flip-flop A with a separate clock input. In normal BCD operation input A is the pulse-count input, and the clock inputs to the B section must be externally connected to the A output (the remaining clocks are internally connected). Following decade counting stage inputs are tied to the D outputs, which function as carry pulses. Note that the following stage count is incremented on the falling edge of the D output, i.e., on the tenth count of the driving stage. When operated as a frequency divider (divide-by-10), the separate stage A is sometimes placed last (input frequency to B clock) so that the output is a symmetric square wave. In this case the unit acts as a decade counter, but it is not BCD-encoded.

Two BCD counters per package are available in CMOS (Fig. 5-19b). Normally pulses to be counted are connected to the enable input since the flip-flops change on the enable falling edge. Counting is turned off (inhibited) by bringing the enable input to **1**. Clocking on the rising edge is possible by interchanging the clock and enable inputs (inhibit on **0**).

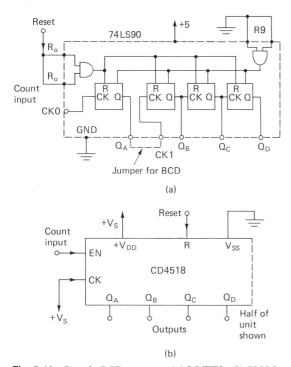

(a)

(b)

Fig. 5-19   Decade BCD counters: (*a*) LS TTL, (*b*) CMOS.

5-9   Type *D* and Related Flip-flops

A variety of flip-flops is available in addition to the *RS* and *JK* flip-flops. The type *D* (Fig. 5-20) is similar to the *JK* except it has only one data input *D*. The *D* input is the same as the *J* input of the *JK* flip-flop with the *K* input at $\bar{J}$ (see truth table of Fig. 5-20). The output changes on the clock rising edge for most types.

A *D* flip-flop can be operated as a binary up counter by tying the $\bar{Q}$ output to the *D* input since the state to which the *Q* output is to be changed is just the state of the $\bar{Q}$ output. In this case the clock is the pulse input. Binary counters, in particular those which are edge-triggered (change state on the rising edge of a pulse), are called *toggle* or *type-T flip-flops* (also see Fig. 5-16).

The standard CMOS type-*D* flip-flop (Fig. 5-21) is recommended for general applications. Because it has a set S and reset R input, it can also be used as an SR flip-flop. Output changes on the rise (positive-going edge) of the clock pulse. In the LS TTL series, the 74LS74 (7474 for TTL) behaves similarly except that the set and reset inputs are inverted (normally held high). Note that the presence of the small circles on the input implies an inversion (see Sec. 5-20 for discussion).

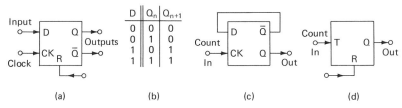

(a)   (b)   (c)   (d)

Fig. 5-20 (*a*) Basic type-*D* flip-flop, (*b*) truth table, (*c*) *D* flip-flop connected as a binary counter, and (*d*) toggle (*T*) flip-flop.

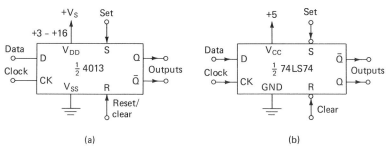

(a)   (b)

Fig. 5-21 Standard type-*D* flip-flops: (*a*) CMOS and (*b*) LS TTL series.

5-10 Special-Purpose Counters

A variety of more complex counters for special applications is available with one or more of the following features: (1) they are presettable to a desired initial value, (2) they can count down as well as up, (3) they can divide by N (a variable), and (4) they can count synchronously. Counters usually have four flip-flops to a package and are encoded in binary or BCD. Specific examples are given in Chap. 17 to illustrate the general type described here.

By presettability is meant that the counter can be set to any given number (0 to 15 in the case of a 4-bit counter) before counting begins rather than just to zero by the reset input. This process is also referred to as *jamming* a number into the counter. The number N in binary (or BCD) form must be applied to the data or preset inputs (Fig. 5-22*a*). When the *preset-enable line* is brought high, the number is transferred to the internal flip-flops and the outputs match the input while the enable is high. When the enable returns to low, normal counting can continue.

An up-down counter can count down as well as up; the direction is controlled by an additional input line (Fig. 5-22*b*). A down counter registering **0000** will change to **1111** on the clock-pulse input (or **1001** for BCD). Carry-pulse generation to subsequent stages cannot utilize the Q_D output as is done in standard up counters; instead internal logic provides a carry when a zero count is detected.

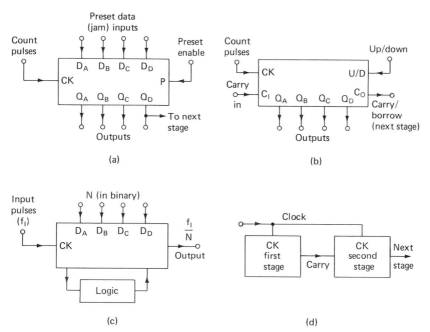

Fig. 5-22   Features of special-purpose counters: (*a*) presettability, (*b*) up-down control, (*c*) divide-by-*N*, (*d*) synchronous.

A divide-by-*N* frequency counter provides one output pulse for each *N* input pulses, where *N* is a number in binary form presented to counter data-input lines. It functions similarly except that *N* is a variable, i.e., programmable. Presettable counters can be connected as divide-by-*N* counters by the addition of external jumpers or logic gates, which depend on the specific device (Fig. 5-22*c*, also see Chap. 17). In this case *N* (or its complement) in binary form is the preset or jam-input data. For example, a down counter preset to *N*=7 by the carry pulse will provide one output (carry) pulse for every seven input (clock) pulses.

Synchronous counters are designed to change the output of all stages simultaneously, a desirable feature in applications where timing is critical. All three digits (all 12 bits) of a counter going from 099 to 100, for example, will change at the same time. An asynchronous, or ripple-type, counter, discussed above, will exhibit a delay (9s will change before the 1 for the example). In order to avoid propagation delay, the clock must be connected to all stages (Fig. 5-22d), but each stage is inhibited unless the carry output of the previous stage occurs. Carry occurs just after the ninth pulse of a BCD counter, thus enabling the subsequent stage on the tenth count of the driving stage.

5-11 Inside TTL Integrated Circuits

The internal amplifier and output driver of various transistor-transistor logic (TTL) circuits are similar in many respects, and it is instructive to analyze a typical circuit such as the NAND gate of Fig. 5-23. A NAND gate

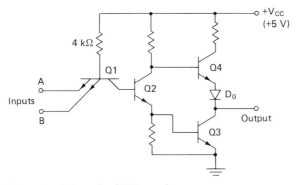

Fig. 5-23 Schematic of TTL two-input NAND gate.

has multiple emitters on the input transistor Q1. It is helpful to think of Q1 as a set of back-to-back diodes, a basic property of BJTs, as illustrated in Fig. 5-24. Suppose input A is made low by grounding A through a low-value resistor R_i. If R_i and thus v_i is sufficiently small, the base current i_B flows through the 4-kΩ resistor, the emitter diode, and R_i to ground. The base voltage V_B is one diode drop above v_i, that is, close to 0.6 V for R_i small. The base current does not flow through the collector diode and into the base of Q2 because that path includes two diode drops to ground and therefore current will not flow unless V_B approaches or exceeds 1.2 V. If both inputs are made sufficiently positive, V_B will rise to the point where current flows into the base of Q2 and the emitter diodes become reverse-biased. As current flows into Q2, the base voltage of Q3 rises and Q3 goes into saturation. The collector of Q3, which is the output of the IC, is then close to zero ($\approx$0.2 V). Under these conditions, the base of Q4 is slightly reversed-biased (the drop across D_0 is larger than the saturation voltage of Q2). If instead the base current into Q2 is zero, Q4 becomes forward-biased and pulls the output up to about +3.8 V. Under these conditions, Q3 is in cutoff. In effect the totem-pole driver shorts the output and acts as a low-impedance emitter follower on high output, thus allowing rapid charging and discharging of any output-line capacitance. The overall result is that if either input is low, the output is high, but if both inputs are high, the output is low, verifying the NAND truth table.

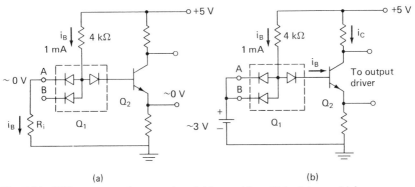

(a)   (b)

Fig. 5-24   TTL NAND-gate input section: (*a*) input *A* low; (*b*) both inputs high.

Low-power Schottky TTL is an improved version of the TTL series. Its internal circuitry is similar in most respects to TTL (Fig. 5-25). The most important difference is the Schottky clamping diode, indicated by the special symbol, which, like a Zener diode, conducts in the reverse direction above a certain voltage. Functionally the clamping diode prevents transistor saturation and decreases recovery and switching time. The diodes also reduce transient oscillation on the inputs, which can occur during switching of poorly terminated lines, thus increasing the allowable speed even under adverse conditions. In Fig. 5-26 the output voltage is plotted as a function of the input voltage.

Some digital devices have an *open-collector output,* as illustrated in Fig. 5-27, rather than the standard totem-pole output, illustrated in Figs. 5-23 and 5-25. The output transistor is simply turned on and off by the driver. An external *pullup* or load resistor R_L connected to a positive power supply is necessary. Since the supply voltages may differ, this circuit can be used for translation or interfacing between different logic voltage levels. Because the voltage is not actively pulled up to the high level (by a saturated transistor), the speed is less than for a standard output. Another application is the summation of output pulses through a common-load resistor. Since this is equivalent to a logical OR, the configuration is referred to as the *wired* OR, and the purpose is often to save gates. Most display drivers have open-collector outputs.

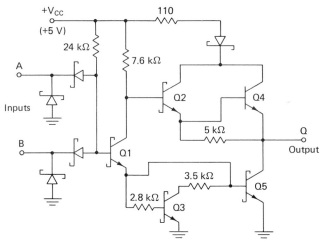

Fig. 5-25 Schematic of a low-power schottky (LS) two-input NAND gate.

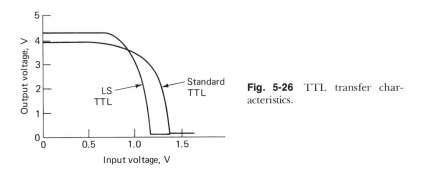

Fig. 5-26 TTL transfer characteristics.

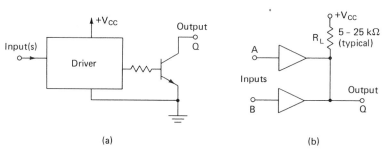

(a)

(b)

Fig. 5-27 Open-collector output: (a) output transistor; (b) output with load resistor as a wired OR.

A common mistake made by the novice is to assume that an open input is equivalent to **0.** Actually with TTL it is usually equivalent to **1** because no ground return or sink for the current flowing out of the input is provided. Logic **0** is provided by a grounded input or by a low-value resistor to ground (under 250 Ω for standard TTL or 1 kΩ for LS TTL). An input connected to the power supply is equivalent to **1.** Some designers prefer to tie fixed **1** inputs to a 3- to 4-V source, e.g., to the output of an inverter with its input grounded.

Nominally **0** (low) is +0.3 ± 0.2 V and **1** (high) is 3.6 ± 1 V. Ground and supply voltages are acceptable **0** and **1** sources, respectively. Negative inputs corresponds to **0,** but most devices have an additional diode from the input to ground (forward-conducting for negative inputs) to suppress negative transients, and thus a voltage more negative than 0.5 to 0.6 V cannot be applied.

5-12   Inside CMOS Integrated Circuits

Complementary metal-oxide-semiconductor (CMOS or COSMOS) IC devices comprise n-channel and p-channel FETs on the same chip. CMOS logic devices are compatible with TTL devices; i.e., they have approximately the same **1** (high) and **0** (low) voltage levels when used with a 5-V power supply, which TTL requires. However, CMOS will operate on any supply voltage between about +3 and +16 V, and the supply current is very small. The somewhat lower speed is usually of minor consequence and is the best choice for instrumentation applications.

The simplicity of a CMOS inverter, as pictured in Fig. 5-28, is notable. Both the n- and p-channel FETs are enhancement type, as implied by the symbol, with cutoff (or cut-in) voltages matched so that when one is on, the other is off. If the gate voltage of an n channel is sufficiently positive, it will turn on while a p channel will turn off. In effect, the pair of FETs acts as a single-pole double-throw switch which can connect the output v_0 to either supply voltage (**1**) or to ground (**0**) according to the magnitude of the input or gate voltage.

In Fig. 5-29, the output voltage v_0 as a function of the input voltage v_i is plotted for two power-supply voltages V_S. Note that with a 5-V supply, the CMOS can be driven by TTL devices because **1** is at least 3.0 V, but there is little margin of error. (See interface discussion, Sec. 5-17.)

During the switching process, when the gate voltage is intermediate between the **1** and **0** levels, both output FETs (Fig. 5-28) are partly on. Appreciable current may flow through the transistors at this point, and the instantaneous power dissipation P_D can be quite high, especially at higher supply voltages (Fig. 5-30). If the switching time is fast, as is usually the case in logic circuits, the power-dissipation pulse is very brief and contributes only slightly to heating the device. If, on the other hand, the switching time is slow compared with the thermal lag (on the order of 0.1 s), internal heating of the device can be high enough to cause damage. Slow switching time or even a static condition occurs in some timing

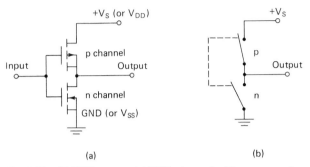

Fig. 5-28 CMOS inverter: (a) FET schematic; (b) output equivalent. Input protective diodes to V_{DD} and V_{SS} are not shown.

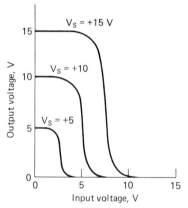

Fig. 5-29 CMOS inverter transfer characteristics.

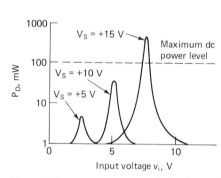

Fig. 5-30 Internal power dissipation of CMOS devices at intermediate logic levels.

circuits or when CMOS is operated as a linear amplifier. Under these conditions the supply voltage must be limited (preferably below 10 V).

Care must be taken to avoid any open input (gates) on CMOS digital circuits because static charges may build up, causing the device to draw excessive current and overheat. Inputs of all unused sections of an IC package should be tied to ground (or $+V_S$).

Nearly all CMOS inputs have internal diodes on the inputs which are tied to V_{SS} (ground) or V_{DD} ($+V_S$). These diodes help protect against damage by static charges during handling or by excessive input voltages (especially if a series protective resistance is present).

5-13   TTL and CMOS Characteristics

A few of the more important characteristics aside from the truth table are listed here.

Supply Voltage V_S, V_{CC}, or V_{DD}   The power supply for TTL logic devices V_{CC} is +5.0 V and must be held within 5 to 10 percent. Voltage surges of a few volts above nominal may destroy the device. The power supply for CMOS V_{DD} may typically be anywhere between +3 and +15 V. The negative side of the CMOS supply, designated by V_{SS}, is usually grounded.

Supply Current I_{CC}   The current drawn by the supply varies with the output state and is typically 1.8 to 5.5 mA per driver for standard TTL, assuming the output is not shorted, and 1 mA for LS TTL.

For CMOS, the equivalent supply current is very small, about 10 pA per gate, but it jumps to around 1 mA per gate during the brief time (0.1 μs) required to change states, primarily because the output-line capacitance must be charged and discharged. A rapidly switched gate might draw an average of 0.1 mA, but in most cases the supply current would be much less. A CMOS gate with the voltage partly on may draw substantial current at higher supply voltages.

Power Dissipation P_D   Power dissipation is a product of the average supply current and voltage. For standard TTL it is 10 to 25 mW per gate, a rather high value, while for LS devices it is appreciably less, about 2 mW per gate. For CMOS, it varies from over 100 mW for an active circuit to 0.01 μW for a quiescent circuit (see Fig. 5-30).

Input/Output Voltage   The maximum range V_{OH} of TTL output voltages at **1** (high level) is typically 2.4 to 4.0 V. The minimum input voltage V_{IH} accepted as **1**, is 2.0 V. The **0** (low) output V_{OL} is typically 0.2 to 0.6 V, while the maximum input voltage V_{IL} accepted as **1** is 0.8 V (minimum −0.5 V). Open input acts as **1**.

For CMOS, the output voltage at **1** is equal to the supply voltage (Fig. 5-28) while the minimum **1** input voltage required is 40 to 70 percent of the supply voltage. The **0** voltage output is 0.0 while any input voltage below 20 to 30 percent of supply is effectively **0**.

All standard CMOS devices have protective diodes tied to the supply lines (V_{DD} and V_{SS}) on all inputs to limit the input voltage to a safe range. While the primary purpose is to prevent or reduce damage due to static-charge buildup during handling, these diodes also allow application of input voltage which exceed the supply-voltage range if a series resistance, for example, 100 kΩ, is added to the input. Input-voltage clamps are also present on TTL and LS TTL devices except that TTL lacks the diode clamp to the positive supply.

Input Current I_I The input current I_{IL} for standard TTL at **0** (low) is typically -1.0 mA (the negative sign indicates an outflow), but the maximum value and thus safe design value is -1.6 mA. At **1** (high) the input current I_{IH} is very small, under 50 μA. Input currents of LS TTL are -0.2 mA.

For CMOS devices the input current is practically zero for both **1** and **0** except for a small capacitive component.

Output Current I_O and Fan-out The output current I_{OH} of TTL at **1** can rise to -18 to -55 mA under transient or short-circuit conditions. At **0** the output current I_{OL} is at least 16 mA (a positive sign indicates an inflow of current, or current sink). Since each TTL input may require -1.6 mA, this output can drive 10 inputs safely. The number of inputs a unit can drive is referred to as the *fan-out*, in this example 10.

The output current for CMOS as a source (**1**) or sink (**0**) is in the range of 0.5 to 3 mA. The fan-out is 50, or larger if less speed is required. Higher-output buffer amplifiers can drive up to 10 mA.

Rise Time t_R and Fall Time t_F Rise time can be defined as the time required for the output voltage to rise from 10 to 90 percent of its final value, assuming a step input. Fall time is similarly defined. Rise and fall times of 5 to 10 ns are typical for TTL, 1 to 5 ns for LS TTL, and 25 to 100 ns for CMOS.

Propagation Delay t_{PD} After a pulse is applied to an input, the time until the output changes is the *propagation delay*. It is in the range of 10 to 30 ns per gate for TTL, 5 to 10 ns for LS TTL, and 20 to 200 ns for CMOS.

5-14 Comparison of Digital Families

Two higher-speed series of TTL are available. The H series is only slightly faster (6 ns) than standard TTL (10 ns) and takes twice the power. It is inferior to the LS TTL series and can be considered obsolescent. The Schottky-clamped TTL high-speed, or H, series is appreciably faster (2 ns) and relatively immune to noise and reflected pulses due to improper terminations, which are troublesome at high speeds. Although not as fast as ECL, this series has a good combination of properties for enough high-speed applications where LS TTL is not quite fast enough.

Emitter-coupled logic (ECL) is the fastest of the widely available logic series. Propagation delays are in the 1- to 5-ns range, and clock speeds up to 400 MHz are possible. High speed is achieved not only by employing emitter followers, which exhibit the best frequency response of the three transistor configurations, but also by biasing the transistors to high current levels and limiting the voltage swings to small values. With a -5.2-V supply (V_{EE} grounded) the output level swings from about -0.8 V (1) to -1.6 V (0), a range which is not compatible with TTL or CMOS without special interface drivers or level translators. Cooling the higher-current (and therefore higher-power) devices often requires a fan or heat sink. Not only are interfacing and cooling more difficult with ECL, but considerable care must be exercised in parts layout, especially the termination of connecting wires which act as transmission lines with a particular characteristic impedance. If this is not done, ECL may perform no better than LS TTL. Use of a printed circuit with all unused areas covered with metal as a ground plane is also desirable to avoid stray coupling. Needless to say, ECL devices should not be employed except where speed requirements leave no other choice.

A lower-power (1 mW) lower-speed (33 ms) series of TTL is available, but in practically all applications for which they are intended CMOS devices are superior.

Resistor-transistor logic (RTL), requiring a 3-V supply, and diode-transistor logic (DTL), requiring a 5-V supply, have characteristics roughly comparable to TTL except for lower speeds. While satisfactory for most instrumentation applications, they have no advantages over the more popular and generally lower-cost TTL or CMOS devices and thus can be considered obsolescent.

A number of large-scale ICs have been developed which utilize current-injection logic (IIL or I^2L) configurations. Advantages include low power dissipation combined with high speed and small area on the IC chip. However, the logic-level change (0 to 1) is small (0.6 V), so that an input and output interface must be built into each line of each chip. Since the inputs and outputs are likely to be TTL- and/or CMOS-compatible, the user can simply treat them as if they were TTL or CMOS throughout.

At the time this is being written, the CMOS or LS TTL series is the best choice for most new equipment design. Generally CMOS is easier to interface and is more versatile while LS TTL has the advantage where higher speed is required. Standard TTL is similar to LS TTL in most respects but requires more power as well as being slower—in other words, has no advantages over LS TTL except perhaps for somewhat wider availability.

5-15   Diode Gates

Simple AND and OR gates can be made from diodes (Fig. 5-31). If a positive voltage of sufficient magnitude is applied to either input of the OR gate, the diode will conduct and the output will rise, i.e., go to 1. The value of the resistor R is unimportant except for loading the input or the output impedances of the driving or following stages. The AND-gate output will be at the supply voltage $+V_S$ or **1** if both inputs are at a positive voltage (**1**) because the diodes are reversed-biased and no current will flow. If either input is at a low voltage or grounded (**0**), the diode will be forward-biased and pulled down to a low voltage (**0**).

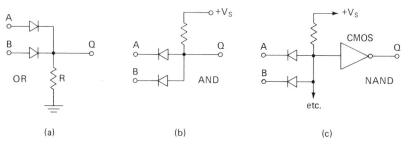

(a)   (b)   (c)

Fig. 5-31 Diode gates: (*a*) two-input OR; (*b*) two-input AND; (*c*) multiple-input NAND utilizing CMOS inverter.

A drawback of the diode gate is that the output-voltage change will be less than the input-voltage change because of the diode drop (0.6 V). Further the rise time of the output is less than that of the input. For many instrumentation applications this degree of attenuation and smearing of the pulse is negligible, especially when CMOS devices are employed. These diode gates cannot be used to construct circuits such as the *RS* flip-flop because they have no internal amplification.

Many of the simpler digital operations can be performed with op-amps. Of course, digital IC circuits (TTL or CMOS) are usually superior, but when only a few simple digital operations are required in an otherwise analog circuit, these op-amp logic circuits may be convenient. Here **0** is defined as negative saturation and **1** as positive saturation (for a +15-V supply, logic **1** is +14 V and **0** is −14 V). An op-amp flip-flop (*RS*) has been described in Chap. 4. Simple gates (AND and OR) which consist of diode gates and a comparator to provide gain are shown in Fig. 5-32. Conversion to NAND and NOR is accomplished by reversing the op-amp input leads.

5-16 Op-amp–TTL–CMOS Interfacing

Since the **1** and **0** voltage levels differ in these series, any interconnection requires an interface circuit to provide proper operation and in some cases to prevent damage. In Table 5-1, the various levels are listed, where the designation +3/+5, for example, means that an input of +3 V is accepted at **1** while the output will deliver +5 V (higher by a substantial safety margin). For op-amps the input must only be slightly positive for **1**, just enough to overcome the offset voltage (<10 mV).

A few of many possible TTL/CMOS interface circuits are shown in Fig. 5-33. If CMOS is operated on +5 V, as required by TTL, CMOS can be driven directly, but a pullup resistor R_p provides a desirable safety margin

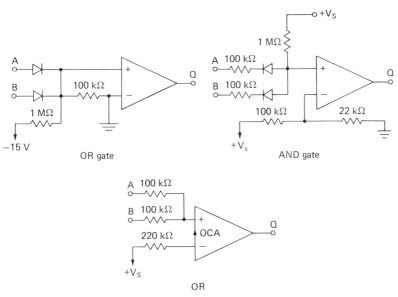

Fig. 5-32   Op-amp versions of OR and AND gates.

TABLE 5-1   Logic Levels, V

Logic	TTL	CMOS		
		$V_S = +5$	$V_S = +15$	$V_S = \pm15$
1 (high)	+2.4/5.0	+3/+5	+8/+15	$0^+/+14$
0 (low)	+0.8/0.0	+2/0	+6/0	$0^-/-14$

by increasing the **1** voltage level to nearly +5 V. If CMOS is operated much above +5 V, the TTL drive will be insufficient but the transistor amplifier will overcome the difficulty. Alternatively a TTL open-collector driver (15 V), line receiver, or a digital comparator (Sec. 5-19) may be used.

CMOS drivers which are designed for higher output currents can sink enough current (3.2 mA) to drive two TTL devices directly, but they must operate on a +5-V supply. However, while CMOS devices (gates, etc.) can drive LS TTL directly without a driver interface, they cannot safely drive standard TTL.

An op-amp comparator can be driven directly by either TTL or CMOS (Fig. 5-34 or Fig. 5-38). If the inverting and noninverting inputs are reversed, the op-amp acts as an inverter.

A CMOS device can be driven directly by an op-amp except that a series resistor must be added, which, together with the internal input diode, prevents the input from swinging negative (Fig. 5-34).

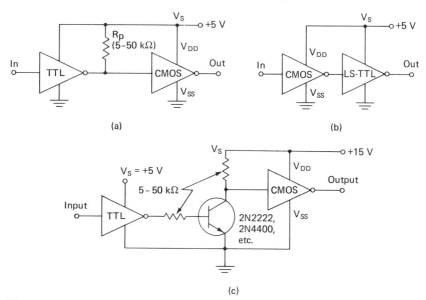

Fig. 5-33 TTL/CMOS interface: (*a*) TTL to CMOS; (*b*) CMOS to LS TTL; (*c*) TTL to high-level CMOS.

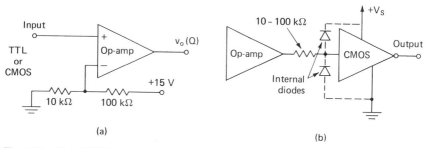

Fig. 5-34 Digital TTL to op-amp interface circuit.

To drive a single TTL input from an op-amp the circuit of Fig. 5-35 is satisfactory. The op-amp current output when negative must be at least 0.4 mA (0.5 mA to be safe) in order to act as a sink for the current flowing out of the LS TTL device. A diode, which most TTL devices include, prevents the input voltage from swinging more negative than -0.6 V. When the op-amp output is positive, no current flows into the TTL device but its input voltage must be between $+2.4$ and $+5.0$ V as provided by the resistor divider. A wide variety of special-purpose interface IC devices are available. Manufacturers' literature should be consulted for details.

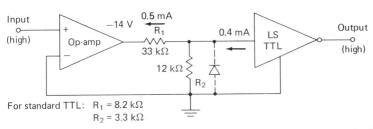

Fig. 5-35 Op-amp–to–LS TTL interface driver. Resistor values for standard TTL are 4 times smaller.

5-17 Schmitt Triggers

A Schmitt trigger (Fig. 5-36) is a logic device with an output which changes state suddenly as a threshold is reached. It is similar to a comparator with hysteresis but is usually operated with logic-level inputs rather than a wider-range analog input. It frequently is employed in analog-to-digital interfacing, especially where a pulse with a slow rise (or fall) is to be changed into a pulse with a fast rise (or fall). Voltage level or time ambiguities, undesirable in digital systems, are reduced or eliminated by the Schmitt trigger.

Various digital devices (drivers, inverters, gates) with built-in Schmitt triggers are available. Alternatively two inverters (Fig. 5-36c) can be connected to provide the same function. Adjustment of the ratio of R_a to R_b controls the positive-feedback fraction and therefore the amount of hysteresis ($H = V_L/V_H$).

5-18 Line Receivers

A digital line receiver is an input port to a digital system designed to convert a noisy, undervoltage, or poor-quality pulse into a standard-level digital pulse. Pulse degradation can occur when signals are sent through long lines between digital systems. Line receivers can be thought of as a Schmitt trigger (Fig. 5-37) with an input low-pass filter (and sometimes an amplifier). The filter reduces high-frequency noise, which may be picked up on the line, but at the expense of response time. Commonly the pulse rate along an external line is comparatively slow to allow for noise filtering. External filter capacitors (and sometimes resistors) are usually chosen so that the time constants are a factor of 3 to 10 times smaller than the pulse period.

Line receivers to match and interface between various logic levels are available. A few devices have enable controls to turn off the input. By the addition of a low-pass filter, the two-inverter Schmitt trigger (Fig. 5-37b) is a satisfactory line receiver for less demanding applications.

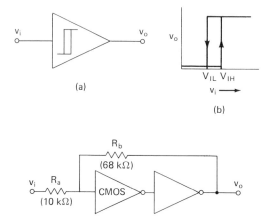

(a)

(b)

(c)

Fig. 5-36   Schmitt triggers: (a) Schmitt trigger driver symbol; (b) input-output transfer function; (c) driver (or two inverters) connected as a Schmitt trigger.

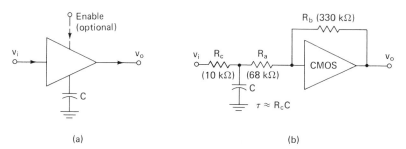

(a)

(b)

Fig. 5-37   Digital line receivers: (a) integrated unit; (b) CMOS driver type.

For very noisy environments, the signal voltage along the line may be higher than normal logic levels (perhaps 40 V) although special high-level transmitters and receivers are required. Transmission of a signal A and its complement $\bar{A}$ along a balanced line (neither grounded) is sometimes required. In this case a differential receiver or optical isolator between the line and receiver is suggested.

5-19   Digital Comparators

A digital comparator is an interface device between analog and digital circuits. The input section is similar to an op-amp while the output section is similar to a TTL driver (usually open collector) and therefore can be connected directly to TTL or CMOS logic devices (Fig. 5-38). Speed

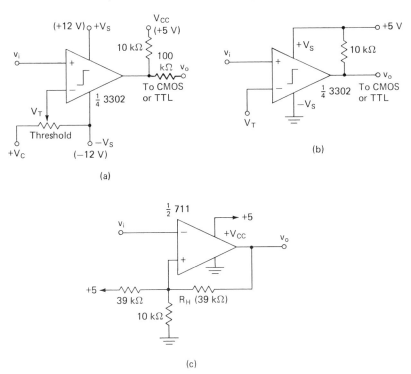

Fig. 5-38   Digital comparators: (*a*) example of use with triple supply; (*b*) single supply grounded, for which the threshold must be positive; (*c*) comparator with hysteresis.

(output rise time or slew rate) is much faster than analog operational amplifiers connected as comparators (Fig. 4-10). Additionally power-supply or voltage-level separation between the input and output is generally possible.

Comparators may require one, two, or three power supplies, depending on the type of application. In Fig. 5-38*a*, the supplies $+V_S$ of the input section are connected to the analog dual power supply (for example, ± 12 V) while the output supply is connected to the digital supply and ground (for example, $+5$ V). The threshold at which the output changes state can be set to a negative, zero, or positive analog input voltage, within the span of the dual analog supply. The output transistor collector is connected through a load resistor R_L to the digital supply $+V_S$ so that the logic level matches the digital devices to be driven. The $-V_S$ input may be grounded, but in this case the threshold must be positive. This type of circuit is useful in level translation, such as the conversion of TTL logic levels ($0 \approx 0.3$ V, $1 \approx 3$ V threshold at $+1.5$ V) to higher voltages ($0 = 0$ and $1 = +15$ V).

Some comparators can be operated from single supplies (Fig. 5-38b and c). In this case the analog input voltage is limited to positive voltages less than the supply voltage. Actually, two supplies can be used if voltage-level translation is desired since the output is open-collector. This type of comparator is convenient for battery-operated applications or level translation.

5-20   Negative-Logic Notation

Standard positive-logic notation, as used throughout this book, refers to 0 V (approximately) as the low logic level and also as logic 0, or **0**. Correspondingly the more positive voltage, for example, $+3$ to $+5$ V, is termed high and logic 1, or **1**. From the boolean algebra or mathematical logic point of view, the assignment of **0** or **1** to one of the two stable states of a particular logic device is, of course, arbitrary. Actually, two conventions are embedded in this notation; the assignment of high and low states to the device voltage states and the assignment of **1** or **0** to the high state. Nearly always the *more* positive state is termed high even if (as for ECL devices) both logic states may involve negative voltage levels. However, it is not uncommon to designate **0** as the high state, a convention termed *negative logic*. Designers often find that mixing negative and positive logic throughout a system reduces the number of gates required or is otherwise more compatible with available hardware. To handle the notational problem, two methods are commonly employed: (1) positive logic is assumed for a signal, for example, A, and the bar added, for example, $\bar{A}$, whenever the signal is inverted or (2) alternate symbols are employed for inverting logic devices, including NAND and NOR gates.

The relation of the inverting symbol (small circle) to negative logic is indicated in Fig. 5-39. Here A represents a positive logic signal ($1 = $ high for A) while $\bar{A}$ is the inverted signal and a is the same signal represented by negative logic ($1 = $ low for a). In all three cases in the example shown, the effective signal inside the device is low when A is high. In Fig. 5-39b this occurs because the applied signal is inverted by an inverter (circle) present on the input. Figure 5-39c shows the same device as Fig. 5-39a, but a negative-logic symbol a is used. Confusion sometimes occurs in relating the truth table to device operation if the truth table lists the signal as A (positive logic) when actually $\bar{A}$ must be applied to the input, as in the example shown.

To see the point in the alternative logic symbols for the NOR and NAND gates of Fig. 5-40, note that a standard positive-logic NOR gate used with negative logic acts as a NAND gate (see truth table). Likewise a standard

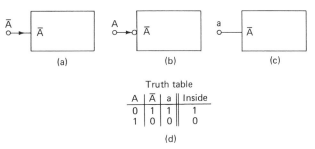

Truth table

A	$\bar{A}$	a	Inside
0	1	1	1
1	0	0	0

(d)

Fig. 5-39 Positive- and negative-logic input symbols.

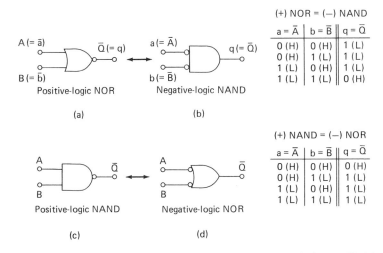

Fig. 5-40 Negative-logic gate symbols: (a) NOR as a negative-logic AND (b); (c) NAND as negative-logic OR (d). Truth table shows interconversion.

positive-logic NAND gate functions as a negative-logic NOR. These boolean algebra relations are a consequence of De Morgan's theorem. Thus by redrawing the device logic symbols, the logical function in the circuit is emphasized. Negative-logic gate symbols may not be found in specification sheets, and the equivalent positive-logic device must be selected.

BASIC CIRCUIT-DESIGN EXAMPLES

Example 1   Inverting Amplifier (High-Z Input)

DESCRIPTION: The circuit provides voltage amplification from dc to 50 kHz.

SPECIFICATIONS

Gain: -10 and -100 (switch selectable)
Frequency response: dc to 50 kHz (minimum)
Input impedance: 10 MΩ (minimum)
Output impedance: 10 Ω (maximum)
Output voltage and current: ±10 V at 10 mA up to 5 kHz (minimum)
Zero offset voltage at output: 0.1 V (maximum)

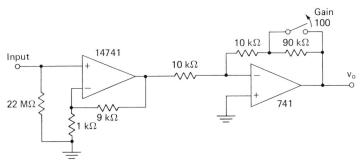

Fig. E-1   A two-stage inverting amplifier.

DESIGN CONSIDERATIONS: The output voltage, current, and impedance are within the capabilities of the standard 741 op-amp (with a ±15-V supply). However, an input impedance of 10 MΩ is difficult to achieve with a BJT op-amp like the 741, primarily because of the bias current, thus indicating that an FET op-amp is needed on the input.

Two stages are desirable because the gain of 100 at a bandwidth of 50 kHz (gain-bandwidth product of 5 kHz) can be obtained with a single stage only with a premium-quality amplifier (318) while two standard amplifiers (741) in cascade can accomplish the same at much lower cost. Another reason for two stages is that the input impedance of the required inverting amplifier is inherently low, suggesting that a noninverting input stage which does have a high impedance would be needed in any case.

At a gain of 100 the input offset voltage must be within ±1 mV in order to achieve an output offset of ±0.1 V. A trimming potentiometer is needed with standard op-amps to achieve this offset. If a gain-bandwidth product of 4 MHz is assumed for both amplifiers, the frequency response is dc to about 200 kHz (at a gain of 100).

Example 2   Integrator with External Zero and Hold Control

DESCRIPTION: The circuit produces a voltage output proportional to the integral of an input voltage over a time interval determined by an external pulse. Output is held until zeroed initially (push button).

SPECIFICATIONS
Input and output voltage range: ±10 V
Input impedance: 100 kΩ (minimum)
Run-hold control amplitude: +2 V (minimum) to run (integrate), below +1 V
to hold
Run pulse time: 1 s at full input amplitude (longer for smaller signals)

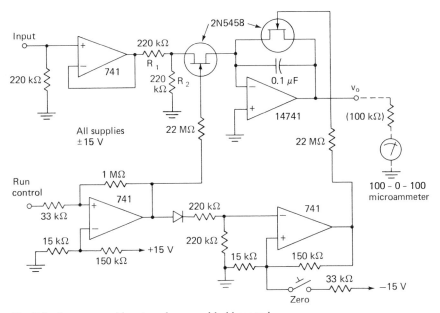

Fig. E-2   Integrator with external zero and hold control.

DESIGN CONSIDERATIONS: A standard integrator (which inverts) is used, to which is
added FET switches to zero the integrator and to switch off the input except when
integration is desired. When the input is off but not zeroed, the integrator is on
hold; i.e., the output remains constant at the level it had at the end of the
integration period (although some drift may occur). An SR flip-flop (op-amp style)
keeps the integrator zeroed to prevent drift, once the zero switch is pushed, until
the run pulse occurs. Digital flip-flops and comparators could be used instead, but
the voltage levels might be insufficient to switch the FETs.

Signal voltage level at the FET is limited to about ±7 V (half op-amp voltage
saturation) by the resistor divider (R_1, R_2), so that the gate-to-source voltage is
sufficient to turn the FET off or on under all conditions. The effective value of the
integrator time constant is RC, where $R = R_1 \| R_2 = R_1/2$. As with all integrators,
the output will go to saturation if the input signal is of one polarity for too long.

Example 3   Decade Counter and Display

DESCRIPTION: The decade counter counts input pulses from an external source.
Output is a seven-segment light-emitting diode (LED) display. TTL and CMOS
logic versions are given.

SPECIFICATIONS

Input voltage: CMOS-compatible (**0** ≈ 0 V, **1** ≈ 5 V)

TTL-compatible [**0** ≈ 0.3 V, **1** ≈ 3.6 V sink 1.6 mA (0.4 mA, LS type)]

Digits: Two decimal digits shown (expandable to three or more)

DESIGN CONFIGURATIONS: Interfacing between sections is simplified by selection of one series either CMOS or TTL (LS or standard). To clear the counter, a momentary **1** level is applied to the clear inputs.

A standard seven-segment common-anode LED driver and display are used (see Chap. 7). Segment resistors limit the current to about 20 mA per segment (140 mA display total).

The CMOS version (Fig. E-3a) uses a half of a dual decade counter (4518). As with any CMOS device, all unused inputs (not output) should be grounded (or $+V_S$) to avoid excessive power dissipation and random switching of open inputs. A supply between about +4 and +15 V is acceptable, but the LED segment resistors were chosen for a +5-V supply (10 mA per segment). Supply current is largely that delivered to the display.

The TTL counter (Fig. E-3b) requires a jumper wire from the first-stage output Q_A to the second input B. Note that the 7490 has two set 9 inputs labeled "S_9" as well as two reset **0**, or clear, lines. A **1** to either reset overrides the counter-pulse input. If reset inputs are left open (ungrounded), the counter will not operate since with any TTL device an open input is equivalent to **1**. Care must be taken to keep the supply between +4.5 and +5.5 V (at 250 mA) at all times. To avoid supply transients and possible false counts, mount the supply bypass capacitor near the counter.

Pulse counters are discussed further in Chap. 17.

BASIC CIRCUIT-DESIGN PROBLEMS

1 Design a noninverting amplifier with a gain of 500 and a frequency response of dc to 2 kHz. What changes would be required if the frequency response were dc to 50 kHz?

2 Design a peak reader to respond to an input pulse with a voltage v_i range of 0.1 to 1.0 V. The input impedance must be 100 kΩ or over, and the output should drive a standard voltmeter (10 V full scale, basic movement of 1 mA).

3 Suggest a circuit for a threshold detector which will turn on a lamp if an input voltage exceeds a fixed level (1.0 V). The 6-V lamp draws 100 mA. Note that the lamp exceeds standard op-amp capacities but not that of discrete transistors.

4 Draw a circuit which will add two signals (v_A and v_B) so that the output v_o is $v_A + v_B$. Next draw a circuit which will subtract two signals ($v_o = v_A - v_B$).

5 Draw a circuit diagram for a pulse generator (astable multivibrator) with a 1-ms-wide pulse and repetition rate of 100 Hz.

6 Suggest a circuit for an amplifier with a gain of 5 and bias voltage adjustable from at least +6 to −6 V.

7 Design an amplifier that will automatically change gain from 100 to 10 if the input signal rises above +0.1 V. Include hysteresis so that the gain will not

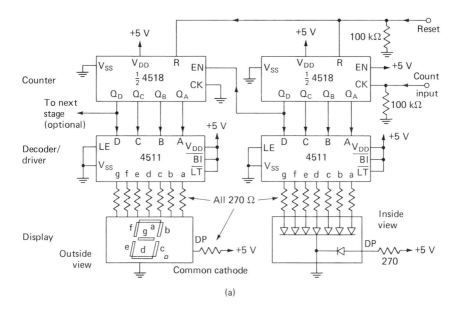

(a)

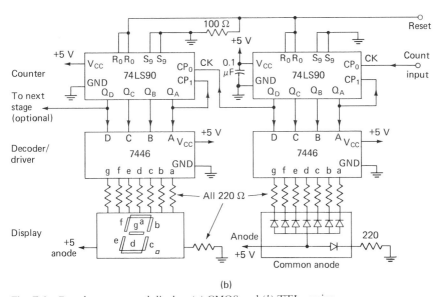

(b)

Fig. E-3 Decade counter and display: (*a*) CMOS and (*b*) TTL version.

change back (from 10 to 100) until the signal level drops below $+0.02$ V. Indicate the range by a lamp (LED). *Hint:* Gain resistors can be switched in or out by an FET or a transmission gate.

8 Design a milliammeter (ac or dc) adapter for an oscilloscope which has an input voltage drop near zero. It must deliver a voltage proportional to the input current. Neither input terminal can be grounded, but the terminal voltages are within 10 V of ground potential. An input range of 0 to ± 1 mA is desired with a sensitivity of 1 V (output) per milliampere (input). *Hint:* Couple a current-to-voltage converter, v_+ input ungrounded, with a differential input amplifier.

9 Design a peak-to-peak reader that will produce an output v_o equal to the difference between the positive and negative peaks of an input signal. The output voltage v_o must be referenced to ground (as the input voltage is). Peak voltage should be held until reset by push button.

10 Design a logic device which will turn on an indicator lamp if an input signal A lies between $+1.0$ and $+1.1$ V provided control signal B is negative and signal C is positive. *Hint:* Use comparators and logic gates.

11 Design a frequency divider with outputs of $f/10$, $f/20$, and $f/100$, where f is the input frequency.

12 Design a frequency divider with outputs of $f/7$ and $f/19$, where f is the input frequency.

13 Two counter outputs (4-bit) are to be compared by an "equal" circuit which is to have an output Q_E of **1** only if the corresponding bits of each counter are equal. Devise a circuit using XOR gates.

14 Design a Schmitt trigger with a high input impedance (>100 kΩ) and TTL-compatible output which produces an output Q_o of **1** if the input voltage v_A is more negative than -6 V and an output of **0** if the input is more positive than -2 V. A response time under 1 μs is desired. This circuit serves as an interface between negative voltage logic and TTL.

15 Design a pulse synchronizer which will change output Q_o state (**0** to **1**) on the negative-going edge of an input clock pulse (CK) *if* an enable pulse E is present (E must be at **1** on the falling edge of the clock pulse). The output, once triggered, should remain at **1** until E returns to **0** (more precisely, $Q_o = \mathbf{0}$ occurs when $E = \mathbf{0}$ and CK $= \mathbf{0}$).

Part Two

Transducers

Chapter Six

Temperature Sensors

Nearly every electrical property of a material or device varies as a function of temperature and could in principle be employed as a temperature sensor. The requirements of operation over a wide temperature range with high sensitivity, reproducibility, and linearity greatly limit the possibilities, especially if cost, size, and ease of readout are also considered. No sensor available has all these desirable properties but may have the proper combination for a particular application.

6-1 Thermistors

One of the most popular thermosensors is the thermistor, basically a resistor with a high temperature coefficient. A thermistor is a semiconductor in various geometrical configurations to which two leads are attached (Fig. 6-1). One of the advantages of the thermistor is the wide variety of shapes and sizes (down to microscopic) possible. Often a coating is applied for electrical insulation and mechanical protection.

Like other semiconductors, the thermistor has a temperature coefficient α which is negative (resistance decreases with increasing temperature) and quite high but nonlinear in resistance-temperature variation.

Over most of its range the resistance decreases exponentially with temperature (Fig. 6-2)

$$R = R_A e^{B/T} \qquad (6\text{-}1)$$

where T is absolute temperature in kelvins and R_A is a constant for the

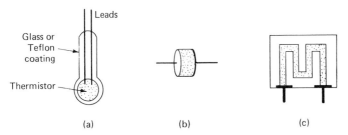

Fig. 6-1 Various forms of thermistor: (a) coated-bead, (b) disk, and (c) film.

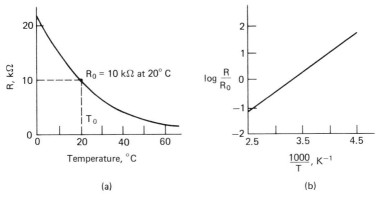

Fig. 6-2 Thermistor resistance vs. temperature (typical): (a) linear scale and (b) logarithmic plot (temperature in kelvins).

device which can be calculated from R_0, T_0, and β. More conveniently the temperature dependence is expressed in terms of a temperature coefficient α, defined as

$$\alpha = \frac{\Delta R / R_T}{\Delta T} \qquad (6\text{-}2)$$

Thermistors with a wide range of resistances are available, but for most commercial units the temperature coefficients are in the range of 3 to 5 percent per degree Celsius. A conversion chart or graph like Fig. 6-2 is generally supplied by the manufacturer. A unit is specified by its resistance at a particular temperature T_0 (e.g., $R_0 = 10$ kΩ at 25°C) as well as its temperature coefficient. For precision units this value R_0 is closely controlled (say ±1 percent) but for standard units only the temperature coefficient is controlled while R_0 may vary by 10 percent or more from unit to unit, thus requiring an individual calibration at one temperature. It is fair to observe that thermistors are not really precision sensors because the resistance-temperature characteristics are not highly reproducible. Also their nonlinearity is a nuisance. Their excellent sensitivity, ease of readout, and small size outweigh these disadvantages for many applications.

Two additional parameters can be important, the time constant τ_T and the internal temperature rise due to power dissipation η. Internal heating is caused by the voltage applied to the thermistor from the readout or converter unit. Both depend on the device thermal insulation and degree of external heat dissipation; they may be 3 to 10 times higher for a thermistor in still air compared with a stirred liquid. Typical values in a liquid are $\tau_T = 0.5$ s and $\eta = 0.1°C/mW$.

Two simple circuits (Fig. 6-3) are considered first for the purpose of

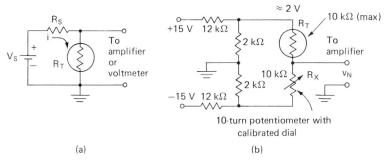

(a)   (b)

Fig. 6-3 Simple thermistor resistance-to-voltage circuit: (*a*) voltage divider; (*b*) bridge.

analyzing resistance-to-voltage conversion and sensor sensitivity. These circuits are practical where accuracy is unimportant, but for more exacting requirements the bridge amplifier discussed later (Fig. 6-6) is preferred. In the circuit of Fig. 6-3*a*, the current through the thermistor i is approximately constant if R_S is made sufficiently large at the highest thermistor resistance (lowest temperature). In this case the voltage across the thermistor V_T is

$$V_T = \frac{V_S R_T}{R_T + R_S} \approx \frac{V_S}{R_S} R_T \tag{6-3}$$

Note that the resistance R_T–to–voltage V_T conversion sensitivity is determined by V_S and R_S. High sensitivity is achieved by increasing V_S, but care must be taken to limit the internal heating effect due to the power dissipation in the thermistor ($P_D = V_T^2/R_T$). Typically a temperature rise of

0.2°C can be tolerated; if we assume $\eta = 0.1$ mW/°C and $R_T = 5$ kΩ, the maximum value of P_D is 2 mW or $V_T \approx 3.3$ V (maximum). It is a good practice to make the sensitivity factor V_S/R_S an even power of 10; for example, if $V_S = 15$ V and $R_S = 150$ kΩ (better, $R_S + R_T \approx 150$ kΩ), then $V_T = 0.10R_T$ and $V_{T,\max}$ is 1 V (which does not exceed $P_D = 2$ mW). In this example, an output voltage of 0.50 V would correspond to $R_T = 5.0$ kΩ, and the temperature would be found from a table. Alternatively a milliammeter (after unity voltage-gain amplification) might be used to read the voltage with the scale recalibrated (nonlinearity) to read temperature directly.

The circuit of Fig. 6-3b, a variation of a bridge circuit, can be operated in either of two modes, a null mode or a small-deviation (limited-temperature-excursion) mode. With the null mode, R_X is adjusted until the output voltage v_N is zero. An analysis of the circuit shows that this will occur when the voltage drops across R_T and R_X are equal; that is, $R_T = R_X$. Since R_X is a precision potentiometer with a calibrated dial, the value of R_T is read off the dial and the temperature found by referring to a calibration table. Although fairly high sensitivity and accuracy at low cost are possible, the indirect readout is tiresome and not very popular.

With the second mode of operation (Fig. 6-3b) R_X is nulled at the lowest temperature of a limited range. For example, if the thermistor were a clinical thermometer with a range of 95 to 105°F, the output voltage v_N would be set to zero at 95°F. At higher temperatures ($R_T < R_X$) the output voltage would be positive; more specifically, circuit analysis shows the following relation for small changes in $R_T = R_X + \Delta R$

$$v_N \approx V_b \frac{\Delta R}{R_T} = V_b \alpha \, \Delta T \tag{6-4}$$

where $R_X \approx R_T$ and $V_b = V_T$ when $v_N = 0$. Suppose α is 2 percent per degree Fahrenheit; then an appropriate choice for V_b would be 5 V because the temperature-to-voltage conversion ratio is an even power of 10, specifically $v_N = 0.10\Delta T$. As above, the internal power dissipation should be limited by choosing R_T sufficiently high (or V_T sufficiently low). To complete the instrument, v_N should be amplified (by say a factor of 10) and the output voltage measured by a milliammeter connected as voltmeter or by a digital voltmeter. The advantage of this circuit is that it can measure small changes in temperature accurately, but it has the disadvantage of a limited range without excessive nonlinearity (for the circuit shown the deviation from linearity is about ±6 percent for $\Delta T = \pm 5$°F).

The nonlinear (exponential) response of thermistors can be compensated by several methods. One method is to employ a nonlinear amplifier with a response characteristic which is the inverse of the thermistor response. The logarithmic amplifier discussed in Chap. 4 will serve, although matching characteristics can be tedious.

A simple and surprisingly effective method of thermistor linearization is to add a fixed resistor in parallel equal to the thermistor resistance R_M at the middle of the temperature range. The circuit (Fig. 6-4) acts as a

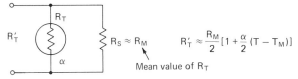

$$R_S \approx R_M \qquad R_T' \approx \frac{R_M}{2}[1 + \frac{\alpha}{2}(T - T_M)]$$

Mean value of R_T

Fig. 6-4   Thermistor linearization by a shunt resistor.

thermistor R_T with effectively half the resistance and temperature coefficient of the actual thermistor but with a substantial linear range (typically a span of 20 to 50°C). The compensation technique is based on the observation that the net resistance of a variable resistor in parallel with a fixed resistor R_S changes nonlinearity with the resistance of the variable resistor. By proper choice of a shunt resistance this nonlinearity just compensates the thermistor nonlinearity over a certain range. This can be shown mathematically by expanding the thermistor response R_T as a power series in $\Delta T = T - T_M$ and retaining terms in $(\Delta T)^2$.

6-2   Other Resistance Thermometers

Various metals make suitable resistance thermometers, but platinum and nickel are the most popular because of their wide temperature range, linearity, and resistance to oxidation and chemical attack. Thin platinum, nickel, or tungsten wire is wound without strain around a mica, ceramic, or other heat-resistant form, as illustrated in Fig. 6-5. Typical room-temperature resistances would be about 100 Ω. At any temperature the resistance is given by

$$R = R_0(1 + \alpha T) \qquad (6\text{-}5)$$

where R_0 = resistance at 0°C
$\quad T$ = temperature, °C
$\quad \alpha$ = temperature coefficient

[see also Eq. (6-2)]. The value of α for platinum is 3.92 mΩ/Ω · °C and for nickel is 6.8 mΩ/Ω · °C. Actually α itself is slightly temperature-dependent, and a polynomial correction factor must be applied for accurate results. Data on resistance thermometers are shown in Fig. 6-6. Platinum thermometers can be used from −250 to 1100°C and nickel from −100 to 300°C.

At low temperatures, below 50 K, germanium resistance thermometers (thermistors) or carbon resistors are good thermometers. Like other thermistors, they exhibit an exponentially increasing resistance as the temperature decreases toward absolute zero.

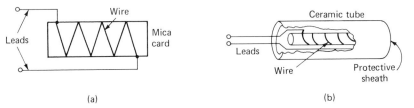

(a)   (b)

Fig. 6-5   Resistance wire thermometers: (*a*) open-wire element; (*b*) protected element (probe).

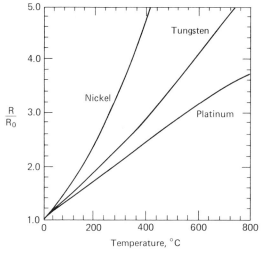

Fig. 6-6   Resistance-thermometer data.

6-3   Bridge Amplifier for Resistance Sensors

The bridge amplifier (Fig. 6-7) discussed in Chap. 4 is well suited as a readout device for a resistance thermometer. Bridge voltage is derived from the standard regulated supply through a voltage divider. The resistance ratio is chosen such that the sensor power dissipation is not excessive. The divider output (Thevenin) resistance must be low enough (R_B low) to ensure that the bridge voltage will not vary appreciably as R_T and/or R_x change.

From the general expression for the bridge sensitivity [Eq. (4-16)] and the thermistor temperature coefficient [Eq. (6-2)] the circuit sensitivity is

$$v_o = S_T \, \Delta T \qquad (6\text{-}6)$$

where

$$S_T = \frac{AV_b\alpha}{4} \qquad (6\text{-}7)$$

and where A is the amplifier voltage gain.

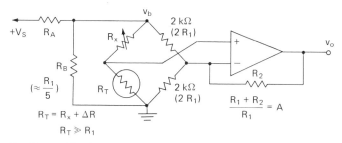

Fig. 6-7 Bridge amplifier for resistance thermometer.

6-4 Thermocouples

A thermocouple is made from two dissimilar conductors, usually metal wires. When the junction is heated, a small thermoelectric voltage is produced which increases approximately linearly with junction temperature. Actually two junctions are always present, a measurement junction and a reference junction, as indicated in Fig. 6-8. At any junction between

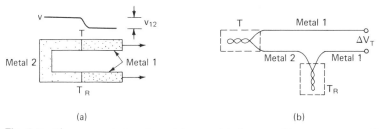

(a)   (b)

Fig. 6-8 Thermocouple junctions: (*a*) potential diagram; (*b*) measurement-reference arrangement.

two metals a potential difference v_{12} exists as a consequence of differences in the effective concentration of electrons in the two metals. The effect is similar to that which produces the internal bias voltage across diode junctions but much smaller. Clearly the two junctions of Fig. 6-8 are connected into the circuit with opposite polarity. When the two junctions are at the same temperature, the two junction potentials cancel ($\Delta V_T = v_{12} - v_{21} = 0$), but otherwise ΔV_T may be positive or negative, depending on whether T is higher or lower than T_R. In physics laboratories an ice bath

(0°C) is frequently used as a reference. An attempt to eliminate the reference junction simply shifts the junction to the terminals of the instrument.

In commercial instruments the bother of providing a constant-temperature bath for the reference junction is circumvented by making the instrument terminals the junction. A thermistor monitors the terminal temperature and provides a temperature-dependent voltage which just compensates for the junction voltage (Fig. 6-9). It is necessary to choose

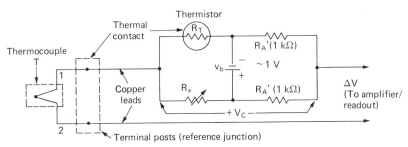

Fig. 6-9 Thermocouple-junction compensating bridge.

the thermistor temperature coefficient and/or battery voltage V_b such that V_C matches the thermocouple temperature coefficient in millivolts per degree Celsius. The zeroing resistor R_x is set so that the output voltage Δv is zero when the thermocouple is at 0°C.

Conversion tables in condensed form for several pairs of thermocouple wires are given in Table 6-1. Examination of these data will reveal that the

TABLE 6-1 **Conversion Factors for Copper-Constantan, Platinum-Platinum plus 10% Rhodium, and Chromel-Alumel Thermocouples**
Potential in millivolts with reference at 0°C

T, °C	Cu–Con	Pt–Pt–10Rh	Cr–Al	T, °C	Cu–Con	Pt–Pt–10Rh	Cr–Al
−60	−2.14	.....	−2.20	60	2.47	0.364	2.43
−40	−1.46	.....	−1.50	80	3.36	0.500	3.26
−20	−0.75	.....	−0.77	100	4.28	0.643	4.10
0.00	0.00	0.00	0.00	120	5.23	0.792	4.92
20	0.78	0.113	0.80	140	6.20	0.946	5.73
40	1.61	0.235	1.61				

temperature-voltage relation is not especially linear but at least is better than thermistors.

Advantages of thermocouples are small size, rapid response, wide temperature range, and simplicity. Disadvantages are low output voltage, need for a reference junction, and nonlinearity.

6-5 Additional Thermosensors

Piezoelectric crystals resonate at a frequency which is temperature-dependent. When they are made a part of an oscillator circuit (Chap. 12), the shift in frequency is approximately proportional to temperature. Often quartz crystals especially cut to *maximize* the temperature dependence are chosen because the frequency shift with temperature is quite linear and reproducible (the temperature dependence is minimized in conventional cuts). A further advantage is the direct digital readout possible. When the circuit is properly designed, the frequency shift (or at least the number of cycles counted) is numerically equal to the temperature in degrees Celsius. The shift with the frequency is rather small (1 kHz/°C at 10 MHz) and therefore a costly stable reference oscillator and frequency counter must be employed. Aside from cost, the crystal sensor has the disadvantage of a relatively large size and long response time.

Temperature-dependent capacitors are sometimes employed as thermosensors. Like the piezoelectric crystals, the capacitor is made a part of an oscillator-resonance circuit, so that the frequency shift is a function of temperature. A Colpitts oscillator (Chap. 12) and ceramic capacitor work well for applications where simplicity is desired but high accuracy is not required. For example, a miniaturized version has been strapped to a bird so that the oscillator frequency, monitored by a radio receiver, indicated its temperature in flight. Available capacitors seem either to have a very small temperature coefficient or are not highly reproducible and therefore are not satisfactory as thermosensors for exacting applications.

Ordinary diodes can be employed as temperature sensors in less exacting applications. When they are operated at constant current I_B, the forward-biased diode voltage varies exponentially with temperature (see Chap. 2). A temperature-measurement circuit is shown in Fig. 6-10. Like the thermistor, it has the disadvantage of an output which is a nonlinear function of temperature except over a short range. Temperature characteristics for ordinary diodes are not highly reproducible and therefore not suitable as sensors, where much accuracy is required. For semiquantitative applications, e.g., detecting the overheating of a heat sink or other component, it would serve well. As an overtemperature detector a switching circuit may be preferred, and in this case the inverting-amplifier configuration shown can be converted into the comparator with positive feedback simply by connecting the feedback resistor R_f to the noninverting input instead of the inverting input.

6-6   IC Temperature Transducer

An improved version of the diode thermosensor is the IC transducer (Fig. 6-11). Linearization of the temperature response in the -55 to $125°C$ range is accomplished by a relatively complex circuit within the block

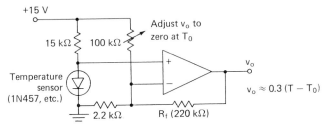

Fig. 6-10   Simple diode thermosensor.

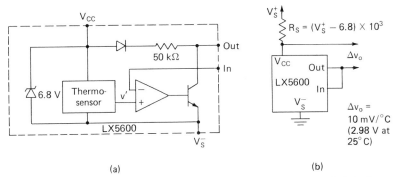

Fig. 6-11   IC temperature transducer: (a) internal block diagram, and (b) basic connection for positive supply.

labeled "thermosensor." The sensor output v' has a sensitivity of 10 mV/°C. When the op-amp is connected as a unity-gain amplifier (Fig. 6-9b), the transducer output v_o is the same as v'. The sensor output v' is 2.98 V more negative than V_{CC} at 25°C and has a voltage range of 2.98 to 3.98 V. External offset is required to set the output to 0 V (at 0°C, for example). Often an emitter follower is added to the output to compensate for the high output impedance (as a current source).

Chapter Seven
Electrooptical Devices

7-1 Optical Spectra and Energy Relations

The choice of optical detectors and sources is strongly influenced by the wavelength of light required. Many detectors, for example, will not respond at all if the wavelength is longer than a certain cutoff wavelength. Except for the insensitive bolometer, all optical detectors involve a quantum effect, in which the photon energy is an important parameter. The energy of the photon E_P is given by

$$E_p = \frac{ch}{\lambda} \tag{7-1}$$

where h = Planck's constant = 6.62×10^{-34} J·s
λ = wavelength of light
c = velocity of light = 3.0×10^8 m/s

It is convenient to express photon energy in electronvolts (eV) and wavelength in nanometers (nm); in this case, the constant ch is equal to 1240 eV/nm. Nearly all electrooptical effects require that a single photon have sufficient energy to free an electron by overcoming the energy of a chemical bond or trap in a solid. Since most bond energies lie in the range of 0.5 to 5 eV, the minimum photon energy is in this range. Actually trapping energies lower than this are common in semiconductors, but the trapped charge carriers are released by thermal energy (0.025 eV at room temperature) unless the detector is cooled and therefore light has little additional effect. In Fig. 7-1 the range of wavelengths, or optical spectrum, is indicated.

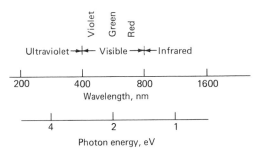

Fig. 7-1 Optical spectrum.

Light intensity referring to the amount of power emitted from a source is commonly expressed in watts. Intensity can also be defined as an energy or radiant flux, expressed in watts per square meter. If, as is usually the case, the intensity varies with wavelength, the power per unit wavelength, or spectral power density, is often specified as a function of wavelength, as in Fig. 7-2.

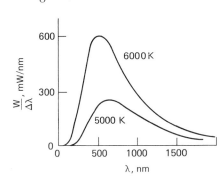

Fig. 7-2 Spectral power density of a blackbody.

Sometimes light intensity is expressed in photometric units, such as candelas (cd) or lumens (lm), which are based on effective brightness as perceived by the eye. Because the sensitivity of the eye depends on the wavelength, no constant ratio between the photometric units and fundamental units such as power exists. At 550 nm (green, where the eye is most sensitive) the ratio (luminous efficiency) is approximately 680 lm/W. A one-candela point source results in a flux of one lumen through a one-square-foot area on a sphere with a radius of one foot. Another way of measuring light intensity is to specify the numbers of photons emitted of incident power per unit time. As indicated by Eq. (7-1), the number of photons per watt depends on the wavelength. At 555 nm, a light beam of 1 W corresponds to 2.8×10^{34} photons per second. Definitions and interconversion of a number of units are indicated in Table 7-1.

TABLE 7-1   Definitions of Optical Units†

Parameter	Symbol	Definition	Units
		Radiometric	
Radiant energy	Q_e		erg, J, cal, kWh
Radiant flux	P	$P = \dfrac{dQ_e}{dt}$	erg/s, W
Radiant emittance‡	W	$W = \dfrac{dP}{dA}$	W/cm², W/m²
Irradiance‡	H	$H = \dfrac{dP}{dA}$	W/cm², W/m²
Radiant intensity	J	$J = \dfrac{dP}{d\omega}$ where ω = solid angle through which flux from point source is radiated	W/sr
Radiance	N	$N = \dfrac{d^2P}{d\omega \, (dA \cos \theta)}$ $= \dfrac{dJ}{dA \cos \theta}$ where θ = angle between line of sight and normal to surface considered	W/sr·cm² W/sr·m²
		Photometric	
Luminous efficacy	K	$K = \dfrac{F}{W}$	lm/W
Luminous efficiency	V	$V = \dfrac{K}{K_{max}}$	
Luminous energy (quantity of light)	Q_v	$Q_v = \displaystyle\int_{380}^{760} K(\lambda) Q_e \lambda \, d\lambda$	lm·h, lm·s (= talbot)
Luminous flux	F	$F = \dfrac{dQ_v}{dt}$	lm

TABLE 7-1   Definitions of Optical Units†   (*Continued*)

Parameter	Symbol	Definition	Units
		Radiometric	
Luminous emittance‡	L	$L = \dfrac{dF}{dA}$	lm/ft²
Illumination (illuminance)‡	E	$E = \dfrac{dF}{dA}$	footcandle (fc = lm/ft²), lux (lx = lm/m²), phot (ph = lm/cm²)
Luminous intensity (candlepower)	I	$I = \dfrac{dF}{d\omega}$	candela (cd = lm/sr)
Luminance (brightness)	B	$B = \dfrac{d^2F}{d\omega\,(dA\cos\theta)}$ $= \dfrac{dI}{dA\cos\theta}$	cd/in², etc., stilb (sb = cd/cm²), nit (nt = cd/m²), footlambert [ft·L = cd/(π ft²)], apostilb [asb = cd/(π m²)], L [= cd/(π cm²)]

†Used by permission of the Illuminating Engineering Society.
Table from the "American National Standard Nomenclature and Definitions for Illuminating Engineering."
‡W and L refer to "emitted from" while H and E refer to "incident on."

7-2   Photodiodes

The principle upon which the photodiode operates is a quantum effect. Consider the reverse-biased *pn* junction sketched in Fig. 7-3 operated as a photodiode. If light is absorbed in a semiconductor, with a certain proba-

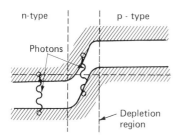

Fig. 7-3 Light interaction in a semiconductor.

bility an electron will be excited into the conduction band. The extra charge carriers, if produced far from the junction, result in a slightly increased conductivity or photoconductivity, which is negligible if the diode is reverse-biased. Minority carriers produced near the junction, in the depletion region, however, have a marked effect because they are drawn by the field through the junction and contribute to the reverse current, which is very small in the dark. The net effect is that the reverse current i_p of the photodiode is proportional to the light intensity W, that is,

$$i_p = KW \tag{7-2}$$

where K is a sensitivity factor. Sometimes the photodiode junction is coincident with the base-collector junction of a transistor, which then becomes a phototransistor. It is equivalent to a photodiode connected to the base of a transistor (Fig. 7-4c) which provides a current gain.

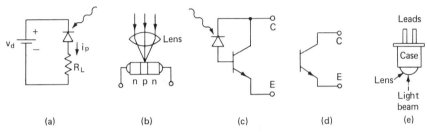

(a)   (b)   (c)   (d)   (e)

Fig. 7-4 Basic photodiodes: (a) single diode; (b) back-to-back diode; (c) phototransistor equivalent; (d) phototransistor symbol; (e) case with lens.

Typical photodiode characteristics are shown in Fig. 7-5. Note that the current is linearly proportional to the incident light intensity over a wide range. It must be emphasized that it is the reverse current, and not the conductivity, of the device that is proportional to light intensity. The sensitivity depends on the wavelength and always exhibits a long-wavelength cutoff which corresponds to the energy needed to excite a carrier across the bandgap (Fig. 7-3). The current at a given light level is nearly independent of diode voltage, and any bias voltage higher than about 1 V will do. It should be pointed out, however, that the diode capacitance C_d decreases, and therefore the frequency response increases, as the reverse bias voltage is increased.

Two circuits suitable for photodiode detectors are shown in Fig. 7-6. In the circuit in Fig. 7-6a the current through the photodiode i_p produces a voltage at the input $v_+ = R_1 i$ which is amplified by the noninverting amplifier. In the circuit in Fig. 7-6b a current-to-voltage converter (see Chap. 4) produces an output $v_o = R_L i_p$ which is proportional to light intensity. An FET-input op-amp is desirable, especially if R_b is high. A zero control which compensates for dark current is useful at low light levels or high gain (R_b large), but if it is not used, the noninverting input should be grounded. For high sensitivity, R_b should be as high as possible although bias current is not a problem for FET op-amps. For good high-frequency response R_I must be kept small since the circuit time constant

$R_I C_d$ limits the response, where C_d is the photodiode capacitance at the existing reverse voltage. The circuit of Fig. 7-6b is much faster than that of Fig. 7-6a.

7-3   Light-emitting Diodes

The principle of the light-emitting diode (LED) is the inverse of the photodiode. When it is forward-biased, charge carriers flow across the

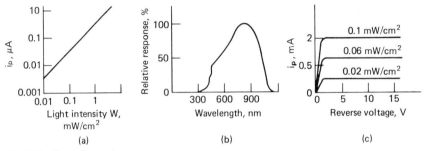

Fig. 7-5   Photodiode characteristics.

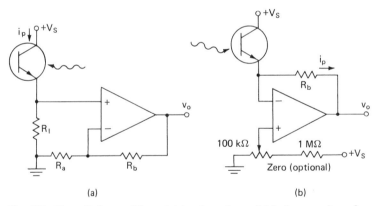

Fig. 7-6   Photodiode amplifiers: (a) low-impedance (high-frequency) configuration; (b) high-sensitivity configuration. Photodiode and phototransistors may be interchanged.

junction, recombine with the majority carriers, and release energy in the form of photons with a photon energy equal to that of the band-gap energy. Semiconductors with higher gap energies result in higher photon energies, i.e., light of shorter wavelengths. For silicon diodes, the wavelength maximum is about 900 nm, which lies in the near infrared. For semiconductors with higher bandgap energies, such as GaAs, the emission is in the visible region. Currently available visible LEDs are red, orange, yellow, and green. Emission spectra are indicated in Fig. 7-7b.

The light output may increase nonlinearly with forward current but is not always noticeable except near overload. Because the efficiency of the LED increases slightly with current, it can be operated in a pulsed mode such that the average power does not exceed the maximum power dissipation of the device. The permissible peak current is usually 5 to 20 times

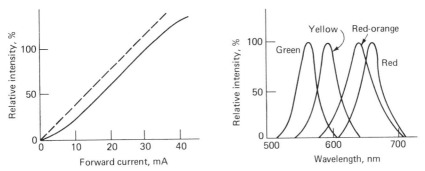

Fig. 7-7 Light-emitting-diode (GaAsP) characteristics: (*a*) light output vs. current and (*b*) emitted spectrum.

the average or direct current, which is typically 10 to 50 mA. The efficiency drops with increasing temperature, including the temperature rise due to internal heating.

Current-voltage characteristics (Fig. 7-8) are similar to those of any

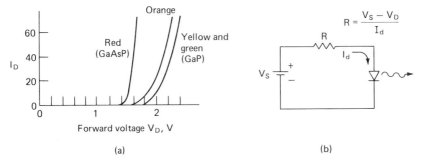

Fig. 7-8 LED electrical characteristics.

forward-biased diode except that appreciable current does not flow until 1.4 to 2.7 V is applied (both conduction threshold voltage and photon energy increase with band-gap energy). In order to limit the current flow to a safe value, a series resistor is ordinarily required. Connection to a fixed voltage source may destroy the diode. The reverse voltage is also rather small, typically 3 to 10 V.

Two common applications are as status or logic-level indicators (Fig. 7-9). Since the current output of most op-amps is rather limited, a transistor driver is desirable unless low light output is tolerable. Further the transistor driver does not apply an excessive damaging reverse voltage to the diode.

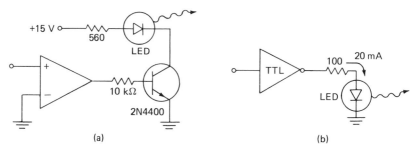

(a)   (b)

Fig. 7-9   LED applications: (*a*) op-amp–transistor–driven; (*b*) TTL-driven.

7-4 Semiconductor Photocells and Related Devices

A semiconductor type of cell is based on photoconductivity. In contrast to the photodiode, which acts as a current source, the photocell increases in conductivity with light intensity (Fig. 7-10). The resistance of a cell may change from 10 MΩ in the dark to under 10 Ω in a bright light.

In the design of photocells there is a trade-off between sensitivity and fast response. In commercial devices the sensitivity has been optimized at

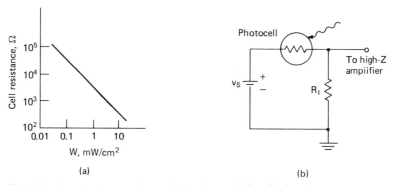

(a)   (b)

Fig. 7-10   Semiconductor photocell: (*a*) characteristics; (*b*) circuit.

the expense of response time. The response time is difficult to define precisely because it varies greatly with light level and the fall time is longer than the rise time. Response at 60 Hz is marginal.

A signal voltage is provided simply by the load resistor R_l of Fig. 7-10*b*. For linear output R_l is selected so that it is one or two orders of magnitude lower than the minimum photocell resistance R_{min} (highest light level) in order to maintain a nearly constant voltage across the photodiode. If linearity is unimportant, as in an application requiring only a discrimination between light and dark, there is a wide latitude in the choice of R_l but an appropriate value might be $10R_{min}$.

Bolometers are identical to photocells in most respects, but the mechanism responsible for the conductivity change depends simply on the temperature rise due to absorption of radiation rather than the direct excitation of charge carriers across an energy gap. Electromagnetic energy absorbed at any wavelength, including microwave or RF, will have the same effect. This broad response is not approached in other optical detectors and is its most important advantage. Sensitivity and response time are rather poor, however.

The vacuum-tube photocell, an obsolete device, is described briefly under photomultipliers.

7-5 Photomultipliers

A photomultiplier is based on the photoelectric effect, i.e., the release of electrons from a metal surface hit by higher-energy photons. Its operating principle is basically similar to that of the vacuum-tube photocell (Fig. 7-11). When a photon strikes a surface, an electron is released if its energy

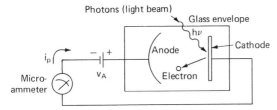

Fig. 7-11 Vacuum-tube photocell schematic.

($E = h\nu$) exceeds that of the work function of the cathode surface. The electrons are collected by a positively charged anode ($v_A \approx 150$ V) resulting in a photocurrent i_p proportional to the number of photons, i.e., to the intensity of light at any particular wavelength. Because the current is small, a sensitive current amplifier is required unless light is very intense. Another disadvantage is that lower-energy photons, specifically the red and infrared end of the spectrum, cannot be detected because the work function of available materials is comparatively high, over 1 eV.

Adding a multistage electron multiplier to a photocell produces a photomultiplier (Fig. 7-12). Photons of sufficient energy striking the cathode release electrons, which are attracted to the neighboring dynode because of its positive potential with respect to cathode. The potential difference is fairly high (50 to 200 V), accelerating the electrons so that as they strike the dynode secondary electrons are emitted. Typically two or three low-energy secondary electrons are emitted for each incident electron, thus producing an electron-multiplying effect. This occurs at each stage or dynode. The result is that the photocurrent emitted by the cathode is multiplied by $A_m = (A_s)^n$, where A_s is the stage current gain

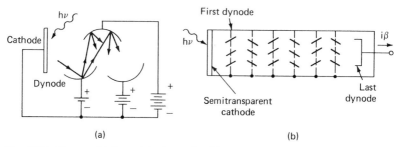

(a)   (b)

Fig. 7-12   Photomultipliers: (a) schematic of first stages; (b) end-on multiplier.

(ratio of secondary to incident electrons) and n is the number of stages, typically 7 to 14. Gains A_m of 10^5 or 10^6 are common and account for the high sensitivity of the photomultipliers.

An end-on photomultiplier has a semitransparent photocathode film on the end window. Light passing through the glass causes an electron to be released from the back side of the film. It is drawn by focusing electrodes (not shown) to the venetian-blind-type dynodes, where secondary electron emission occurs as before.

Voltage is applied to the dynodes and cathode from a negative high-voltage source V_H through a resistor chain (Fig. 7-13). A positive ground is

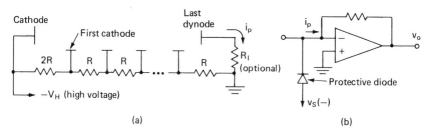

(a)   (b)

Fig. 7-13   Photomultiplier: (a) dynode voltage divider; (b) amplifier.

required because the load resistance R_l and/or amplifier common must be connected to the power-supply positive terminal. The secondary electron emission is voltage-dependent, and the total gain A_m depends on the nth power of stage gain, so that the total gain is strongly dependent on the applied voltage V_H. As in the photodiode, the photomultiplier output is a current source, and thus the input amplifiers described for the photodiode are suitable here. A diode to protect against high-voltage surges (due to exposure to high light intensities) is suggested.

The photomultiplier is the fastest and most sensitive photodetector available. It will respond to single photons with high efficiency. At low light levels, where the single photon pulses do not overlap, pulse-counting techniques can be used to measure light intensity. If the load resistance R_l is low (50 to 1000 Ω), the pulse width is typically 5 to 50 ns. More specifically, $R_l C_p$ is small, where C_p is the last dynode capacitance to ground. The photomultiplier will respond to ultraviolet if the envelope is ultraviolet-transparent (quartz), but the inherent lack of red and infrared response is often a severe practical limitation. The photomultiplier cathode can be damaged by exposure to higher light intensities when the high voltage is on. The fragility, size, and cost of the tube and regulated high-voltage supply are substantial disadvantages, and the photomultiplier is usually used only where the ultimate in speed and/or sensitivity is required.

7-6   Optical Isolators

The most popular electrooptical couplers or isolators consist of an LED and photodiode mounted in the same package. The photodiode is mounted to capture the maximum amount of light from the LED; i.e., the two diodes are optically coupled but electrically isolated. These devices are useful in high-voltage power supplies and as a means of isolating grounds in larger systems. Most isolators have a built-in transistor, which can be operated in either the common-collector mode or the common-emitter (phototransistor) mode. Typical characteristics are shown in Fig. 7-14. Because of the slightly nonlinear LED characteristics, the current gain ($\beta = i_o/i_i$) varies with the current. Current gains of 0.05 to 0.50, frequency response beyond 1 MHz, and insulation of 1 to 3 kV are typical.

A voltage-to-current amplifier (Chap. 4) is a convenient driver for the LED isolators (Fig. 7-15), and the current-to-voltage converter is a convenient receiver. Only positive signals are allowed. Resistor R_b and diode D1 limit the driver amplifier voltage and current swing. Because no feedback exists here between the receiver and driver, the overall voltage gain will vary somewhat with temperature and with the individual isolator current gain. Of course the power supplies for the driver and receiver sections must be separate if ground isolation is required. If batteries are impractical, an isolated dc-to-dc power supply (Chap. 21) can be used.

Remote digital systems are often coupled with isolators, especially in electrically noisy environments, where substantial voltage differences may exist between local "grounds." It must be recognized that simply running a wire between grounds is usually ineffective because either the resistance required is too low for a wire of practical size and/or a voltage is actually being induced along the wire, the ground wire or loop acting as a transformer secondary.

The coupling arrangement of Fig. 7-16 is suggested for digital data transmission with CMOS logic. Sometimes a low-pass filter (time constant

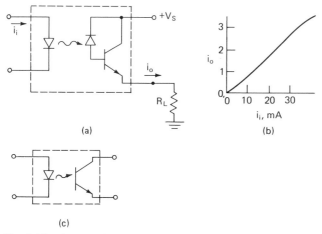

(a)

(b)

(c)

Fig. 7-14 Electrooptical isolator: (*a*) standard phototransistor connection, (*b*) characteristics, and (*c*) alternate symbol.

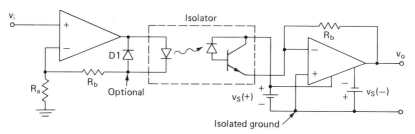

Fig. 7-15 Isolator driver and receiver.

R_LC) is added to reduce noise pickup but at the expense of response time or data transmission rate.

Optical couplers or isolators which combine an incandescent (or neon) bulb and photocell in one package are also available. They are similar in function to the LED photodiode devices discussed above, but their response time is much slower. While in most applications the LED type is superior, for others, e.g., automatic-gain-control circuits, the linear current-voltage characteristic of the photocell (a resistance, not a diode) is important.

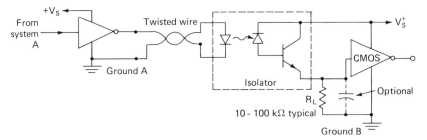

Fig. 7-16   Optical isolators for digital coupling (CMOS).

7-7   Digital Displays

Perhaps the most common digital display is the seven-segment LED display pictured in Fig. 7-17a. Each segment or bar can be individually illuminated. Any digit from 0 to 9 can be displayed by turning on the

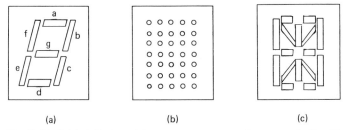

(a)   (b)   (c)

Fig. 7-17   Digital displays: (a) seven-segment array for numerals (0 to 9); (b) 5 × 7 dot array for alphanumeric readout; (c) 16-segment alphanumeric display.

proper combination of segments. One lead is brought out for each segment in a standard display (without a built-in decoder). Most units also contain a decimal point.

In some devices each segment is formed from a row of illuminated dots, and a few of these can also display the alphabetic characters A through F as needed to read out hexadecimal numbers. Special arrays of dots are needed to display the full set of alphanumeric characters available on a typewriter or keypunch (see Sec. 19-5). Because it is impractical to bring out one lead per dot, these units either have a built-in decoder or, more commonly, bring out one lead for each row and column (12 leads for a 5 × 7 array). A dot at the intersection of a specific row and column is illuminated by applying a voltage between the specific row and column leads. To produce a complete character, the row or column is scanned (rapidly to avoid flicker).

Although some displays contain built-in decoders (and perhaps latches and counters as well), most require an external decoder/driver. Displays are divided into two types, common-cathode and the more popular common-anode. Decoders must match the display type. Decoder input is supplied in 4-bit BCD form. The decoder output is seven lines, one for each segment. A transistor driver within the decoder/driver is switched on to turn a segment on. For common-anode displays the anode is connected to the positive power supply (Fig. 7-18*a*). To turn a segment on, the

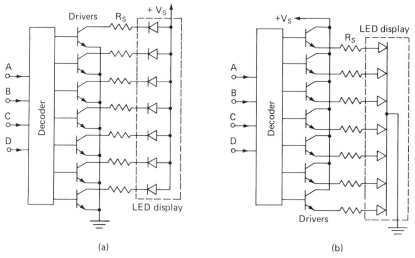

(a)   (b)

Fig. 7-18   Decoder-display wiring: (*a*) common anode and (*b*) common cathode. Decimal point not shown.

cathode is grounded through a current-limiting resistance R_S. The open-collector driver transistor is driven to saturation to switch a segment.

For the common-cathode displays, the display cathode is grounded and the transistor collectors (within the IC) are connected to the positive supply. As with other LEDs, the series resistance is chosen to limit the current to a specified value, typically 20 mA.

A BCD-to-seven-segment decoder/driver and seven resistors are required for each digit with the standard circuit arrangement. Display multiplexing or scanning techniques which require only one segment decoder are discussed in Chap. 17.

Neon or incandescent-lamp seven-segment displays are similar to the LED displays except for (*a*) the current and voltage requirements on the drivers and (*b*) the lamp segments are electrically unpolarized so that either the common-anode or common-cathode drive circuits will work with any display. Neon or gas-discharge displays are bright, large, and low-cost but require a high voltage drive (about 100 V at 1 mA). IC drivers are generally inadequate, and a high-voltage-drive transistor per segment is the common solution (Fig. 7-19). A decoder equivalent to that of the LED common-cathode type is required. Segment current is limited by R_S. While the cost of the display is low, the cost of the high-voltage supply and drivers often makes the total cost higher than that of the LED display.

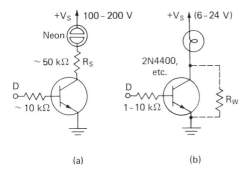

Fig. 7-19 Segment drive circuits for (*a*) neon lamp, (*b*) incandescent lamp. D = decoder output (one segment).

Incandescent displays have roughly the same advantages and disadvantages as the gas-discharge displays except that the special power-supply requirement is a relatively high current, usually at a lower voltage. Here also individual transistor segment drivers are usually required. Sometimes a resistor R_W is added to keep the filament warm, thus reducing the warmup shock. These displays require a fair amount of power but can be

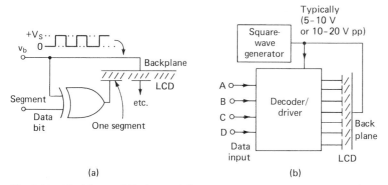

Fig. 7-20 Liquid-crystal displays and drivers: (*a*) driver (one segment); (*b*) driver connection.

made as large as desired so that the display can be read some distance away.

In principle liquid-crystal displays (LCD) can use the same decoder/drivers as the common-cathode LED type, but usually the drive-voltage requirements are often higher and no series resistor is needed. These displays are based on the electric field–dependent (polarized) light absorption of the liquid crystals which make up each segment. Ambient light is reflected or absorbed (white or black segment) depending on the voltage (either polarity) applied between the back plane and individual segment. Both the back plane and segment electrodes are transparent conductors. Because the current drawn is very small, LCDs are desirable for battery-operated devices.

Display life is greatly extended by ac rather than dc drive. The drive circuit of Fig. 7-20a applies a square wave (typically 30 to 100 Hz) between the segments and back plane. The peak value is equal to the amplitude of the supply voltage. The XOR gate reverses the truth table, thus reversing the polarity of the drive signal across the display with the frequency of the square wave. The voltage across an activated segment at any instant is equal to the supply voltage V_S, but since both the back plane and segment voltage reverse simultaneously, the effective peak-to-peak voltage is $2V_S$. This is done to increase the electric field strength as well as to generate an ac voltage in the crystal.

Chapter Eight

Displacement Transducers

Displacement transducers sense the change in position or displacement of an object. A wide array of sensors is based on measurement of displacement. Linear-displacement transducers sense the displacement Δx or movement along a line. Angular-displacement transducers sense rotation about an axis θ such as the turning of a wheel. Proximity or contactless sensors do not require a mechanical connection to the object being measured. Other sensors may use a shaft to transmit the movement from the object to the sensor, or, like strain gages, be fastened or cemented to the object. For these contact sensors, care must be taken to choose the sensor so that the elastic or frictional force needed to drive the sensor is negligible compared with other forces on the body. In other words, the sensor should not disturb the system being measured. In addition the range of the sensor must roughly match the displacement expected since most sensors become nonlinear or limit beyond some maximum or full-scale value and are insensitive if operated far below this value.

8-1 Strain Gages

A strain gage is intended to measure the small displacements or strains associated with the stress on solid objects such as the bending of a beam.

Strain ε is defined as

$$\varepsilon = \frac{\Delta L}{L} \tag{8-1}$$

where L is the length of an object, or more generally the distance between two reference points fixed in the object. Most gages consist of a metal foil or wire bonded to a plastic sheet or other insulating base, as illustrated in Fig. 8-1. The base is cemented to the surface of the object under test, and

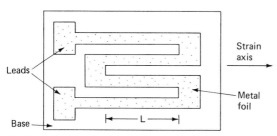

Fig. 8-1 Strain gage diagram.

the gage therefore experiences the same strain as this surface. As the length of the foil increases, its electric resistance R increases proportionally, at least for small displacements, as

$$\frac{\Delta R}{R_0} = G_f \varepsilon \tag{8-2}$$

where ΔR = small resistance increase
R_0 = unstressed resistance
G_f = gage factor

A resistance increase as the metal strip is stretched occurs both because the strip becomes longer and because its cross-sectional area decreases, according to the relation $R = R_0 + \Delta R = \rho(L_0 + \Delta L)/(A_0 + \Delta A)$, where A is the foil-strip cross-sectional area (strip thickness times width) and ρ the resistivity of the conductor. Since $-\Delta A/A_0$ is proportional to strain, this relation reduces to Eq. (8-2) for small displacements with the aid of the binomial expansion. For the ideal case of an incompressible metal $\Delta A/A_0$ = ε, and thus $G_f = 2$. Gages made from semiconductor films usually have much larger gage factors because the resistivity of the semiconductor film ρ itself varies with the strain in the semiconductor.

Two output circuits are shown in Fig. 8-2. Figure 8-2a is a minimal circuit suitable for such applications as the analysis of the vibration frequency of a beam. Here R_s is chosen to set the voltage across the gage and therefore limits the power dissipation of the gage to an acceptable level. Calculation of the break frequency of this high-pass amplifier is discussed in Chap. 11.

The readout circuit of Fig. 8-2b is a bridge amplifier (Chap. 4), which has been shown [Eqs. (6-61) and (4-16)] to have an output voltage of

$$v_o = \frac{V_b}{4}\frac{R_b}{R_a + R_b}G_f\varepsilon = S_g\varepsilon \tag{8-3}$$

For some applications, in particular those involving small dc signals, the temperature coefficient of the gage is important since a resistance change

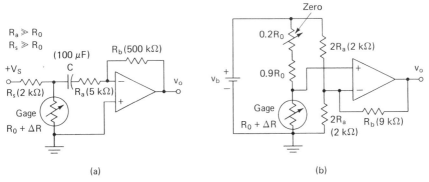

Fig. 8-2 Strain-gage amplifiers; (a) ac type; (b) bridge type.

due to temperature is indistinguishable from that due to strain. For these applications, in addition to choosing a gage with a low temperature coefficient, a second gage may be added, as indicated in Fig. 8-3. The gages should be mounted close to each other so that they are at the same temperature. Electrically they are connected into the two arms of the bridge circuit such that no bridge unbalance or output is produced if the resistance change in the two gages is identical, as it will be if the ambient temperature changes. Two methods of mounting are common. In Fig. 8-3b the second gage is mounted where it will not be strained but close enough to the first or sensing gage so that it is at the same temperature. Sometimes, as in Fig. 8-3c, the second gage can be mounted in a position where the strain is equal in magnitude but opposite the strain in the first gage. In this case not only is temperature compensation provided, but the bridge output voltage is twice that of the single-active-gage circuit.

The high-frequency response is ordinarily limited by the mass of the body to which the gage is attached. For optimum signal-to-noise ratio at dc or low frequency, a more elaborate carrier-amplifier system, in which the bridge excitation is ac instead of dc, may prove superior to the above circuits.

8-2 Electromagnetic Velocity Sensors

A common motion sensor is based on the emf generated by a coil moving in a magnetic field. Two versions of the electromagnetic pickup are shown in Fig. 8-4. The output voltage v_e is proportional to the time derivative of displacement dx/dt, more specifically to the velocity of the coil with respect to the magnet (either may move)

$$v_e = K\frac{dx}{dt} \tag{8-4}$$

where K is a constant for the device proportional to the number of turns

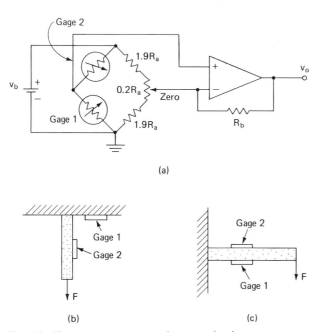

(a)

(b)   (c)

Fig. 8-3 Temperature-compensating gage circuits.

in the coil and the magnetic field strength. Sensors of the first type (Fig. 8-4a) are available commercially, or they can be fabricated without difficulty. For many applications, an ordinary loudspeaker can be operated as a motion pickup transducer.

An output proportional to displacement x can be obtained by integrating the amplified output of the transducer (Fig. 8-5). The low-pass amplifier version of the integrator (Chap. 11) was selected here; the resettable version could also be used (Chap. 4). Because the sensor voltage approaches zero as the frequency approaches zero, this sensor cannot be used for direct current or very low frequencies. With the circuit shown the circuit time constant RC determines the low-frequency limit. High-frequency response is limited by mechanical properties, in particular the mass of the magnet or coil. Mechanical mounts often exhibit a resonance near their high-frequency limit, which is rarely above 10 kHz and may be much lower.

Accelerometers can be made from velocity transducers by simply processing the velocity-sensor signal with a differentiator circuit (Fig. 11-8). The resulting response is

$$v_o = K_A \frac{d^2x}{dt^2} \tag{8-5}$$

where $K_A = -R_bCK$ is a sensitivity factor. Alternatively the force can be

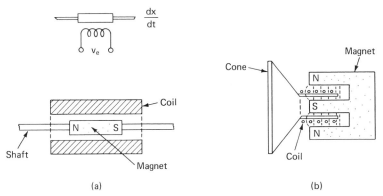

(a)   (b)

Fig. 8-4 Electromagnetic sensors: (a) linear velocity transducer; (b) loudspeaker.

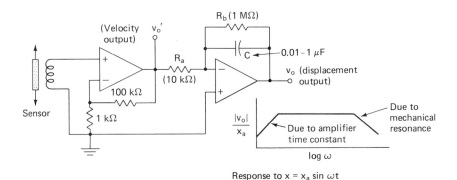

Response to $x = x_a \sin \omega t$

Fig. 8-5 Electromagnetic sensor amplifier with velocity and displacement outputs.

measured and converted into acceleration from Newton's second law ($f = M\, d^2t/dt^2$). A piezoelectric crystal (Fig. 8-15) with masses attached to the faces is a satisfactory transducer since the force on the elastic crystal produces a proportional displacement and thus the output voltage is proportional to acceleration.

Numerous devices based on the electromagnetic sensor exist, including the dynamic microphone and magnetic phonograph pickup, but because their applications are rather specialized, they will not be covered here.

8-3 Inductive Transducers

Inductive displacement transducers utilize the change of inductance of a coil (inductor) in the vicinity of a ferromagnetic material, as indicated in Fig. 8-6. The inductance increases as the material is inserted (Fig. 8-6a) or

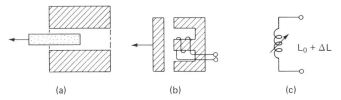

(a)   (b)   (c)

Fig. 8-6 Inductive transducers: (a) contacting type; (b) proximity type; (c) electrical equivalent.

comes closer (Fig. 8-6b). In order to read out the signal, the inductor is made a part of a bridge (Fig. 8-7) driven by an ac generator (perhaps 60 or or 400 Hz). A variable inductor, which is commonly a second (dummy) sensor, is required in one arm of the bridge. Sometimes the second

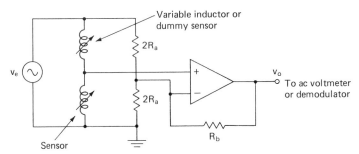

Fig. 8-7 Inductance-sensor bridge amplifier.

inductor is built into the transducer. The output of the amplifier is an ac voltage proportional to displacement which can be read on an ac voltmeter or converted into direct current by an ac-to-dc converter or a phase-sensitive detector

$$v_o = \frac{V_b}{4} \frac{R_b}{R_a} Kx \tag{8-6}$$

where K = sensitivity = relative inductance change, $\Delta L/L_0$ per unit distance

V_b = exciting voltage

x = displacement

A linear variable differential transformer (LVDT), Fig. 8-8, is similar to the inductive transducer in construction, but the principle and electrical characteristics differ. The LVDT has a primary connected to an ac source

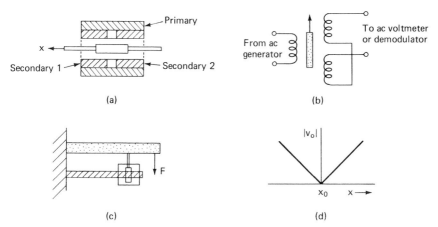

(a)   (b)

(c)   (d)

Fig. 8-8 LVDT displacement transducer: (*a*) cross-sectional drawing; (*b*) electrical equivalent; (*c*) typical mount; (*d*) output voltage.

(typically 60 Hz to 20 kHz) and two secondaries, one on each side of the center, connected so that their output voltages are in opposition. Coupling to the secondaries is determined by the position of the movable magnetic core. When the core is in the center, the coupling is equal, the voltages are equal, and the output voltage is zero. As the core is displaced, coupling to one secondary increases and the output voltage increases. Over a certain range the output is linear with displacement of the core x

$$v_o = v_e K x \qquad (8\text{-}7)$$

where v_e is the excitation voltage and K is the sensitivity.

In practice an exact null, as measured by an rms meter, is not obtained at center because a small out-of-phase (quadrature) component is present due to capacitive feedthrough and nonideal transference properties. Further, an rms meter does not register the change in phase of the output as the core moves from one side of center to the other. To overcome these difficulties and to reduce the noise level as well the use of a phase-sensitive detector (Chap. 15) is desirable. A block diagram of a readout circuit is shown in Fig. 8-9, and a detailed circuit is presented in Example 5 (at the end of Chap. 10).

8-4   Capacitive Transducers

Capacitive displacement transducers are based on the change of capacitance between two electrodes as the separation of the plates changes or the dielectric between them is moved. Three of many versions of this trans-

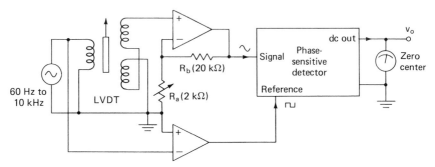

Fig. 8-9   LVDT with phase-sensitive-detector readout.

ducer are suggested in Fig. 8-10. Few capacitive transducers are available commercially as separate units because the electrodes are easily incorporated into the body being displaced or mounted on it. Usually one plate is grounded. Capacitances are small, typically 2 to 100 pF, depending of

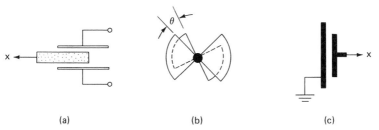

Fig. 8-10   Capacitive transducers: (a) movable dielectric (linear), $C = C_0 + KX$; (b) rotating plate (angular), $C = C_0 + K'\theta$; and (c) movable plate (linear).

course on electrode area and separation. For a parallel-plate capacitor, the capacitance neglecting fringing fields is

$$C = 0.0885 \frac{A}{d} \qquad (8\text{-}8)$$

where C = capacitance, pF
   A = area, cm^2
   d = electrode separation, cm
   K = dielectric constant ($K = 1$ for air; $K \approx 3$ for many plastics)
   Advantages of the capacitive transducers are their high sensitivity, the

fact that contact is not required, and simple versatile (electrode) mounting. Disadvantages are nonlinearity (unless careful attention is paid to the electrode geometry and fringing fields), the fact that capacitance change is small so that connecting cables are a problem, and requirement of a relatively complex readout, also a consequence of the low capacitance.

A bridge amplifier may be used, similar to that shown in Fig. 8-7 but with the inductors replaced by the capacitive transducer and an adjustable capacitor. A high-value resistor, that is, 100 MΩ, from the v_+ input to ground is also required. For most applications an FET input op-amp is required because of the high input impedance of the capacitors. Care must be taken in the bridge wiring to minimize stray capacitance, especially changes in stray capacitance, which are equivalent to a drift or noise at the output. Rigid wiring and shielding as well as mounting of the bridge and op-amp close to the electrodes is desirable.

A second type of readout is based on the frequency shift of an oscillator circuit in which the capacitive transducer is a frequency-determining element. In Fig. 8-11, a Colpitts oscillator (Chap. 12) version is given.

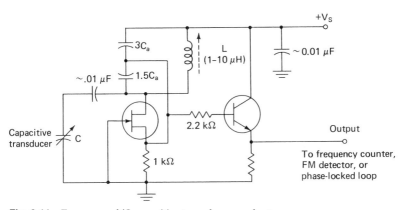

Fig. 8-11   Frequency-shift-capacitive-transducer readout.

Although a wide range of operating frequencies is practical, commonly a center frequency in the region of 1 to 10 MHz is chosen. Calculation of the frequency shift due to changes in the transducer capacitance ΔC is straightforward. When ΔC is small, the following relation is obtained by binomial expansion:

$$\frac{\Delta f}{f_0} = \frac{-\Delta C}{2C_0} \tag{8-9}$$

where $f_0 = 2\pi\sqrt{LC_0}$ and $C_0 = C_a + C'$, C' being the sum of the stray capacitance and the minimum capacitance of the transducer. Note that the frequency range of sensitivity is determined by the choice of C_a and L.

If a frequency meter is used to measure the output, it is convenient to adjust L and C_a so that the frequency shift for a unit displacement is an integral amount. For example, if $f_0 = 5000.0$ kHz, Δf might be chosen to be 100.0 kHz for a 1.00-mm displacement with the nominal full-scale range set at 5999.9 MHz. In this case the first digit (5) would not be displayed and the decimal point on the display would be set so that a frequency of 5100.0 kHz would register as 1.000; that is, the counter would read out directly in millimeters.

8-5   Digital Displacement Transducers

When relatively large displacements are to be determined, a digital encoded transducer is often the best choice, especially if the readout is digital. Two types of transducers will be discussed, incremental and absolute. In Fig. 8-12 the incremental type of pickup is illustrated for the

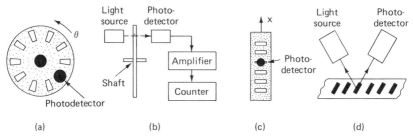

Fig. 8-12   Basic incremental digital displacement transducers: angular-transmission type (a) front view and (b) side view; linear (c) front view; (d) reflection type, side view.

range subdivided into 10 intervals. For many practical applications, subdivision into more intervals, perhaps 100 or 1000, would be required, while for others one pulse per turn (half opaque, half transparent) would do. As the wheel rotates, light pulses converted by the light detector into electric pulses are amplified and counted by a decade counter. The digital reading of the counter is thus equal to the angular displacement of the wheel and/or number of revolutions. Linear transducers can be constructed in a similar fashion. Drawbacks to incremental transducers are the lack of a zero reference and directional sensitivity.

It should be noted that the frequency of the pulse train at the output of the comparator is proportional to velocity. A frequency counter connected to the output in place of the pulse counter will register velocity directly.

Absolute digital displacement transducers, both angular and linear, use several optical pickups to provide an encoded output directly, as illustrated in Fig. 8-13. Both transmission type (opaque-clear) and surface-reflection type (black-white) are popular. Each bit requires a separate photodiode pickup as well as a separate strip on the encoding card. With this arrangement there is no ambiguity with respect to card position or direction of movement. Note that the outputs can be sent directly to a BCD-to-seven-segment decoder and display. The main disadvantage is the multiple pickups required. Encoding in binary or codes other than BCD are common.

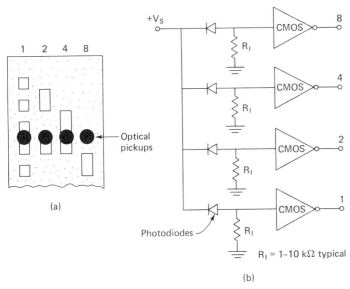

Fig. 8-13 Absolute digital displacement transducers: (*a*) movable (4-bit) BCD-encoding card (linear); (*b*) output circuit.

8-6 Other Displacement Transducers

Potentiometers are convenient displacement transducers in applications where the movement is of large amplitude and slow and friction is not a problem. A 10-turn potentiometer is best because of its good electrical linearity and resolution. A servo type with ball bearings may be required for heavy use. A friction (clutch) connection is usually needed to prevent the potentiometer from being driven beyond its range. Its applications as an angular transducer (limited to 10 revolutions) is obvious and, as Fig. 8-14 indicates, the extension to linear displacement measurements requires only the addition of a simple cable drive. Readout in terms of a voltage proportional to displacement requires that a fixed voltage be applied across the potentiometer. Where low output impedance is required, a unity-gain amplifier can be added.

Resolution of the potentiometer is typically limited to 0.02 to 0.5 percent of full scale, due to its mechanical construction, and is not suited for measuring small displacements. This point and the possible mechanical difficulties (friction, backlash) are often overlooked by those who are focusing on the selection of a displacement transducer with the simplest electrical readout.

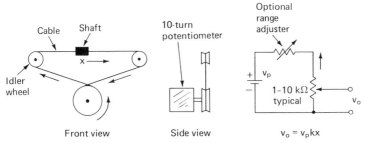

Fig. 8-14   Potentiometer as a linear displacement transducer: (*left*) front view; (*middle*) side view; (*right*) output circuit.

Piezoelectric transducers take advantage of the voltage developed across certain crystals when they are strained. Ionic crystals within which the repeating arrangement of atoms (unit cell) is asymmetric, i.e., falls in a crystallographic space group which lacks a center of symmetry, develop a separation of charge across the crystal faces because the positive and negative ions separate slightly when the crystal is strained. Two common piezoelectric crystals are quartz and barium titanate. The magnitude of the voltage and the direction of strain sensitivity depend on the cut of the crystal with respect to the crystallographic axis, but the response is always of the form

$$v_x = K\varepsilon \qquad (8\text{-}10)$$

where v_x = ac voltage developed across crystal
K = sensitivity
ε = strain

Because the crystal is a nonconductor, however, the generator voltage is coupled to the electrodes at the crystal faces via the crystal capacitance C_x, as indicated in Fig. 8-15. Since any practical amplifier has a finite input resistance R_i, the result is that the voltage v_x must pass through a high-pass network (Chap. 11), where $f_x = 1/2\pi R_i C_x$, and thus the transducer will not respond to direct current. A high-input-impedance FET op-amp is desirable at lower frequencies although at high frequencies lower-impedance RF amplifiers are adequate. Each crystal has one or more sharp mechanical-resonance points, typically in the 20-kHz to 10-MHz range, at which point the output is several orders of magnitude higher than off resonance.

Optical levers consisting of a light source, reflecting surface, and photodetector can be used to measure small displacements. Several arrangements are illustrated in Fig. 8-16. Displacement is detected by the move-

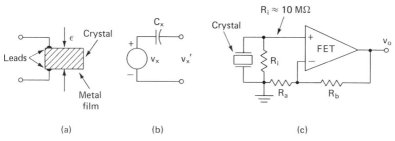

(a)   (b)   (c)

Fig. 8-15   Piezoelectric transducers: (*a*) configuration; (*b*) equivalent circuit; and (*c*) amplifier.

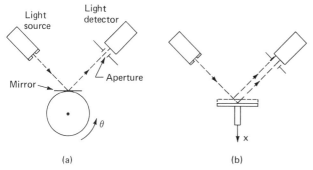

(a)   (b)

Fig. 8-16   Optical level-displacement detector: (*a*) angular; (*b*) linear.

ment of the edge of the light beam across the photodetector aperture, which causes a variation in detected intensity. Since the light intensity does not fall off linearly with distance from the center of the beam, the overall response of the transducer is quite nonlinear except for very small displacements, and the response range is also quite limited. These transducers are most useful in detecting small vibrations or as a displacement null detector.

8-7   Magnetic Proximity Detectors

In some cases it is convenient to determine when an object has rotated or translated into a predetermined position by sensing the presence of a small magnetic field. In Fig. 8-17, for example, a magnet attached to a wheel is detected if it moves close to either proximity device (*A* or *B*). One method of detecting the static field of the magnet is by a flexible magnetic reed, which makes an electric contact as it is bent by the force of the nearby magnet. The device is rather small (1 to 3 cm long) and usually has

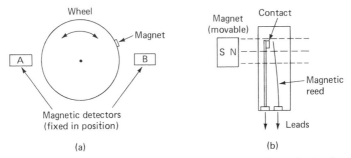

Fig. 8-17 Magnetic proximity detectors: (*a*) example of use; (*b*) sketch of magnetic reed detector.

a sealed construction similar to the reed relay. Note that the reed detector need not contact the magnet or rotating object. Unlike dynamic pickups, e.g., the induction coil, the reed detector has no minimum speed of approach of the magnet for detection to occur.

A more sensitive static magnetic detector, and one which in some devices exhibits a degree of proportional response, is the *Hall detector*,

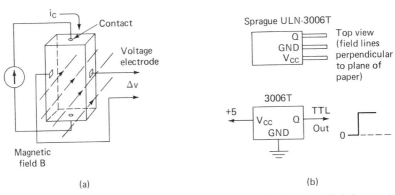

Fig. 8-18 Hall-effect magnetic detectors; (*a*) principle; (*b*) IC device (digital output).

based on the Hall effect, which all semiconductors possess to a greater or lesser degree. Current i_c is passed through the semiconductor (Fig. 8-18*a*) in a direction perpendicular to the magnetic field (at least a component must be perpendicular). Voltage leads are placed perpendicular to the current-flow direction, as close as possible on an equipotential line when the magnetic field B is zero, so that $\Delta v = 0$ for $B = 0$. For maximum response, the plane of the voltage and current, i.e., surface of the semiconductor slab, must be perpendicular to the magnetic field. When the field is applied, the charge carriers giving rise to the current flow are deflected slightly by the magnetic field and drift to one voltage electrode or the other (depending on current and magnetic field direction). The net result is that the voltage Δv between the pickup arms is proportional to B, the strength of the magnetic field at the semiconductor.

In the IC version of the Hall detector (Fig. 8-18b), the small-signal voltage Δv is amplified by a difference amplifier and converted into a logic (TTL) level with a comparator with hysteresis. A small magnet (cube, $\frac{1}{2}$ cm each side, say) would be expected to produce the required field of a few hundred gauss at a distance of 2 mm from the surface. As indicated above, the response is a function of orientation of the detector and magnet. Care must be taken to avoid an orientation where the output is nearly zero. A linear version is also available (3008).

Chapter Nine

Fluid-Measurement Gages

9-1 Pressure Gages

Pressure transducers are basically displacement or strain sensors attached to flexible elements which move as pressure is applied. Discussion of the numerous types and variations of pressure transducers will not be attempted here because the electrical-readout techniques have already been discussed in Chap. 8 and the mechanical aspects of the flexible elements which translate pressure into displacement are beyond the scope of this text.

Perhaps the three most common types of flexible elements are the diaphragm, bellows, and bourdon tube, sketched in Fig. 9-1. A strain-gage, LVDT, capacitive, piezoelectric-crystal, or other transducer is fastened to the point indicated to measure the displacement Δx. Calibration of the readout is made in terms of pressure rather than displacement, but otherwise the electronics are identical to that described in Chap. 8.

Two examples of pressure transducers are shown in Fig. 9-2. The first is a diaphragm type intended to measure absolute pressure since one side of the diaphragm is evacuated rather than exposed to atmospheric pressure.

A differential pressure gage is identical in construction except that a second pressure-inlet port is provided on the opposite side of the diaphragm. The second example is a relative gage utilizing a C-type bourdon tube and a LVDT sensor.

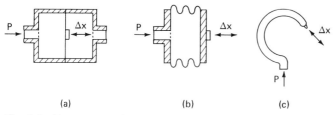

(a)   (b)   (c)

Fig. 9-1 Pressure-transducer elements: (*a*) diaphragm; (*b*) bellows; and (*c*) bourdon tube.

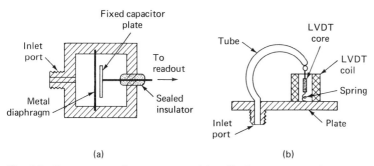

(a)   (b)

Fig. 9-2 Pressure transducers: (*a*) capacitive, diaphragm type and (*b*) LVDT bourdon tube.

9-2 Flow Gages

Devices which measure the rate of flow of a fluid (in units such as liters per minute) are called *flow gages*. Most are based on the pressure drop along the tube through which the fluid is flowing (Fig. 9-3). Constrictions in the tube or other aerodynamic mechanisms are employed to produce the pressure drop ΔP, and the difference in pressure is measured with a differential pressure gage. In most cases the pressure drop is proportional to the square root of fluid velocity u and therefore also to the net flow rate through the pipe.

A second type of flow gage depends on the increase in cooling rate produced by a flowing fluid. One version is the *hot-wire anemometer*. A thermistor version of the gage is pictured in Fig. 9-4. The thermistor is deliberately operated at a fairly high applied voltage V_t. As a result of the internal power dissipation $P_D = V_t^2/R_T$, the temperature of the thermistor rises substantially above ambient. In other words, the thermistor is a heater as well as a temperature sensor. The temperature rises until the power dissipated is equal to the loss through the leads (relatively small) and the convective loss to the fluid, which increases with fluid velocity u. The rate of heat loss and therefore the temperature change of the sensing

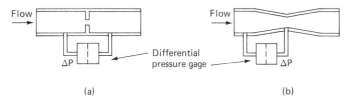

Fig. 9-3 Flow-gage sensing elements ($\Delta P = K\sqrt{u}$): (*a*) orifice type and (*b*) venturi type.

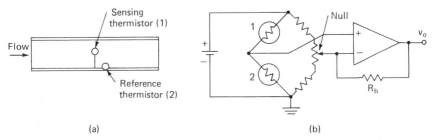

Fig. 9-4 Thermoconductive flow gage: (*a*) thermistor placement; (*b*) bridge readout.

thermistor (thermistor 1 of Fig. 9-4) with respect to its temperature in the static fluid is approximately proportional to the square root of velocity. In order to compensate for changes in ambient temperature a second thermistor is placed on the wall of the pipe where the fluid velocity is zero. Good thermal contact is required so that this thermistor does not self-heat appreciably. Electrically this thermistor is connected into the second arm of the bridge so that equal resistance changes in the two elements caused by changes in the ambient temperature result in no change in the bridge output. Changes in thermistor temperature result in a voltage output v_o from the bridge amplifier which is proportional to the temperature difference, as discussed in Chaps. 4 and 6. The bridge is balanced ($v_o = 0$) when the fluid is static, and not at equal thermistor temperatures. A flowing fluid of velocity u causes a temperature decrease of the sensing thermistor, resulting in an output, for small velocities and small temperature changes, of

$$v_o \approx K_f \sqrt{Q} \tag{9-1}$$

where Q is the flow rate and K_f is a sensitivity factor. It takes into account the aerodynamic flow, the conversion from fluid velocity u to flow rate Q, the power dissipation in the thermistor, bridge sensitivity, and amplifier gain. Calculation or estimation of K_f is possible, but an experimental calibration is usually required.

9-3 Vacuum Gages

In the high-vacuum region, the pressure gages discussed above, based on the force produced on an elastic element and the consequent measurement of displacement, are too insensitive. Pressures are usually expressed in torrs, where 1 torr is the pressure required to raise a column of mercury 1 mm (760 torr = 1 atm). With care, diaphragm gages which can detect pressure changes as small as 10^{-5} torr are possible, but special-purpose vacuum gages are generally used below about 1 torr. High-vacuum techniques commonly result in residual pressures in the range of 10^{-5} to 10^{-8} torr. Pressures below 10^{-10} torr can be achieved by special techniques, e.g., cryogenic pumping.

Some knowledge of high-vacuum techniques is required if serious measurement errors are to be avoided. Among these errors are the gases produced by the gage itself (outgassing), pressure differentials along the tube which connects the gage to the system, and differences in the gage sensitivity between various constituent gases (especially between air and condensible gases like water).

At higher pressures, in the 10^{-3} to 1 torr range, a popular sensor is the thermocouple gage. It is not very accurate, but it is inexpensive and conveniently covers the lower operating range of mechanical vacuum pumps. Usually the gage is a monitor of proper pump operation where only a rough idea of the pressure is needed. The gage consists of a thermocouple attached to a heater, as illustrated in Fig. 9-5a. The amount of power dissipation in the heater is a constant, but the temperature depends on the degree of cooling, which in turn depends on the gas pressure. High vacuum is an insulator, and thus the thermocouple heats up (the main heat loss is through the lead wires and mount). When gas is present, convection cooling takes place and the thermocouple temperature is only slightly above ambient. The voltage produced by the thermocouple is read by a microammeter. Scale calibration is nonlinear with the electrical zero at atmospheric pressure. To adjust the meter, a good vacuum ($<<10^{-3}$ torr) is desirable, at which point the heater power is adjusted so that the meter reads full scale electrically (note that electrical full scale corresponds to zero pressure).

Another thermoconductive gage uses a resistance thermometer as the temperature sensor. In the version shown (Fig. 9-5b) the thermistor is also the heater. Basically this arrangement is identical to the thermoconductive flow gage pictured in Fig. 9-4, and the electrical readout is identical.

Practical versions of this gage have a somewhat wider pressure range and are more accurate than the thermocouple gage.

At lower pressures, the glow discharge or cold-cathode ionization gage, shown in Fig. 9-6, is employed where accuracy in the pressure measurement is not important. It is based on the fact that the current of a high-

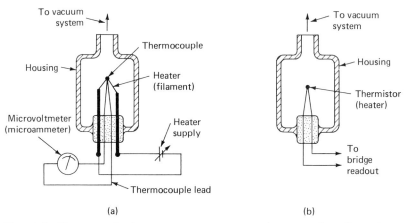

(a)   (b)

Fig. 9-5 Thermoconductive vacuum gases: (*a*) thermocouple; (*b*) resistance thermometer.

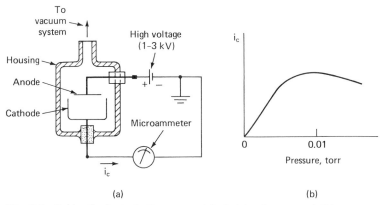

(a)   (b)

Fig. 9-6 Cold-cathode ionization gage: (*a*) sketch of gage and (*b*) pressure response.

voltage discharge varies approximately linearly with pressure at low pressures. As indicated in Fig. 9-6*b*, at higher pressures the linearity no longer holds, and the discharge current eventually drops to zero. Because the current is affected by electrode cleanliness, type of gas, and other parameters difficult to control, high accuracy cannot be expected. The device works best in the range of 10^{-5} to 10^{-3} torr.

Perhaps the most accurate pressure gage at high vacuum is the hot-cathode ionization gage (Fig. 9-7), generally intended for use below 10^{-3} torr, where the mean free path of electrons is fairly long. The cathode or filament is heated to produce electrons, as in a vacuum tube. Emitted electrons are accelerated by a screen with a positive potential (typically 100

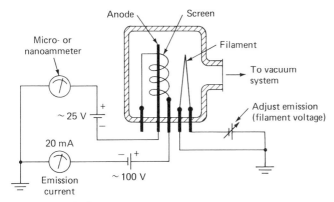

Fig. 9-7 Ionization-type vacuum gage.

V). Emission currents of 20 mA are typical. The energetic electrons collide with neutral gas molecules, causing ionization. A second, negatively charged plate is placed to collect the positive ions. For a fixed electron current and path length, the number of ions produced, and therefore ion current hitting the plate, will be proportional to the pressure. Since ion currents are below 10^{-6} A at lower pressures, sensitive high-input-impedance amplifiers (electrometers) are required to measure the current. The gage has a wide range at lower pressures. The lower limit is well below 10^{-10} torr. A drawback is that the gage will not work at higher pressures, and indeed if it is accidentally exposed to air at higher pressures, the filament may burn out.

9-4 Moisture and Humidity Sensors

Moisture sensors measure the amount of water in a solid or liquid, usually in terms of percent by volume. Humidity sensors measure the amount of water in a gas, especially air, in terms of mass fraction (absolute humidity) or in terms of percentage of vapor saturation (relative humidity at a given temperature). Most sensors rely on the absorption of water onto a surface coated with a hygroscopic chemical in contact with the gas, liquid, or solid phase to be measured. Some physical property of the surface, e.g., resistance, capacitance, or mass, is determined as a function of the moisture content or humidity of the adjacent phase.

Two humidity sensors which depend on electric properties are shown in

Fig. 9-8. The one in Fig. 9-8a depends on the decrease in resistance of a porous film impregnated with a hygroscopic chemical (LiCl or KH_2PO_4). Water is adsorbed to form a conducting salt solution as a film (1 to 5 percent water). As the humidity increases, the adsorbed water increases and thus the resistance between the electrodes decreases. The second

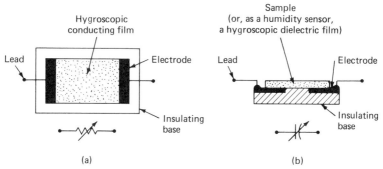

(a)   (b)

Fig. 9-8 Electrical humidity and moisture sensors: (a) conducting film; (b) capacitive (fringing-capacitor) type.

gage is based on the increase in the dielectric constant of a material as water is adsorbed. This approach is practical because the dielectric constant of water is much higher (80) than that of most materials (2 to 4). When used as a gage, the material, e.g., paper for paper manufacturing, is placed in contact with the surface containing the electrodes so that the electrostatic (fringing) field passes through the material. The increase in capacitance is approximately proportional to the percentage of water (by volume) in the material. When the gage is operated as a humidity gage, a hygroscopic material is coated on the surface so that the amount of adsorbed water in equilibrium with the gas increases with the humidity. Both sensors can be protected from mechanical damage by an additional film permeable to water vapor. Readout methods for the resistance sensors are the same as those discussed in Chap. 6 in connection with thermistors. Readout methods for the capacitance sensors were discussed in Chap. 7 in connection with capacitance-type displacement sensors.

Two humidity sensors based on changes in mechanical properties with adsorbed moisture are shown in Fig. 9-9. Figure 9-9a shows a piezoelectric crystal (quartz) connected to an oscillator circuit (discussed in Chap. 12). A hygroscopic polymer coats one surface. As water is adsorbed, the mass of the coating increases, mechanically loading the crystal and decreasing the resonance frequency. Of course, the readout is electrical even though this is classified as a mechanical (mass-dependent) sensor. Although the frequency shift is small, it can be measured accurately with a frequency counter. Readout circuits have been discussed in connection with the crystal as a temperature transducer (Chap. 6).

A variation of the oscillating-crystal sensor can be used to detect the formation of ice, e.g., on an airplane wing. In this case the hygroscopic coating is omitted and the crystal placed in the area where ice formation is

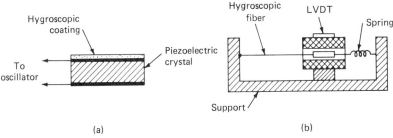

Fig. 9-9 Mechanical humidity sensors: (*a*) oscillating-crystal type; (*b*) stretched-fiber type.

expected. As ice deposits on the crystal, the mass increases and the frequency decreases approximately proportional to the thickness of the ice layer.

A popular, although rather inaccurate, humidity sensor is based on the change in length of fibers or membranes containing collagen (human hairs, animal skins) which occurs as water is adsorbed (Fig. 9-9*b*). The change in length is small and nonlinear (about 1 percent for a 50 percent change in relative humidity). Often a mechanical linkage and pointer with dial is the readout, but sometimes an LVDT displacement gage is used, as shown in Fig. 9-9*b*. Readout circuits for the LVDT were discussed in Chap. 8.

9-5 Liquid-Level Gages

Liquid-level gages measure the height or amount of liquid in a container, tank, or natural body of water. They are classified as discrete and continuous (Fig. 9-10). A discrete level sensor simply indicates whether the liquid is above or below the position of the sensor, while a continuous type measures the depth of the liquid. Discrete sensors are often used in pairs in high- and low-level indications which use control valves or pumps to maintain the liquid level between these two positions. Because many physical properties of a liquid differ from the air above, a wide variety of sensors is possible, and the choice depends on the particular liquid and size of container (different sensors would be appropriate for sulfuric acid in a chemical plant, molasses in a storage tank, liquid nitrogen in a physics experiment, and the level of Lake Erie).

Several discrete level sensors are illustrated in Fig. 9-11. An internally heated thermistor (or a thermocouple mounted on a heater) mounted to the container wall at the desired level is shown in Fig. 9-11a. Because the

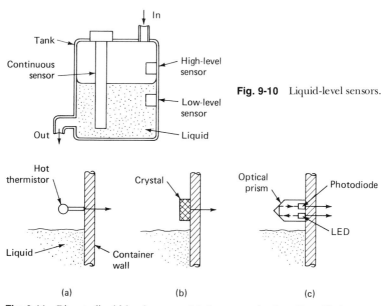

Fig. 9-10   Liquid-level sensors.

Fig. 9-11   Discrete liquid-level sensors; (a) thermoconductive; (b) oscillating crystal; and (c) optical reflection.

thermal conductivity of the liquid is better than a gas, the thermistor will be much cooler when immersed in the liquid. The principle is the same as used in fluid-flow and vacuum gages, discussed above. Readout circuits are given in Chap. 6.

The second sensor (Fig. 9-11b) is a piezoelectric crystal which is made a part of an oscillator. When the crystal is immersed in the liquid, its frequency will shift downward slightly because of mass loading and it will also be damped to a greater degree. In most cases the small but reproducible frequency shift is detected, but the drop in amplitude or even the loss of oscillation of a marginal oscillator can be detected instead.

The third discrete sensor (Fig. 9-11c) relies on the internal reflection of light in a prism. Complete internal reflection will occur if the difference of index of refraction between the prism and outside is sufficiently great and the angle of incidence of the beam with respect to the reflecting surface (surface of prism) is sufficiently small. The prism angle (about 45°) is chosen so that light is internally reflected in air. Light from the LED will be twice reflected into the photodiode detector. Since the index of refraction in liquids is substantially higher than that of air (or any gas), the difference in the indexes will be small when the prism is in the liquid and therefore the light beam will be transmitted instead of reflected and the light hitting the photo detector will drop markedly. Readout circuits are discussed in Chap. 7.

A generally applicable continuous liquid-level gage is the capacitance type, illustrated in Fig. 9-12. It is based on the difference in dielectric

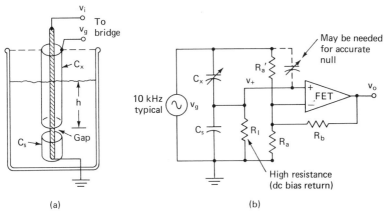

Fig. 9-12   Capacitance liquid-level gage: (*a*) dual coaxial gage immersed in liquid; (*b*) ac bridge with amplifier.

constant of the liquid and gas above. One of several geometric forms of the capacitor is the hollow coaxial cylinder into which the liquid can rise. The capacitance increase C_x in the cylinder is proportional to the height of the liquid h. In order to compensate for possible variations in the dielectric constant of the liquid, a reference capacitor C_s may be added. Electrically it is connected into the opposite arm of an ac bridge, as shown in Fig. 9-12*b*. If the liquid dielectric constant is fixed, an ordinary capacitor can replace C_s. Note that the outside cylinder of the coaxial cylinder acts as a shield for C_x (as well as C_s) which has a low impedance to ground, assuming that the ac generator has a low internal impedance. The bridge amplifier is discussed in Chap. 6. A simpler geometric configuration for C_x for metal containers is a rod or wire partly immersed in the liquid, the container acting as the shield. In this case C_x and C_s (fixed capacitors) are interchanged in the bridge amplifier because one side of C_x is grounded. Care must be taken to avoid errors due to liquid conductance. Since the conductivity of tapwater may be in the range of 1 to 30 mS/cm, the conductive component may dominate the capacitive component at frequencies below the megahertz region. Insulation of the electrodes may suffice.

A number of other liquid-level sensors are suggested by Fig. 9-13. If the liquid is conducting, a simple metal dipstick *A* connected to an ohmmeter

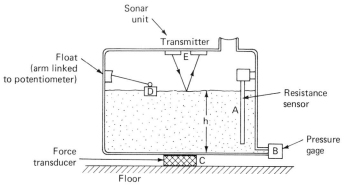

Fig. 9-13   Several continuous-type liquid-level gages.

will work. A reference resistor (pair of electrodes fully immersed in the liquid at the bottom) is necessary if the conductivity varies, and in this case a bridge readout similar to those discussed above can be used. A pressure gage *B* mounted at the bottom of the tank measures the liquid level by the pressure level produced by the height *h* above the gage. Alternatively the weight of the tank can be measured by a force transducer at the bottom *C*. Another method is to use a float *D* connected through a mechanical linkage to a potentiometer so that the resistance is proportional to height, but as with other potentiometer-type displacement transducers, providing a trouble-free mechanical linkage can be difficult.

For larger tanks, lakes, and the like, a sonar method *E* is best. A burst (pulse) of sound waves is sent by a transmitter, reflected by the liquid surface, and returned to a detector. The time delay is proportional to the distance between the sonar unit and the liquid surface, so that change in delay is proportional to the height of the liquid.

Chapter Ten

Chemical and Biological Electrodes

The purpose of a chemical or biological electrode is to make good electrical contact with a solution, tissue, or part of a cell. Practically all chemical electrodes of interest involve contact with an ionic aqueous solution typified by a 0.1 N KCl or NaCl solution. Contact with biological systems likewise is via the ionic solution surrounding cells inside them. Both chemical and biological electrodes must act as an interface between a salt solution and the metal lead-in wires from an electronic instrument. A good contact will have a low or known constant voltage across the interface, and the resistance will be as low as possible. With a number of exceptions, this is not accomplished successfully simply by sticking metal wires into a beaker or tissue. The problem usually encountered is that the voltage difference between the wire and solution is not constant unless the electrode is reversible. For some applications a dc drift in interfacial potential is tolerable, but for most cases a reversible electrode is required.

10-1 Reversible Electrodes

Reversible electrodes will be discussed here in terms of a specific example, the silver–silver chloride (Ag|AgCl) electrode pictured in Fig. 10-1. Silver metal (wire) coated with AgCl is immersed in a water solution containing chloride ions. A reversible chemical reaction occurs at the interface between the solution and metal. In the forward reaction Cl⁻ in solution reacts with Ag (metal), AgCl (solid) is deposited, an electron is transferred to the metal, and a corresponding current flows through the wire to the external surface. In the reverse direction, electrons flow from the metal,

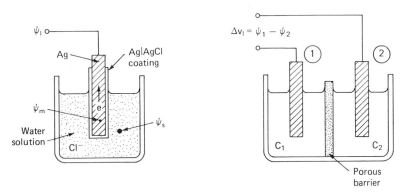

Fig. 10-1   Reversible electrodes; Ag + Cl⁻ ⇌ AgCl + e^-: (*a*) potentials at one electrode (theoretical); (*b*) potential between a pair of electrodes.

AgCl is removed, and Cl⁻ is transferred into the solution (Ag metal is deposited). The process can be visualized as an interchange between the charge carriers in water (Cl⁻) and the charge carriers in metal (electrons) by a specific, reversible process at the interface.

As expected for any chemical reaction, the standard free energies of the charged species (Cl⁻ and e^-) are unequal in general, and therefore the Cl⁻ deposit or removal involves an energy for the reactants and products

$$\mu_{Ag} + \mu_{Cl} = \mu_{AgCl} - \mathfrak{F}\psi_i \qquad (10\text{-}1a)$$

where ψ_i is the potential difference between the solution and metal and $\mathfrak{F}$ is the faraday (electron charge times Avogadro's number). Chemical potentials of solids (μ_{Ag} and μ_{AgCl}) are constant at a given temperature, but that of the solution depends on concentrations C, here that of the chloride ion. Reexpressing the above equation in terms of concentrations, we have

$$\mu_{Ag} + \mu_{Cl}^\circ + RT \ln C = \mu_{AgCl} - \mathfrak{F}\psi_i \qquad (10\text{-}1b)$$

where μ_{Cl}° = standard free energy† of chloride ions, a constant at a given temperature

T = absolute temperature, K

R = gas constant

After dividing by $\mathfrak{F}$ and collecting the constant terms into a single constant

---

†The usual approximation of expressing μ in terms of concentrations rather than activities has been made for the sake of simplicity.

ψ_0, we obtain an expression for the potential difference between solution and metal ψ_i as a function of chloride-ion concentration C

$$\psi_i = -\frac{RT}{\mathcal{F}} \ln C + \psi_0 \qquad (10\text{-}2)$$

This relation is not very practical in this form because the constant ψ_0, which is a kind of contact potential, is not accurately known or experimentally measurable directly although it can be estimated theoretically. Of course since two electrodes must be present for a complete circuit, ψ_0 simply cancels in the expression for the potential difference between the two identical electrodes Δv_i. For Δv_i to be nonzero at least one of the following conditions must be true: (1) the two electrodes are not identical (one might be Ag|AgCl and the other Hg|HgCl, perhaps); (2) the concentrations of chloride ions in which the two electrodes are immersed are not identical (a porous barrier or membrane will maintain a concentration difference); or (3) there is voltage difference in the solutions in which the electrodes are immersed due to membrane or other diffusion potential. An expression for the potential difference Δv_i between a pair of electrodes immersed in solutions of different concentrations C_1 and C_2 and separated by a membrane with a potential across it Δv_m is

$$\Delta v_i = 2.30 \frac{RT}{\mathcal{F}} \log \frac{C_1}{C_2} + \Delta v_m + \psi_0^{(1)} - \psi_0^{(2)} \qquad (10\text{-}3)$$

where $2.30 RT/\mathcal{F} = 59.12$ mV at 25°C. Usually the experimental conditions are such that only one term of this equation is nonzero. For example, if no membrane is present ($C_1 = C_2$), Δv_i represents the difference contact potential between pairs of electrodes of different metals (and metal chloride coatings). Tables listing these half-cell potentials with the hydrogen electrode as reference can be found in texts on electrochemistry.

An Ag|AgCl reference electrode can be made simply by electroplating chloride onto a silver wire, as illustrated in Fig. 10-2. Hydrochloric acid diluted with distilled water is put into a small beaker. Current densities of about 1 to 10 mA/cm² are best, the lower currents producing a finer deposit (uniform purple-brown color). Any two electrodes made in this manner should have a potential difference between them of less than 1 mV when placed in a chloride solution, although for many applications differences up to 10 mV can be tolerated.

A popular reversible electrode is the calomel electrode, which is based on the reaction between mercury (metal) and chloride ions in solution to form mercury(I) chloride (solid)

$$\text{Hg} + \text{Cl}^- \rightleftharpoons \text{HgCl} + e^-$$

The reaction is similar to that of the Ag|AgCl electrode, and the same equations apply except that the value of ψ_0 is different. Construction details for the calomel reference electrode are given below.

10-2 Reference Electrodes

An obvious drawback to the Ag|AgCl and other reversible electrodes is that they require Cl$^-$ of a fixed concentration to be present in the solution

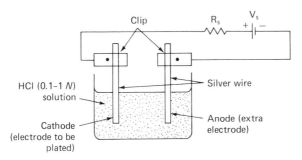

Fig. 10-2 Ag|AgCl electroplating.

adjacent the electrode. In order to make contact with solutions which contain no chloride ion or chloride ion with an unknown concentration, a reference electrode like that pictured in Fig. 10-3 is almost universally used. It consists of a reversible electrode in contact with potassium chlo-

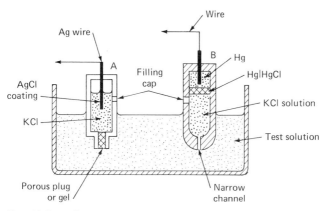

Fig. 10-3 References electrodes: (*a*) Ag|AgCl type; (*b*) calomel type.

ride (KCl) solution at a fixed concentration, which in turn is in contact with the test solution through some sort of barrier. The barrier must be such that a true liquid junction exists but the bulk flow of the KCl solution into the test solution (due to gravity, etc.) is extremely slow. A narrow channel or porous plug (the equivalent of many channels) suffices.

At any liquid junction where bulk flow is negligible, the diffusion of ions can still take place along their concentration gradients. At the reference electrode, K^+ and Cl^- diffuse across the junction into the test solution (assuming, as is ordinarily the case, that the concentrations of K^+ and Cl^- are higher inside the reservoir). When an ion diffuses across a junction, it produces a junction potential (for a cation, the lower concentration side is positive). When both cation and anion diffusion is present, their effects tend to cancel and the faster-diffusing ion determines the sign of the junction potential. By coincidence the diffusion coefficients of Cl^- and K^+ are almost equal and therefore the potential at a KCl junction is nearly zero. For this reason KCl is almost always chosen as the salt for a reference electrode. Thus it can be seen that the potential difference between the lead-in wire and test solution is always a constant, independent of the chloride-ion concentration of the test solution, in effect eliminating the concentration term of Eq. (10-3). When identical reference electrodes are used to make contact with the solution, $\psi_0^{(1)} - \psi_0^{(2)}$, Eq. (10-3) reduces simply to $\Delta v_i = \Delta v_m$.

10-3 Membrane Potentials

Practically all biological potentials are associated with membranes, especially the membranes surrounding cells. Most chemical transducers also are based on membrane voltage. Potentials develop across membranes when the concentrations of a permeable ion are unequal on either side of a membrane. A positive ion (cation) diffusing from the higher- to lower-concentration side leaves the anion (assumed incapable of penetrating the membrane) behind, and the resulting charge separation produces the membrane potential, which can be measured by a pair of reference electrodes (see Fig. 10-4). The voltage difference continues to build up

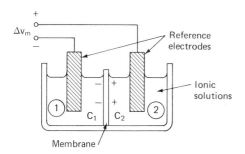

Fig. 10-4 Membrane-potential test chamber; sign of potential is shown for a cation.

until at equilibrium the electric field within the membrane is strong enough to inhibit further diffusion (no external current flow is assumed). This process is similar to that which occurs at (diode) *pn* junctions except that here cations and anions replace holes and electrons. An expression for the potential can be derived from the electrochemical potential for an ion. Since at equilibrium the chemical potentials of the two sides of the membrane are equal, $\mu_1 = \mu_2$ or

$$\mu_0 + RT \ln C_1 + n\mathfrak{F}\psi_1 = \mu_0 + RT \ln C_2 + n\mathfrak{F}\psi_2 \qquad (10\text{-}4)$$

where C_1 and C_2 are the concentrations of the (single) permeating ion of valence n [also see Eq. (10-1)]. Solving for $v_m = \psi_1 - \psi_2$, we obtain

$$v_m = 2.30 \frac{RT}{n\mathfrak{F}} \log \frac{C_1}{C_2} \qquad (10\text{-}5)$$

where $n = 1$ for univalent cations, that is, K^+, and -1 for univalent anions, that is, Cl^-. This equation is known as the *Nernst equation*. Note that at 25°C, Δv_m is 59 mV for a tenfold difference in concentration and 118 mV for a hundredfold difference. If the membrane is permeable to more than one ion, Eq. (10-5) must be replaced by a more complicated expression involving the relative ion permeabilities (or diffusion coefficients or mobilities).

10-4   pH Meters

Measurement and control of pH is important in numerous chemical and biological studies. It is a measure of hydrogen-ion concentration C_H defined as

$$\text{pH} = \log C_H \qquad (10\text{-}6)$$

A pH meter consists of a glass electrode, reference electrode (usually calomel), and a high-input-impedance dc amplifier (Fig. 10-5). The glass

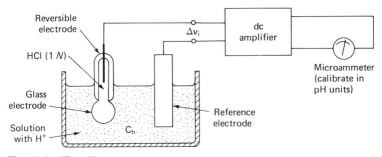

Fig. 10-5   The pH meter.

electrode is made from a special glass which is permeable to hydrogen ions (only), and the thin bottom acts as a membrane. Hydrochloric acid (1 N) inside the sealed electrode is in contact with the glass membrane as well as with a reversible electrode at the top. Both the reversible electrodes inside the glass electrode and reference electrode are identical, and therefore the potential differences due to the reference electrodes are zero. The net potential across the leads connected to the amplifier is just the membrane voltage.

The potential across the membrane is given by Eq. (10-5) with $C_1 = C_H$ (test solution) and $C_2 = 1.0$ since the H^+ concentration is fixed at 1 N inside.

Substituting into Eq. (10-6), we get the pH-voltage relation

$$\Delta v_i = \frac{-2.30RT}{\mathfrak{F}} \quad \text{pH} \tag{10-7}$$

As might be expected, the resistance of the glass membrane is quite high (10^6 to 10^8 Ω), and therefore a high-resistance dc amplifier or electrometer is required to make the potential measurement. A suitable amplifier is shown in Fig. 10-6. It is just a standard noninverting amplifier

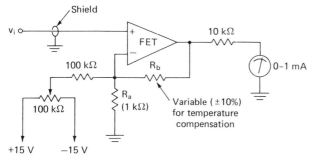

Fig. 10-6 Electrometer amplifier as a pH meter.

with an FET op-amp (for high impedance) and meter calibrated in pH units. A zero offset control to compensate for inequalities in the reference-electrode contact potentials, etc., is present. Usually it is adjusted by placing the electrodes in a standard buffer solution (say pH = 7.00) and setting the meter to that pH. Often the gain is made adjustable over a narrow range through an adjustment of R_b in order to provide temperature compensation [see Eq. (10-7)].

10-5 Specific-Ion Electrodes

Electrodes to measure the concentration of specific ions besides H^+ have been developed. Basically they are identical to the pH meter except that the membrane utilized is permeable to specific ions to be measured. Ideally the membrane should be completely impermeable to all other ions, and the chemical engineering problem is to design such membranes.

Glass electrodes specifically permeable to Na^+ and K^+ have been developed and are interchangeable with the pH electrodes. Since interference between Na^+, K^+, and H^+ occurs to some extent, it is not possible, for example, to measure a low concentration of Na^+ accurately with a Na^+-specific electrode if the K^+ and H^+ concentrations are high.

Another example of a specific-ion electrode is one intended to measure fluoride-ion concentration. The construction of the electrode, shown in Fig. 10-7a, is based on the observation that the conductivity in lanthanum

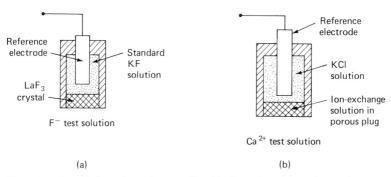

Fig. 10-7   Specific-ion electrodes: (a) fluoride ion; (b) calcium (ion-exchange) electrode.

fluoride crystals involves the transport of F^-. It can be thought of as a thick membrane permeable to F^- and mounted between a test solution and a standard solution of F^- (0.01 N KF). When reference electrodes are in contact with both the standard test solutions, the potential developed is given by Eq. (10-5) with $n = -1$.

Calcium-ion concentrations can be determined with an ion-exchanger phase separating the test solution containing Ca^{2+} and a standard solution of KCl (Fig. 10-7b). In effect the exchanger phase acts as a membrane. The liquid ion-exchanger phase consists of a porous plug filled with a standard $CaCl_2$ solution (connected to a reservoir). The common anion is Cl^-, and the cations which exchange are Ca^{2+} and K^+. For each Ca^{2+} which diffuses out of the exchanger phase into the test solution, two K^+ ions diffuse into this phase from the KCl region and a potential builds up across the exchanger "membrane." The potential is given by Eq. (10-5) with $n = 2$, where C_2 is the calcium concentration of the test phase and C_1 the calcium-ion concentration of the exchanger phase ($CaCl_2$ concentration in the reservoir). While enjoying the advantages of other specific-ion electrodes, this electrode has the disadvantages of a comparatively large size, easy chemical contamination, and Ca^{2+} leakage into the test solution.

10-6 Microelectrodes

Microelectrodes are used in biological experiments to measure voltage inside cells, especially nerve cells. The most common version is basically a glass reference electrode (discussed above) with a hollow tip tapered to a very fine point. To make an electrode, a section of glass tube (~1 mm in diameter) is heated until it is soft and then suddenly pulled (by a solenoid, perhaps) to produce the taper. Tip diameters below 1 μm (10^{-4} cm) can be made. The inside of the tube is filled with KCl solution (3 N) by capillary action, and a reversible Ag|AgCl electrode is inserted into the large end (Fig. 10-8). In an experiment the electrode is held by a micromani-

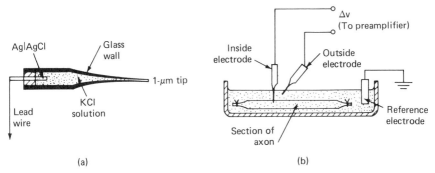

Fig. 10-8 Glass microelectrodes: (*a*) electrode-construction schematic; (*b*) test setup.

pulator and pushed into the cell while the process is being watched under a microscope. A second electrode with a coarser tip is positioned outside the cell in the vicinity of the first electrode contact point. As the electrodes break through the cell wall (plasma membrane), the potential difference between the electrodes jumps from 0 to about 50 to 100 mV (resting potential, inside negative) for a nerve cell. Skill is required in making and using microelectrodes, as can be imagined.

Usually a differential-input preamplifier (see below) is employed to measure the potential difference, and a general ground to the solution is added both to provide a ground return for the signal paths and to provide some degree of shielding. A metal shielded enclosure is also almost indispensible.

Microelectrodes have a high internal resistance (typically 1 to 100 MΩ) due to their long taper and small tip. To avoid a signal reduction or a large dc voltage offset due to bias current, an FET input preamplifier is almost essential (Fig. 10-9). Even with an FET preamplifier the high electrode resistance R_e coupled with the unavoidable shield capacitance to ground C_s results in an additional problem if rapidly changing potentials (action potentials) are to be recorded. This occurs because R_e and C_s act as a low-pass network (Chap. 11), which reduces the high-frequency components of the signal. For example, the action-potential pulse height is reduced and its width widened by a low-pass filter. To negate the effect of

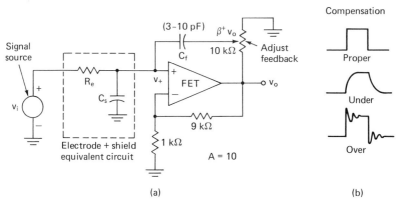

(a)   (b)

Fig. 10-9   Microelectrode equivalent circuit and negative-capacitance preamplifier.

C_s, and thus raise the frequency response, a negative-input-capacitance preamplifier (Fig. 10-9) is effective. Positive feedback through C_f provides the negative-input-capacitance effect.

The negative capacitance input can be analyzed by summing the currents into the node at the noninverting input v_+

$$\frac{v_i - v_+}{R_e} + C_s \frac{dv_+}{dt} - C_f \frac{d(\beta^+ v_o - v_+)}{dt} = 0 \tag{10-8}$$

where β^+ is the fraction of the voltage feedback as determined by the potentiometer setting. Recognizing that the inverting and noninverting inputs are equal and that $v_- = v_+ = v_o/A$ (where A is the dc voltage gain, here 10), we see that the above equation reduces to

$$v_o = Av_i + R_e(C_s + C_f - \beta^+ C_f) \frac{dv_o}{dt} \tag{10-9}$$

The term involving dv_o/dt is the transient effect which is to be minimized. The term is zero when $C_s = \beta^+ C_f$, that is, by proper setting of the feedback potentiometer (preferably C_f is chosen so that this occurs at $\beta^+ \approx 0.5$).

In practice the feedback is adjusted for each experiment before taking data with the microelectrode in place. A test square wave is applied to the solution, and the preamplifier output is observed on the oscilloscope. With proper adjustment the output is also a square wave; otherwise a sluggish rise or ringing is observed (see Fig. 10-9b). When too much positive feedback is present (extreme overcompensation), the circuit will oscillate.

Ordinarily two microelectrodes are employed in experiments, and in this case two negative-capacitance preamplifiers are desirable. Their outputs are connected to a differential amplifier, as discussed below. Often it is more convenient to locate the reversible electrode away from the end of the glass microelectrode. In this case a salt-bridge extension is used. A thin plastic tube filled with KCl and connected to a syringe with extra solution works well. A KCl gel (1 to 2 g agar in 100 mL boiling water plus 3.7 g KCl), which does not drip as connections are made, is preferred.

Metal microelectrodes are often preferred in applications where only an ac signal is needed, in particular where the nerve action potential and not resting potential is recorded. They consist of a thin (0.002- to 0.020-in) stiff wire insulated except for the tip. Metals used include tungsten, platinum, platinum–10% rhodium, stainless steel, and silver. They are chosen for their stiffness (except silver) and nontoxicity. A stiff wire is desirable because it can be guided into position more easily. Teflon-insulated wire is manufactured for this purpose. Often the tip is tapered to a point by dipping the wire tip into a solution containing the metal ion, and the metal is removed by electrolysis (HNO_3 is satisfactory for silver, for example, because $AgNO_3$ is soluble). Except for silver, these metals do not make reversible electrodes, and therefore the dc potential between the metal and solution or tissue is quite variable. A low-resistance contact is formed, however, and that is why they are suitable for ac applications. They are also suitable as stimulation electrodes driven by a constant-current source (usually a short pulse) since in this case the electrode contact potential is immaterial. Because metal electrodes have relatively low resistance, a negative-capacitance preamplifier is not usually necessary but an FET op-amp may still be desirable because the bias current of a BJT amplifier flowing through the electrode into the cell can be intolerable.

10-7 Skin Electrodes

Surface electrodes placed on the skin measure potentials originating from the electrical activity of the heart (electrocardiogram, ECG or EKG), brain (electroencephalogram, EEG), or muscle (electromyogram, EMG). The electrode must make a reasonably stable ac contact with the skin to pass these small ac signals (0.01 to 1 mV). Low resistance and low contact potentials are desirable but not essential.

A typical skin electrode, sketched in Fig. 10-10a, consists of a metal disk (or screen) held against the skin by a suction cup or strap. Usually a conducting (electrolytic) paste is smeared on the skin or electrode to provide better contact. The disk may be made of silver (plate), on which AgCl has been deposited to make a reversible electrode; or the disk may be a compressed tablet of silver and silver chloride powder. Stainless steel

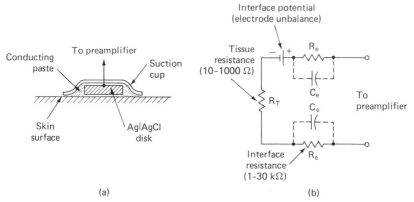

Fig. 10-10   Surface or skin electrodes: (*a*) construction details; (*b*) electrical equivalent circuit for a pair of electrodes.

and other irreversible electrodes work well if they have a large enough contact area. Numerous variations of this type of electrode are commercially available. Electrode wires placed below the surface of the skin by hypodermic needle provide a more permanent method of attachment if an invasive technique can be tolerated.

The contact resistance between the electrode lead wire and the tissue underlying the surface of the skin is comparatively high and variable compared with the tissue resistance. It can be modeled by the equivalent circuit of Fig. 10-10*b*. At higher frequencies (>1 kHz) the capacitive coupling between the electrode and tissue becomes important, but for practically all bioelectric signals which involve lower frequencies, the capacitive component is negligible. Actually electrode impedance is not important if the preamplifier input impedance is sufficiently high, as discussed below.

Electrode noise varies with the type of electrode material, its area, and the care with which the surface is prepared. The lowest noise is obtained with the Ag|AgCl disk. Movement of the electrodes, like that associated with muscular activity, can cause a type of noise which is in part a rapid variation in the electrode-skin contact potential. It too can be minimized by careful application of the electrode.

10-8   Differential Electrode Preamplifiers

A preamplifier suitable for amplifying electrode potentials in nearly all cases must have a high input impedance since the membrane and electrode (signal-source) resistances are generally rather high. The basic noninverting amplifier discussed in Chap. 4 is quite satisfactory provided the op-amp selected has an FET input (see Fig. 10-6). A requirement for many biological applications is that the preamplifier have a differential input. The reason for this requirement is indicated in the example below.

In EKG determinations a small signal (~ 1 mV) appearing across two electrodes (Fig. 10-11) is amplified by a differential-input preamplifier. A

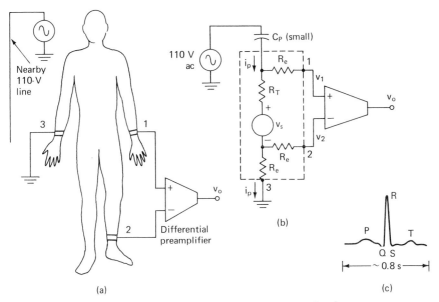

Fig. 10-11   EKG: (*a*) measurement and (*b*) equivalent circuit; (*c*) signal.

third electrode acts as a ground. As indicated previously, the EKG signal v_s, tissue resistance R_T, and electrode resistance R_e can be represented by the equivalent circuit of Fig. 10-10*b*. Unless the measurement is made in an electrically shielded room, pickup of unwanted signals, especially hum (60-Hz line), is unavoidable.

The human body and the power lines (ungrounded side) act as two widely separated plates of a capacitor C_p with air as the dielectric. Due to this capacitive coupling (~ 10 pF), a small current i_p flows into the body, through the tissue, and through electrode 3 to ground. Practically no current flows through the preamplifier electrodes 1 and 2 because the preamplifier input resistance is very high. The voltage drop across tissue resistance R_T due to i_- is quite small ($R_T \ll 1/2\pi 60 C_p$, $R_T \ll R_e$) although the voltage drop across the grounding electrode may be substantial. In other words, the hum voltage across leads 1 and 2 is small, assuming a third electrode ground is present to provide a path for i_p, but the pickup voltage with respect to the ground lead (1 or 2 and 3) is relatively high. The necessity of a high-impedance differential-input preamplifier which responds only to the difference voltage ($v_1 - v_2$) and not the voltage to ground v_1 is apparent.

The differential-input precision-gain voltage amplifier or instrumentation amplifier was discussed in Chap. 4. Specific requirements are high input impedance (FET op-amps) and good common-mode rejection (high m). Care must be taken to limit the dc gain of the input section to avoid amplifier overload due to the electrode potential unbalance v_j. A satisfactory fixed-gain ($A = 10$) amplifier is shown in Fig. 10-12. To minimize the

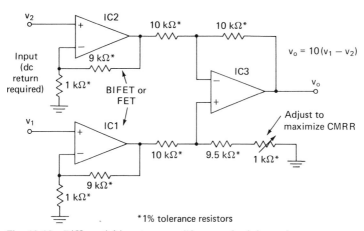

Fig. 10-12   Differential-input preamplifier, standard dc version.

common-mode response (maximize m), the test circuit of Fig. 4-13b is suggested. Negative-capacitance input (Fig. 10-9) may be added to either or both input amplifiers IC1 and IC2.

Requirements for electrocardiogram EKG, EEG, and EMG preamplifiers are rather similar. Differential input is almost essential, and only the ac component is amplified. A high-pass filter is present to block the dc electrode voltage v_j. A low-pass filter to limit the frequency response to no higher than required is also very helpful in reducing noise and other unwanted signals. Sometimes a notch filter at the line frequency (60 Hz) is added, although in the EKG signal some distortion occurs because the desired signal has frequency components in this range. Appropriate filters are discussed in Chap. 11.

An EKG preamplifier must amplify the signal, typically 1 mV (R wave), to a convenient level (say 1 V) without distortion and noise. Frequency-response requirements differ depending on whether the purpose is diagnostic, in which case no noticeable distortion is tolerated, or monitoring, in which case all that is needed is the determination of the presence of the signal (heart-rate measurements) or perhaps the detection of missing or grossly abnormal R waves. For diagnostic purposes, a frequency range of 0.2 to 100 Hz is required with a high-frequency rolloff of 6 dB/octave above 150 Hz. It is best to roll off at 12 db/octave below 0.1 Hz. For monitoring purposes a range of 0.5 to 50 Hz or sometimes even 2 to 20 Hz is satisfactory. Because of the limited range, electrode preparation and shielding against hum pickup is much less critical.

As with any amplifier with extended low-frequency response, there is a long delay before the amplifier baseline stabilizes. Large dc excursions, like those occurring when the electrodes are first connected, are especially annoying. Circuits which automatically zero the amplifier or provide a faster time constant for large-signal excursions are desirable.

Amplifiers for the EEG must be capable of amplifying signals in the 1- to 10-μV range. Fortunately the frequency response is limited to 3 to 30 Hz, and rejection of line-frequency (60 Hz) pickup by sharp cutoff and/or notch filters is not a major problem. The alpha wave occurs at about 12 Hz. Gains of 10^4 to 10^5 are typical, but otherwise the design is rather similar to the monitoring-type EKG preamplifier.

EMG preamplifiers are also similar to EKG preamplifiers except that the required frequency range is 5 to 500 Hz and because the amplitude of the signal varies considerably (0.1 to 10 mV) depending on electrode placement (muscle mass being monitored), a gain control is desirable. Often only the presence or count of impulses is required. For this purpose a differentiator (Chap. 11) followed by a comparator is satisfactory (Fig. 10-13). The main problems are choosing the differentiator time constants to match the signal being detected and setting the threshold V_T. Actually the differentiator or active high-pass filter is usually adjusted so that it removes most of the low-frequency components of the signal itself, with baseline removed, and the derivative seems to provide the most unambiguous input signal to the comparator v_0.

The electrode amplifiers discussed here are not recommended for use on human subjects unless ground isolation is provided on the output to eliminate the shock hazard (see next section).

10-9   Electric-Shock Hazards

Attachment of electrodes to human subjects may involve a shock hazard. Prudence and strict federal laws dictate that precautions be taken, particularly in the isolation of subject "ground" electrodes from equipment

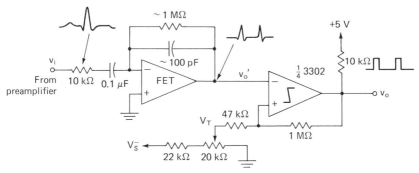

Fig. 10-13   EKG-EMG signal detector.

grounds. Approval of both experimental and commercial biomedical equipment, however innocuous, is required before it can be used on people.

Primary adverse physiological electrical effects are interference with the action of excitable membranes of nerve and muscle, and local heating. Although at the cellular level, specific effects are explained in terms of membrane voltage, gross effects are expressed in terms of currents flowing through a tissue mass or even the whole body. Of course there are wide variations in response, depending on the path of current flow as well as the build and tolerance of individuals. The most life-threatening current path is through the thorax since even moderate current levels can produce cardiac or respiratory failure. A low resistance at the skin-electrode interface implies that a particular current level can be produced by a relatively low voltage.† Heating effects are important only at higher current levels (Fig. 10-14).

Most people begin to sense current flow (as a tingle) above about 1 mA. When the current through the thorax reaches 100 mA, the heart is likely to undergo ventricular fibrillation (muscular quivering rather than rhythmic beating). This potentially fatal condition often persists after the current flow stops. Still higher currents produce complete heart stoppage and paralysis of the respiratory muscles. Obviously this condition is fatal if continued, but if the current ceases, recovery is likely. In other words, if

†Electrocution has been caused by only 40 V.

the current flow is brief, a higher current level can be less likely to be fatal than a lower level. Muscular paralysis implies that the victim cannot voluntarily move away from the "hot" wire, a widely recognized problem. It is less widely recognized that these physiological effects other than heating become negligible at higher frequencies (Fig. 10-14*b*). Sinusoidal

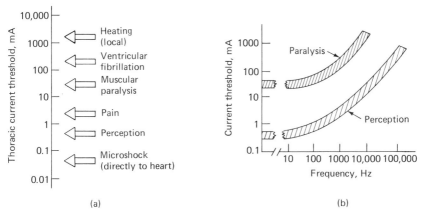

(a)   (b)

Fig. 10-14 Physiological effects of a current through the thorax: (*a*) as a function of current magnitude and (*b*) as a function of frequency.

currents over 1 A at 1 MHz have been passed through a subject without sensation or ill effects. Local heating and tissue damage will occur if the current flows through a constricted area, producing a high local current density (this is the basis of electrosurgery). Unfortunately, at line frequency (60 Hz) the adverse physiological effects are comparable to those at dc. Asymmetric pulse trains, transients, or nonsinusoidal high-frequency excitations with a low-frequency component are a potential hazard.

It is not immediately evident that an instrument like the EKG, which only monitors biopotentials and is not a current or voltage source to the electrodes, is any hazard at all. The problem is the ground lead or any current path to ground connected to the subject through the amplifier input leads under possible overload conditions. Should the subject accidentally come in contact with a grounded voltage source such as the hot lead of the line, current will flow through the subject and to ground through the grounded lead. In other words, a grounded electrode provides one of the two contacts to the body necessary for shock, thus greatly increasing the probability of an accident. It may be rashly assumed that the chances of a subject coming in contact with a hot lead is negligibly small, but sad experience indicates that this is not the case. Various metal objects such as recording instruments, appliances, hospital bed frames, and water pipes are electrically connected to a ground. If by accident the ground becomes shorted to the hot side of the line or a high current flows through the ground return wire due, say, to a short circuit in a neighboring room, the ground becomes hot, at least momentarily. Touching this seemingly (and usually) innocuous object provides the possibly fatal second contact. Any instrument which is grounded directly or indirectly to the power-line ground should have ground isolation such as provided by optical isolators (Chap. 7).

TRANSDUCER DESIGN EXAMPLES

Example 4 Sensitive Differential Thermometer

DESCRIPTION: The thermometer provides an output voltage proportional to the temperature difference between a main and reference thermistor.

SPECIFICATIONS

Common-mode range: 10 to 40°C

Differential range: 0 to 5.0°C

Thermistor required: R ≈ 10 kΩ (25°C), α = 0.04/°C (internal temperature rise of 0.1°C/mW)

Sensitivity: 1 V/°C

Linearity: ±2 percent

DESIGN CONSIDERATIONS: A thermistor with a room-temperature resistance R_0 of 10 kΩ and a sensitivity of 4 percent per Celsius degree is arbitrarily selected. The voltage across the thermistor is set at 1 V in order to limit the power dissipation to 0.1 mW and internal temperature rise to a negligible 0.01°C. Bridge voltage v_b is

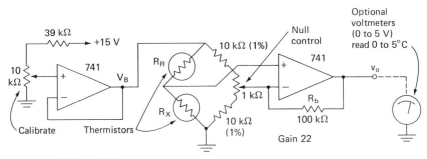

Fig. E-4 Differential thermometer.

derived from a regulated source and can be adjusted from the nominal 2-V value for sensitivity or calibration purposes. Bridge-feedback resistances are arbitrarily chosen to be 10 kΩ. The effective value of the bridge resistors is thus 10.5 kΩ, and the parallel combination which determines the amplifier gain is 5.45 kΩ.

From Eq. (4-15) the bridge-amplifier (Fig. 4-10) gain must be 22 in order to achieve a sensitivity of 1 V/°C. This can be met by setting R_b to 100 kΩ. Null adjustment R_4 compensates for inequalities in the thermistors, bridge resistors, and dc offset of the op-amp. It is adjusted such that $v_o = 0$ when both thermistor temperatures are equal.

Example 5   An LVDT Readout

DESCRIPTION: The readout provides an output voltage proportional to the displacement of the core of a LVDT. Positive displacement produces a positive voltage, and negative displacement produces a negative output voltage since the unit incorporates a phase-sensitive detector.

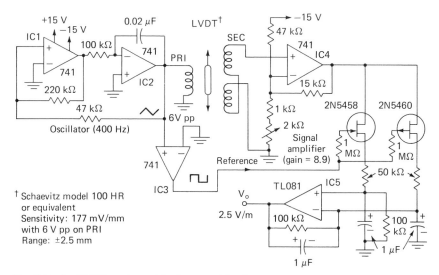

Fig. E-5   An LVDT readout with phase-sensitive detector.

SPECIFICATIONS

Sensitivity: 10 V per millimeter of displacement (with selected LVDT)

Range: ± 1 mm

Resolution: ± 0.001 V ($\pm 10^{-3}$ mm), limited by short-term drift

DESIGN CONSIDERATIONS: Assume that the LVDT has an ac differential secondary voltage of 0.18 V for a primary voltage of 6 V (400 Hz, primary impedance of 100 Ω) at a displacement (from zero) of 1 mm. Excitation by a triangular-wave source (~6 peak to peak) was chosen rather than a sine-wave source because a constant amplitude control is easier to implement with a triangular-wave source (see Fig. 12-10).

With 6 V peak to peak on the primary, the amplified secondary voltage V' will be 4 V peak to peak with the amplifier of gain 8.9. The phase-sensitive detector described in Chap. 15 has a dc output v_o of +2.5 for this input. Thus the overall response corresponds to 10 V output for a 1-mm displacement from center (zero). Note that the phase-sensitive-detector output v_o reverses polarity when the displacement is in the opposite direction, an important feature of this circuit.

Frequency stability is not important, but the amplitude stability of the source is. Variations in sensitivity from the design center are compensated by the gain or calibration control. A bias voltage on the amplifier output (IC4) prevents reverse polarity on the phase-sensitive-detector capacitors.

TRANSDUCER DESIGN PROBLEMS

16 A thermostat is needed which indicates digitally (TTL or CMOS) when the temperature is over a set point (25°C). The output should change to **1** if the temperature exceeds 25°C and change back to **0** when it drops below 23°C; that is, the hysteresis is 2°C. Use a thermistor of 10 kΩ at 25°C with a temperature coefficient of 4 percent per degree Celsius.

17 Draw a circuit which will turn on a relay (12 V, 50 mA) when a light beam is interrupted. Assume a phototransistor is available which delivers a current of 50 μA with the light on.

18 A vibration detector based on a strain is to be made. Only frequencies above 10 Hz (3-dB point) need be passed. Sufficient gain should be provided so that a strain of 10^{-4} (peak to peak) results in a 1-V output. A gage with a dc resistance R_0 of 100 Ω, a maximum power dissipation of 20 mW, and a gage factor of 2.0 is available. Do not neglect offset-voltage effects.

19 A crystal microphone has a sensitivity of 1 mV rms at a sound level of 100 dB and capacitance of 100 pF. Design an amplifier which will result in an output of 1 V rms at a sound level at 100 dB. The frequency response should be 20 Hz to 100 kHz.

20 Design an electronic readout for a pressure transducer based on a variable inductor sensor. Assume that the inductor varies from 5 mH at zero pressure to 7 mH at 1 kg/cm² (full scale). The maximum recommended operating frequency is 10 kHz. A digital voltmeter (0 to 1 V dc) is available as the indicator. A phase-sensitive detector or precision rectifier as an ac/dc converter is suggested.

21 A peak-reading photometer is required to measure flash intensity for photography. Assume a phototransistor with a sensitivity of 1 μA/lm is provided. An output of 100 lm (full scale) is desired, indicated by a microammeter-type (panel-type) voltmeter. The reading should hold following the flash until reset by a push-button switch.

22 Design a readout circuit for a conducting-film type of moisture sensor. Assume for simplicity that the conductance G_s of the sensor increases in proportion to the relative humidity such that 0 percent of humidity corresponds to $G_s = 0$ and 100 percent corresponds to $G_s = 0.01$ mS.

23 An accelerometer based on the electromagnetic sensor is required. The sensor (Fig. 8-4a) sensitivity K is 10 mV · s/cm. An output voltage of 1 V for an acceleration of 980 cm/s², that is, acceleration of gravity, is desired. The maximum frequency response required is 1 kHz. Consult Chap. 11 for differentiator design.

24 Design a device which will read liquid level in a storage tank of gasoline using a capacitance-type gage. The tank is cylindrical (2 m in diameter, 5 m tall), and a single dipstick sensor in the center is specified (not separate coaxial cylinder as in Fig. 9-12). The dielectric constant of gasoline is 2.2. An accuracy of 5 percent is adequate, and a digital voltmeter 0 to 1 V dc full scale is available. Consult Chap. 15 for an ac-to-dc converter.

25 Design a battery-operated EKG monitor which will flash an LED indicator each heartbeat. Differential input is not essential with battery operation. Allow for an unpredictable electrode-polarization (dc) voltage offset up to 100 mV.

Part Three

Signal Amplification and Processing

Chapter Eleven

Signal Filtering

Analog signals are filtered according to frequency to a greater or lesser degree in most electronic instruments. Sometimes filtering is introduced as a result of an amplifier nonideality, e.g., limited frequency response or the necessity of blocking a dc offset. In other cases the signal itself is not in the form desired and must be shaped or otherwise processed.

This chapter is primarily concerned with active filters which affect the frequency and time response of amplifiers. Integrators, differentiators, and simple (single-pole) RC filters fall into this category, as do the sharp-cutoff-frequency filters more commonly associated with active-filter design. While the analysis of simple filters is not difficult mathematically, an understanding of higher-order filters requires a good background in circuit analysis. To make it easier for the reader uninterested in mathematical details, the higher-order filters are described in terms of summary design equations. Derivations of equations are deferred until the end of the chapter.

11-1 Ideal Filters

Perhaps the most common form of signal processing is the decomposition of a signal into its frequency components by means of a filter. Ideal-filter response is illustrated in Fig. 11-1. An ideal high-pass filter will pass all frequencies above a chosen break frequency f_x without attenuation but allow no frequencies below f_x to pass. A low-pass filter does the reverse, and a bandpass filter allows only frequencies within $\Delta f_0/2$ of a center frequency f_0 to pass. A notch, or band-reject, filter is the complement of a

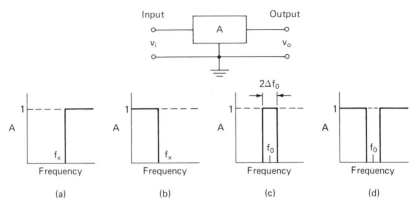

Fig. 11-1 Frequency response of ideal filters: (*a*) high-pass; (*b*) low-pass; (*c*) bandpass; and (*d*) notch.

bandpass filter. Nonideal filters do not show an infinitely sharp transition at the break frequency, but real filters approaching the ideal are achievable. Actually for most purposes a broad transition, perhaps covering a decade or more in frequency, is quite adequate.

Phase shifts are associated with all real filters. According to an electric-network theorem, the slope of a passive filter gain response on a log-log scale is related to filter phase shift ϕ in radians

$$\phi = \frac{\pi}{2} \frac{d(\log |A|)}{d(\log f)} \tag{11-1}$$

where $|A|$ is the magnitude of the response, $|v_o|/|v_i|$. If the frequency response of a filter (or any network) is known, the phase shift can be inferred. Phase response is important in determining the stability of control systems and other feedback networks, but since for many instrumental applications only the signal amplitude is of interest, the phase response can often be ignored.

Filter response can be expressed in three ways: in the time domain as a differential equation, in the frequency domain as a frequency response A, or in the s domain as a Laplace transform or transfer function H. In the following sections, alternate expressions are given where appropriate.

Sharp cutoff filters can distort signals which have Fourier frequency components in the vicinity of the break frequency, an inherent drawback. An example is given in Fig. 11-2. In this respect an ideal filter does not produce an ideal result. The tailoring of filters to minimize distortion of various classes of signals is a well-developed subject but outside the scope of this text.

Input   Output

Fig. 11-2   Response to a step function of a sharp (near-ideal) low-pass filter.

11-2   Single-Stage High- and Low-Pass Circuits

Single-section RC high- and low-pass filters (Fig. 11-3), which separate the higher frequencies from the lower-frequency or dc signal components,

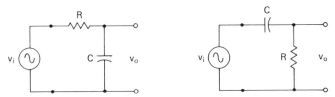

Fig. 11-3   Single-section RC filters: (a) low-pass; (b) high-pass.

abound in electronic instruments. Analysis of these circuits is uncomplicated if the output is unloaded. In either circuit the input current is $v_i/(R + Z_c)$, where $Z_c = -j/\omega C$. The output voltage is iZ_c for the low-pass and iR for the high-pass, for which the gain ($A = v_o/v_i$) is

$$|A_L| = \frac{1}{\sqrt{1 + (\omega/\omega_x)^2}} \qquad \text{low-pass} \qquad (11\text{-}2a)$$

$$|A_H| = \frac{1}{\sqrt{1 + (\omega_x/\omega)^2}} \qquad \text{high-pass} \qquad (11\text{-}3a)$$

where $\qquad\qquad \omega_x = \dfrac{1}{RC} \quad \text{or} \quad f_x = \dfrac{1}{2\pi RC} \qquad\qquad (11\text{-}4)$

Here $|A|$ indicates the gain magnitude. If the phase shift is required, Eqs. (11-2a) and (11-3a) become, in phasor form,

$$A_L = |A_L|e^{j\phi_L} \qquad (11\text{-}2b)$$
$$A_H = |A_H|e^{j\phi_H} \qquad (11\text{-}3b)$$

where $\qquad\qquad \tan\phi_L = -\dfrac{f}{f_x} \quad \text{and} \quad \tan\phi_H = \dfrac{f_x}{f} \qquad (11\text{-}5)$

Plots of gain are shown in Fig. 11-4*a* and *b* for $|A|$ expressed in decibels, where $|A \text{ (dB)}| = 20 \log A|$.

Interpretation of f_x as a break frequency is apparent from the logarithmic frequency plots but less obvious on a linear scale (Fig. 11-4*d*). At the breakpoint $|A_H|$ or $|A_L|$ is at -3 dB, which on a linear scale is $1/\sqrt{2} =$

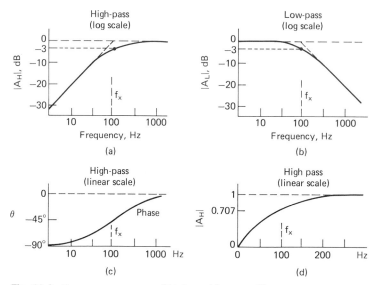

(a)

(b)

(c)

(d)

Fig. 11-4   Frequency response of high- and low-pass filters.

For $\omega_x = 100$ rad/s or $f_x = 17$ Hz

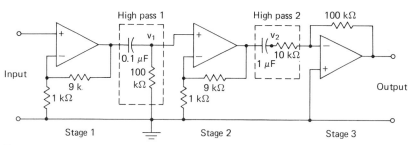

Fig. 11-5   Examples of high-pass filters in interstage coupling.

0.707. Note that the capacitor impedance and resistance are equal in magnitude at $f = f_x$.

A common application of the high-pass circuit is an interstage coupling of an ac amplifier, as illustrated in Fig. 11-5. It eliminates the need for compensating dc offset because the direct current is blocked between stages. In this example each stage has a gain of 10, and both high-pass filters have a break frequency of 17 Hz. For the first filter it is clear that v_1 corresponds to v_o of Fig. 11-2, but it may not be obvious that v_2 of the second filter corresponds to the filter v_o until it is recognized that the input v_- of the third op-amp is at virtual ground.

An alternative interpretation of the high and low pass, particularly in

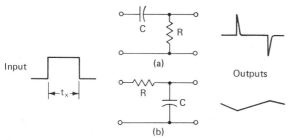

Fig. 11-6 (*a*) Differentiating networks and (*b*) integrating network.

pulse applications, is that of differentiating and integrating networks, respectively. When the filter is used as a differentiating network, the break frequency f_x is usually set so that the time constant $\tau = RC$ is shorter than the pulse width t_w, as shown in Fig. 11.6*a*. In this case only the edges of the pulse are passed. If τ is sufficiently small, the output signal approaches the derivative of the input signal but this simple circuit is inferior to the differentiator (see Sec. 11-4) in this application. When it is operated as an integrating network, the time constant $\tau = RC$ is made much larger than the pulse width. It is not much used because the active integrator circuit (see below) is superior for this application.

11-3 Integrator-Averager (Active Low-Pass)

A single-section active low-pass circuit is shown in Fig. 11-7. It is identical to the simple integrator except for the addition of the resistor R_b, normally a high value, which can be thought of as a reset mechanism with a long time constant. It acts as an amplifier at sufficiently low frequencies or dc. It can be analyzed as follows, again under the infinite-gain approximation:

$$i = \frac{v_a}{R_a} = \frac{-v_o}{R_b} - \frac{C\,dv_o}{dt} \tag{11-6a}$$

$$I(s) = \frac{V_i(s)}{R_a} = \frac{-(1 + sCR_b)}{R_b}\,V_o(s) \tag{11-6b}$$

In this case the gain for a sinusoidal input, ignoring the phase shift, is

$$A = \frac{|v_o|}{|v_i|} = \frac{R_b/R_a}{\sqrt{1 + (f/f_x)^2}} \tag{11-7a}$$

which corresponds to a transfer function $H(s)$ of

$$H(s) = \frac{V_o(s)}{V_i(s)} = \frac{-1/R_aC}{s + 1/R_bC} \tag{11-7b}$$

It can be seen from Eq. (11-7a) that the circuit acts as an inverting amplifier (Chap. 4) and low-pass filter (Sec. 11-2) in series.

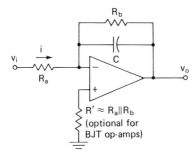

Fig. 11-7 Active low-pass filter amplifier.

Its frequency response is identical to the low-pass filter response of Fig. 11-4a except that the gain axis is shifted to take into account the gain of the amplifier (R_b/R_a at dc).

The low-pass filter acts as a time averager, where the time over which the average is taken is roughly the time constant of the circuit R_bC. By contrast an integrator ideally averages over an infinite time.

11-4 Differentiator (Active High-Pass)

The inverse equivalent of the integrator is the differentiator circuit, shown in Fig. 11-8. Its output v_o is proportional to the derivative of the input signal v_i, which, assuming the infinite-gain approximation, can easily be derived as

$$v_o = -R_bC \frac{dv_i}{dt} \tag{11-8}$$

Note that the factor R_bC acts as a gain and the output is inverted.

If v_i is sinusoidal ($v_a \sin \omega$), the output will be proportional to ωv_a; that is, the gain increases linearly with frequency. The implication in terms of signal quality is that high-frequency noise, ever present in the signal input, will be highly amplified and perhaps overwhelm the desired signal. To minimize noise in the practical version of the circuit, a series input resistor R_a is added to limit the high-frequency gain to $-R_b/R_a$. Sometimes a small capacitor C' is also added, as shown in Fig. 11.8b, to further reduce the high-frequency gain. Proceeding as with the active low-pass filter, we see that the gain analysis (without C') yields for a sinusoidal input

$$|A| = \frac{|v_o|}{|v_i|} = \frac{2\pi f C R_b}{\sqrt{1 + (f/f_x)^2}} \quad \text{where } f_x = \frac{1}{2\pi R_a C} \quad (11\text{-}9)$$

In terms of a transfer function this becomes

$$H(s) = \frac{-R_b C s}{1 + R_a C s} \quad (11\text{-}10)$$

In Fig. 11-9 the frequency response implied by Eq. (11-9) but with C' added is plotted.

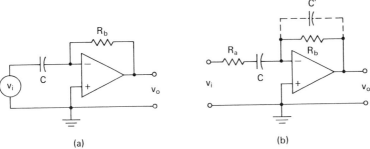

Fig. 11-8   Differentiators or active high-pass filter: (a) basic circuit; (b) practical version.

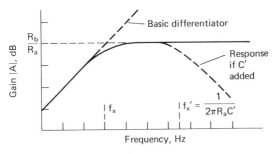

Fig. 11-9   Differentiator frequency response.

11-5 LC **Bandpass Filters**

Bandpass filters based on inductor-capacitor resonance are widely employed in high-frequency communication circuits. At lower frequencies LC networks are less satisfactory and not often used, largely because the higher-value iron-core inductors needed in the lower-frequency range are bulky, costly, and nonlinear. For these reasons, the discussion of LC

networks will be brief. The most common means of coupling a signal to an LC parallel resonance filter is via a drive circuit approximating a current source (Fig. 11-10). At resonance ($f = f_0$) the impedances of the inductor and capacitor are equal in magnitude but opposite in sign, so that the net impedance of the circuit is R_p and the voltage at resonance v_0

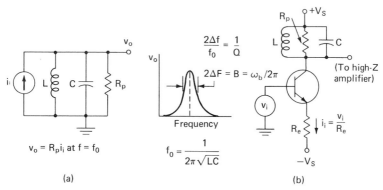

Fig. 11-10 LC resonance-bandpass filter: (*a*) equivalent circuit; (*b*) transistor version.

is then $R_p i_i$. Away from resonance impedance drops rapidly, and consequently so does v_0. It can be shown that the bandwidth ($2\Delta f/f_0$ or -3 dB points) is the inverse of the Q factor, which here is equal to $\omega L/R_p$. If the resistance in series with the inductance (R_L, not shown) is nonzero, in a parallel resonance circuit R_p should be replaced by $R_p \| R_L'$, where $R_L' = (\omega L)^2/R_L$. Since a nonideal current source is equivalent to an ideal source in parallel with a resistor R_i, the effect of a nonideal source is to reduce R_p to $R_p \| R_i$.

A transistor is a good approximation to a constant-current source. In Fig. 11-10 the alternating current is $i_i = i_c = v_i/R_e$, and thus the gain is R_p/R_e. A high-input-impedance amplifier must be connected to the output to avoid loading the resonance circuit. The parallel resonance LC network is commonly used as the collector load of tuned high-frequency amplifiers (Chap. 13).

11-6 Notch and Bandpass Filters

A twin-T notch filter is an RC network which preferentially replaces the LC network in low-frequency applications. The filter has the response indicated in Fig. 11-11. The components must be carefully adjusted to the values shown, and the network must be driven by a low-impedance source and followed by a high-input-impedance amplifier, e.g., by unity-gain amplifiers. If these conditions are not met, the gain $|A|$ at resonance f_0 will not be zero, but with 1 percent matching of the capacitors and resistors $|A|$ will be below 0.01 at f_0.

Perhaps the easiest notch filter to adjust, of those which utilize RC elements, is that based on the Wien bridge (Fig. 11-12). At the notch frequency f_0, the series impedance Z_s is equal (in magnitude and phase) to twice that of the parallel branch Z_p, and thus $v_2 = v_i/3$. Resistors on the inverting side are chosen so that the inverting and noninverting gains

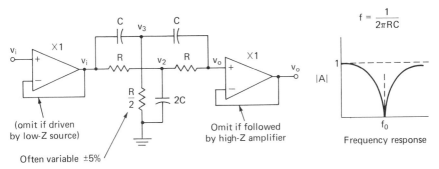

Fig. 11-11 Twin-T notch filter.

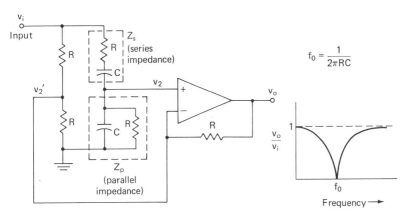

Fig. 11-12 Wien-bridge notch filter.

match (at f_0), giving an output voltage v_o of zero. Close-tolerance components are desirable but not critical since minor adjustment in the notch frequency can be made by inserting a resistor in the RC branch. By varying one of the resistors on the inverting side the transmission at the notch frequency can be set very close to zero.

Adjustment of the center frequency is possible with a comb filter, which can be considered a type of digital filter. A comb filter will reject a fundamental f_0 and its odd harmonics, thus producing a characteristic frequency-response plot (Fig. 11-13), from which the name is derived. It is based on the analog delay line usually implemented with a charge-coupled device, as discussed in Chap. 18.

Operation is based on signal cancellation as a result of phase delay. The delay-line output v_i' simply reproduces the input signal with a time delay τ equal to the sum of the delays $1/f_c$ of each of N identical sections for a total delay of N/f_c. Suppose that a sine wave of frequency f_i is applied to the input and the delay is adjusted so that $2\tau = 1/f_0 = 1/f_i$. In this case the phase of the delayed signal v_i' is 180°, which when summed with the

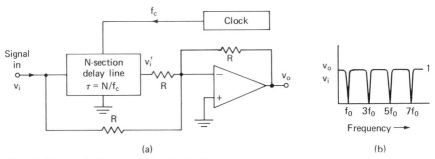

Fig. 11-13 Comb filter: (a) basic circuit; (b) frequency response.

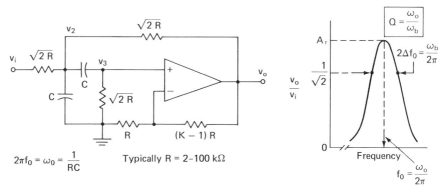

Fig. 11-14 Active bandpass filter.

undelayed signal v_i results is a null output. For a given value of τ, the null output or notch occurs for $f_i = f_0, f_i = 3f, f_i = 5f_0$, etc. Should the delay circuit not be unity-gain, the ratio of the summing resistors can be varied to obtain null. Note that the notch frequency is a multiple of the clock frequency f_c and that control of f_c provides a convenient means of varying f_0.

One of a number of excellent active bandpass filters is shown in Fig. 11-14. It is a combination of the active high- and low-pass filters discussed in the next sections. Tuning to a specific frequency f_0 requires first the selection of the (equal-valued) capacitors and then the calculation of R (and other resistors) according to the relation given. Usually C is chosen as some convenient standard value subject to the restriction that R not be excessively large or small. Close-tolerance components are required to avoid oscillation. The amplifier gain K is chosen to obtain the desired Q, that is, the reciprocal of bandwidth. At peak ($\omega = \omega_0$) the stage gain is A_r (see Table 11-1).

TABLE 11-1   Bandwidth $2\Delta f/f_0$ and Stage Gain A_r of the Active Bandpass Filter as a Function of Amplifier Gain K

K	$\dfrac{2\,\Delta f}{f_0}$	Q	A_r
(2.00)		0.7	(1.0)
3.00	0.71	1.4	3.0
3.50	0.35	2.8	7.0
3.70	0.21	4.7	12.3
3.80	0.14	7.1	19.0
3.85	0.12	9.4	25.7
3.90	0.07	14.1	39.0

11-7   Butterworth Filters

The multiple-section active filters designed by Butterworth are high- or low-pass filters which exhibit a sharper cutoff than simple filters. In general the sharpness of the cutoff and the approach to the ideal (Fig. 11-1) increase with the number of section. For brevity, the discussion here will be restricted to certain popular even-order Butterworth filters.

A key design requirement for Butterworth filters (and many other types of filter as well) is that the magnitude of the gain $|A|$ at the crossover (angular frequency ω_x) be -3 dB or $1/\sqrt{2}$, just as it is for a simple single-section RC filter. Butterworth filters are composed of quadratic active filters which have this property singly or in cascade. A quadratic filter has a transfer function, or gain, expressed by

$$A = \frac{1}{1 + jb\omega/\omega_x - (\omega/\omega_x)^2} \tag{11-11}$$

If b is set equal to $\sqrt{2}$ for one quadratic filter, it is shown in Sec. 11-10 that

$$|A| = \frac{1}{\sqrt{1 + (\omega/\omega_x)^4}} \tag{11-12}$$

The response of a quadratic filter, which is identical to that of a second-order Butterworth filter, is plotted in Fig. 11-15.

Two of many op-amp realizations of the quadratic filter are shown in Fig. 11-16. Both exhibit the response indicated by Eq. (11-12) except for a scaling factor or dc gain. Various combinations of capacitors and resistors will produce the same filter characteristics. While it is possible to set the dc gain A_{dc} to an even value (greater than unity) for the noninverting configuration (Fig. 11-16a), the specification of equal-valued resistors and capacitors was considered more desirable. For the inverting configuration (Fig. 11-16b) the unity gain was the chosen criterion even though odd-

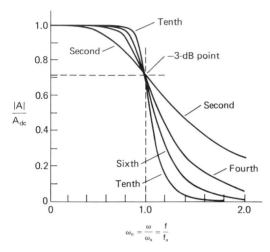

$$\omega_n = \frac{\omega}{\omega_x} = \frac{f}{f_x}$$

Fig. 11-15 Frequency response of Butterworth low-pass filters of several orders.

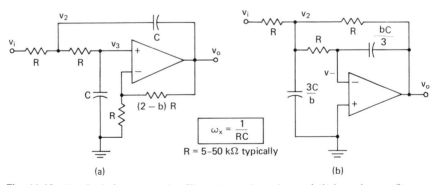

$$\omega_x = \frac{1}{RC}$$

R = 5–50 kΩ typically

(a)   (b)

Fig. 11-16 Quadratic low-pass active filters: (a) noninverting and (b) inverting configurations. Values are given for Butterworth filters ($a = 1$ and b chosen from Table 11-2).

valued capacitors are then required. Note that the parameters ω_x and b are independently adjustable. Since capacitors are available in only a limited range of values, the capacitor value is usually selected first and then R calculated from $R = 1/\omega_x C$. Close-tolerance components are generally required because the filter characteristics and stability may depend critically on several components.

Higher-order Butterworth filters consist of two or more quadratic filters connected in cascade, as indicated in Fig. 11-17. While each filter is identical in form and ω_x is the same, the value of b is different for each filter. The total gain function becomes

$$A = \frac{1}{1 + jb_1\omega/\omega_x - (\omega/\omega_x)^2} \frac{1}{1 + jb_2\omega/\omega_x - (\omega/\omega_x)^2} \cdots \quad (11\text{-}13)$$

If the values of b are selected according to Table 11-2, the gain (magnitude) can be expressed as

$$|A| = \frac{1}{\sqrt{1 + (\omega/\omega_x)^{4n}}} \quad (11\text{-}14)$$

where n is the number of quadratic filters and $2n$ is the order of the Butterworth filter. Although the procedure is tedious, the value of b for

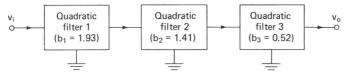

Fig. 11-17 Cascaded quadratic filters as higher-order Butterworth filters. A sixth-order filter is shown.

any even-order Butterworth filter can be derived by substituting Eq. (11-13) into $|A|^2 = AA^*$, where A^* is the complex conjugate (Sec. 11-10), and solving for b as the roots of a polynomial equation. However, it is not often that a filter sharper than sixth or at most tenth order is needed, and therefore Table 11-2 will suffice.

TABLE 11-2 Butterworth Filter Constants $(a = 1)$

Order	b_1	b_2	b_3	b_4	b_5
2	1.414				
4	1.845	0.7654			
6	1.932	1.414	0.5176		
8	1.962	1.663	1.111	0.3896	
10	1.976	1.783	1.414	0.9081	0.3128

High-pass Butterworth filters can be designed with the same design criteria except that the variable ω/ω_x is inverted, i.e., replaced by ω_x/ω in equations for A [Eq. (11-14)]. The values of b in Table 11-2 still hold. Of course the circuit configuration of a high-pass filter differs, as indicated in Fig. 11-18. The first circuit has a high-frequency gain A_{HF} greater than unity. The second (Fig. 11-18b) has unity high-frequency gain but has the disadvantage of low input impedance at high frequencies.

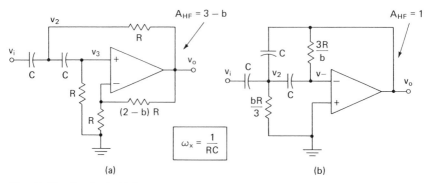

Fig. 11-18 Quadratic high-pass active filters: (*a*) noninverting and (*b*) inverting configurations.

11-8 Chebyshev and Other Active Filters

It must be recognized that multiple-section or higher-order *RC* filters have many degrees of freedom and there are many configurations and sets of component values which will satisfy the minimum design requirement of sharp cutoff. The experienced active-filter designer may appreciate and take advantage of the design flexibility, but the occasional user is likely to be confused by the multiple design criteria. Here various criteria have been fixed (arbitrarily chosen) to simplify the filter design procedure even though the result may be somewhat suboptimal for a particular application. The general parameters to be chosen are the break frequency ($\omega_x = 2\pi f_x$) and the degree of sharpness of the cutoff, which is determined by the order of the filter and the choice of parameters *b* (and *a*) according to some frequency-response criteria. Once these main criteria are satisfied, there are still degrees of freedom left, and then lesser criteria are added, e.g., equal circuit resistances, even capacitor values, or unity stage gain. Not all the lesser criteria can be satisfied simultaneously.

Another active filter, the *Chebyshev filter,* is similar to the Butterworth filter except that it has a sharper cutoff for a given order. It is also formed from cascading quadratic filters. The characteristics of the filter can be specified by the value of ω_x and *b*, but the choice of *b*'s and ω_x differs. Instead of specifying different values of ω_x for each filter, it is better to define ω_x as identical for each filter and generalize the previous quadratic transfer function [Eq. (11-11)], which, written in terms of the Laplace transform, becomes

$$H(s) = \frac{H_0}{a + bs/\omega_x + (s/\omega_x)^2} \tag{11-15}$$

where a = second adjustable constant
H_0/a = dc gain
$s = j\omega$ for sinusoidal response

A disadvantage of the Chebyshev filter is the appearance of gain maxima and minima, i.e., a gain ripple below the break frequency (see Fig. 11-19). A Butterworth filter, by contrast, has a monotonic decrease in gain

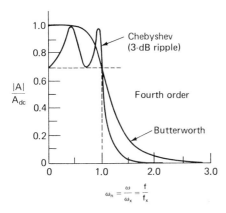

Fig. 11-19 Comparison of a fourth-order Chebyshev filter (3-dB ripple) with a Butterworth filter.

with frequency and a maximally flat response below the cutoff. The deviation from flat response is an adjustable parameter in Chebyshev filter design. It is usually expressed as gain ripple in decibels. Values of b needed to design various Chebyshev filters are listed in Table 11-3.

TABLE 11-3   Chebyshev Low-Pass-Filter Constants (1-dB Ripple)

Order	a_1	b_1	a_2	b_2	a_3	b_3
2†	1.000	1.414				
4	0.2794	0.6737	0.9883	0.2791		
6	0.1247	0.4641	0.5577	0.3397	0.9908	0.1244

†Same as Butterworth.

Both the Butterworth and Chebyshev filters exhibit large phase shifts, especially near and above the break frequency. For applications where phase is important, a minimal phase-shift filter, e.g., Bessel filter, may be preferred even though the frequency cutoff is not very sharp. Some filters, in particular the *all-pass filter*, are intended only to shift phase while allowing all signal frequencies to pass. Detailed discussions of suitable filters can be found in a number of texts on active filters.†

†J. Wait, L. Huelsman, and G. Korn, "Introduction to Operational Amplifier Theory and Applications," McGraw-Hill, New York, 1975; A. Budak, "Passive and Active Network Analysis and Synthesis," Houghton Mifflin, Boston, 1974.

11-9 Gyrators

A gyrator is an active circuit which converts an impedance into its reciprocal. The most common application of the gyrator is a simulated inductance which is equivalent to the reciprocal of the impedance of a capacitor. The gyrator circuit utilizes only resistors and capacitors, in addition to the amplifiers, and therefore a simulated LC (or LR) network can be formed without inductors. As pointed out above, inductors as components have many practical drawbacks especially for miniaturized low-frequency circuits.

Strictly speaking, a gyrator is a two-port network (Fig. 11-20). The

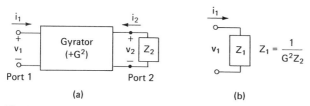

Fig. 11-20   The gyrator (a) as a two-port network; (b) equivalent input impedance.

control impedance Z_2 is connected to port 2, and the reciprocal impedance appears at port 1. The terminals of port 1 act as the terminals of the reciprocal impedance. Transfer of the impedance from ports 2 to 1 is accompanied by a proportionality constant of G^2, where G is a conductance. Often G can be identified with one or more resistors in the actual circuit, but this is not necessarily the case. For the usual case where port 2 is loaded by a capacitor ($Z_2 = -j/\omega C_2$) the impedance at port 1 is

$$Z_1 = \frac{1}{G^2 Z_2} = \frac{+j\omega C_2}{G^2} = +j\omega L_E \qquad (11\text{-}16a)$$

where

$$L_E = \frac{C_2}{G^2} \qquad (11\text{-}16b)$$

If a capacitor C_1 is connected across the port 1 terminals, an equivalent LC resonance circuit will be created which resonates at an angular frequency ω_0 of $1/\sqrt{L_E C_1}$ or

$$\omega_0 = \frac{G}{\sqrt{C_1 C_2}} \qquad (11\text{-}17)$$

One of several op-amp configurations which act as a gyrator is shown in Fig. 11-21. Circuit analysis (Sec. 11-10) reveals that it has a proportionality constant of G^2. Limitations on the amplifier frequency response confine the range of operation to lower frequencies (typically under 100 kHz). Capacitor C_3 is sometimes included to prevent oscillation. It is desirable to choose R so that R and $X_c = 1/\omega C$ are roughly comparable.

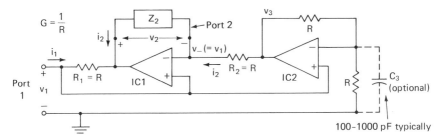

Fig. 11-21   A gyrator op-amp circuit.

11-10   Derivation of Active-Filter Equations

Derivations of the transfer functions $H(s)$ and gains $A(\omega)$ of various filters discussed in this chapter are given below. A familiarity with circuit analysis by nodal and Laplace-transform methods is assumed.

Twin-T Filters

The twin-T filter (Fig. 11-11) is best solved by nodal analysis. It is assumed that the input is connected to voltage generator v_i and the output v_o is unloaded. Summing the currents out of the v_2, v_3, and v_o junctions in turn, we get

$$sC(V_3 - V_i) + sC(V_3 - V_o) + \frac{2V_3}{R} = 0 \qquad (11\text{-}18a)$$

$$\frac{V_2 - V_i}{R} + \frac{V_2 - V_o}{R} + 2sCV_2 = 0 \qquad (11\text{-}18b)$$

$$sC(V_o - V_3) + \frac{V_o - V_2}{R} = 0 \qquad (11\text{-}18c)$$

Variables V_2 and V_3 are eliminated from the three equations, and v_o is expressed in terms of v_i. The method suggested is to multiply all equations by R, solve (11-18c) for V_2, solve (11-18a) for V_3, and substitute V_2 and V_3 into (11-18b). The result is

$$H(s) = \frac{V_o(s)}{V_i(s)} = \frac{1 + (sCR)^2}{1 + (sCR)^2 + 4sCR} \qquad (11\text{-}19a)$$

or, in terms of ω (setting $s = j\omega$),

$$|A(\omega)| = \frac{|y|}{\sqrt{y^2 + (4\omega RC)^2}} \qquad \text{where } y = 1 - (\omega RC)^2 \qquad (11\text{-}19b)$$

Note that $y = 0$ at the notch frequency.

Analysis of the Wien-bridge notch filter (Fig. 11-12) is simplified by treating the series branch impedance Z_s as a factor α times that of the parallel branch Z_p. In general (away from the notch frequency) α is a complex number or $\alpha = \alpha(s)$. The formulation obtained by summing currents at the v_2 and v_2' nodes, assuming the infinite-gain approximation $(v_2 = v_2')$, is

$$\frac{V_2 - V_i}{\alpha Z_p} + \frac{V_2}{Z_p} = 0 \qquad (11\text{-}20a)$$

$$\frac{V_2 - V_1}{R} + \frac{V_2}{R} + \frac{V_2 - V_o}{R} = 0 \qquad (11\text{-}20b)$$

We eliminate V_2 [the suggested method is to solve for V_2 in (11-20b), and substitute into (11-20a)] to get the transfer function

$$H(s) = \frac{V_o(s)}{V_i(s)} = \frac{2 - \alpha}{1 + \alpha} \qquad (11\text{-}21a)$$

where α is obtained by dividing Z_s by Z_p

$$\alpha = \frac{1 + 2sCR + (sCR)^2}{sCR} \qquad (11\text{-}21b)$$

From Eq. (11-21a) it can be seen that $\alpha = 2$ at the notch frequency ($H = 0$). Substituting $\alpha = 2$ and $s = j\omega$ into (11-21b), we get the relation $\omega = 1/RC$. In general ($\omega \neq \omega_0$) for $A(\omega)$ we get

$$|A(\omega)| = \frac{|(\omega\tau)^2 - 1|}{\sqrt{(3\omega\tau)^2 + (1 - \omega^2\tau^2)^2}} \qquad \text{where } \tau = RC \qquad (11\text{-}21c)$$

Active Bandpass

The bandpass filter of Fig. 11-14 can be considered to be made up of an RC network and a noninverting amplifier of gain K. Solving the circuit by nodal analysis, assuming v_i is a voltage generator, we obtain the following relations by summing the currents out of the v_2 and v_3 junctions:

$$V_2 sC + \frac{V_2 - V_i}{\sqrt{2}\,R} + \frac{V_2 - V_o}{\sqrt{2}\,R} + (V_2 - V_3)sC = 0 \qquad (11\text{-}22a)$$

$$(V_3 - V_2)sC + \frac{V_3}{\sqrt{2}\,R} = 0 \qquad (11\text{-}22b)$$

$$V_o = KV_3 \qquad (11\text{-}22c)$$

Variables V_2 and V_3 are eliminated from the three equations, and V_o is expressed in terms of V_i. The suggested method is to multiply (11-22a) and (11-22b) by $\sqrt{2}\,R$, replace RC by τ, substitute (11-22c) into the V_3 terms of (11-22b) and (11-22a) and then (11-22b) into (11-22a). As a result we have

$$H(s) = \frac{V_o(s)}{V_i(s)} = \frac{s\tau\,K/\sqrt{2}}{1 + (s\tau)^2 + s\tau(4 - K)/\sqrt{2}} \qquad (11\text{-}23a)$$

where $\tau = RC$. In terms of $A(\omega)$

$$A(\omega) = \frac{V_o(\omega)}{V_i(\omega)} = \frac{j\omega\tau\,K/\sqrt{2}}{1 - (\omega\tau)^2 + j\omega\tau(4-K)/\sqrt{2}} \qquad (11\text{-}23b)$$

At resonance ($\omega = \omega_0 = 1/\tau$)

$$A(\omega_0) \equiv A_R = \frac{K}{4 - K} \qquad (11\text{-}24)$$

where A_R is the stage gain at resonance. Near resonance ($\omega = \omega_0 + \Delta\,\omega$) the gain is approximately

$$A(\omega) = \frac{A_R}{1 - j\sqrt{2}y/(4 - K)\omega\tau} \approx \frac{A_R}{1 - j2\sqrt{2}(\Delta\omega/\omega_0)/(4 - K)} \qquad (11\text{-}25)$$

where $y = 1 - (\omega RC)^2$. When we set $\Delta\omega = \omega_b/2$, the relative bandwidth (between half-maximum power points) is

$$\frac{\omega_b}{\omega_0} \equiv \frac{1}{Q} \approx \frac{4 - K}{\sqrt{2}} \qquad (11\text{-}26)$$

Here ω_b is the full width of the band (between the -3-dB points at $\omega_0 - \omega_b/2$ and $\omega_0 + \omega_b/2$). Examination of the circuit shows that the stage gain at high Q is the difference of two large quantities, in effect a balance between negative and positive feedback. The gain is thus very sensitive to the exact value of the components. Close-tolerance components are necessary to avoid oscillation.

Quadratic Low-Pass Filters (Noninverting Configuration)

The transfer function of the quadratic low-pass filter of Fig. 11-16a is derived here. It can be seen that part of the circuit is a standard noninverting amplifier configuration of gain $K = 3 - b$, so that $v_o = Kv_3$. Because b lies between 0 and 2, K is between 1 and 3. By summing currents at the v_2 and v_3 junctions we get

$$\frac{V_2 - V_i}{R} + \frac{V_2 - V_3}{R} + (V_2 - V_o)sC = 0 \qquad (11\text{-}27a)$$

$$\frac{V_3 - V_2}{R} + V_3 sC = 0 \qquad (11\text{-}27b)$$

$$V_o = V_3 K = (3 - b)V_3 \qquad (11\text{-}27c)$$

Variables V_2 and V_3 are eliminated from the three equations, and V_o is expressed as a function of V_i. The suggested procedure is to multiply (11-27a) or (11-27b) by R, substitute (11-27c) into (11-27a) and (11-27b), eliminate V_3, and then substitute V_2 from (11-27b) into (11-27a). The transfer function is

$$H(s) = \frac{V_o(s)}{V_i(s)} = \frac{3 - b}{1 + b(s/\omega_x) + (s/\omega_x)^2} \qquad \text{where } \omega_x = \frac{1}{RC} \qquad (11\text{-}28)$$

Expressed in terms of H_{dc} for $a = 1$ (Table 11-2, Butterworth filters), this equation becomes

$$H(s) = \frac{H_{dc}}{1 + bs/\omega_x + (s/\omega_x)^2} \qquad (11\text{-}29)$$

Substituting $s = j\omega$ gives $A(\omega)$. It is identical to Eq. (11-11) except for a multiplying constant of dc gain of H_{dc}. The more general form of the transfer function, where $a \neq 1$, is obtained by multiplying numerator and denominator by a, and redefining b and ω_x; thus

$$H(s) = \frac{H_0}{a + b's/\omega_x' + (s/\omega_x')^2} \qquad (11\text{-}30a)$$

where now

$$\omega_x' = \frac{1}{\sqrt{a}\, RC} \qquad (11\text{-}30b)$$

$$b' = b\sqrt{a} \qquad (11\text{-}30c)$$

$$H_0 = aH_{dc} = 3a - b'/\sqrt{2} \qquad (11\text{-}30d)$$

Quadratic Low-Pass Filters (Inverting Configurations)

The transfer function of the inverting low-pass filter of Fig. 11-16b is derived here. As usual, the infinite-gain approximation will be assumed, so that $v_- = 0$. By assuming the currents out of the v_2 and v_- junctions we get

$$\frac{V_2 - V_i}{R} + \frac{V_2 - V_o}{R} + \frac{V_2 3sC}{b} + \frac{V_2}{R} = 0 \qquad (11\text{-}31a)$$

$$-\frac{V_2}{R} - \frac{V_obsC}{3} = 0 \qquad (11\text{-}31b)$$

To get the transfer function we eliminate V_2 from the two equations; the suggested procedure is to multiply both equations by R, solve (11-31b) for V_2, and substitute into (11-31a). Then we have

$$H(s) = \frac{V_o(s)}{V_i(s)} = \frac{-1}{1 + bs/\omega_x + (s/\omega_x)^2} \qquad \text{where } \omega_x = \frac{1}{RC} \quad (11\text{-}32)$$

With the substitution of $j\omega$ for s this equation becomes identical to Eq. (11-11) except for the negative sign (as expected for an inverting amplifier).

The more general form of the transfer function, where $a \neq 1$, as required for Chebyshev filters [see Eq. (11-15)], is obtained by multiplying numerator and denominator by a and redefining b and ω_x so that the transfer function becomes

$$H(s) = \frac{H_0}{a + b'(s/\omega_x') + (s/\omega_x')^2} \qquad (11\text{-}33)$$

where now $\qquad H_0 = -a \qquad \omega_x' = \frac{1}{\sqrt{a}\,RC}$

and $b' = \sqrt{a}\,b$ replaces b in the circuit of Fig. 11-16b as well as Eqs. (11-31) and (11-32).

Butterworth Low-Pass-Filter Transfer Functions

The magnitude of the low-pass Butterworth filter transfer function $|A(\omega)|$, expressed in terms of frequency, is derived here in terms of one or more quadratic-filter transfer functions [see Eqs. (11-11), (11-29), and (11-32)], that is, Eqs. (11-12) to (11-14) are derived with the results plotted in Fig. 11-15.

Consider a single quadratic filter which is equivalent to a second-order Butterworth filter for which $b = \sqrt{2}$ (Table 11-2). To obtain the magnitude, first the square of the magnitude $|A|^2$ is found, using the relation $|A|^2 = AA^*$, where A^* is the complex conjugate

$$|A|^2 = AA^* = \frac{1}{1 - (\omega/\omega_x)^2 + jb(\omega/\omega_x)} \frac{1}{1 - (\omega/\omega_x)^2 - jb(\omega/\omega_x)} \qquad (11\text{-}34a)$$

$$|A|^2 = \frac{1}{[1 - (\omega/\omega_x)^2]^2 + b^2(\omega/\omega_x)^2} \qquad (11\text{-}34b)$$

Expanding, setting $b = \sqrt{2}$, canceling terms, and taking the square root, we obtain Eq. (11-12).

Consider next a fourth-order Butterworth filter consisting of two quadratic filters in series. The total gain is the product of the individual gains, A_1 and A_2, or

$$A = A_1A_2 = \frac{1}{[1 - (\omega/\omega_x)^2 + jb_1\omega/\omega_x][1 - (\omega/\omega_x)^2 + jb_2\omega/\omega_x]} \quad (11\text{-}35a)$$

For simplicity, this can be rewritten in terms of a normalized frequency ω_n as

$$A = \frac{1}{(1 - \omega_n^2 + jb_1\omega_n)(1 - \omega_n^2 + jb_2\omega_n)} \quad \text{where } \omega_n = \frac{\omega}{\omega_x} \quad (11\text{-}35b)$$

Again multiplying by the complex conjugate to obtain $|A|^2$ gives

$$|A|^2 = \frac{1}{[(1 - \omega_n^2)^2 + b_1^2\,\omega_n^2][(1 - \omega_n^2)^2 + b_2^2\,\omega_n^2]} \quad (11\text{-}36)$$

Expanding, we get

$$|A|^2 =$$

$$\frac{1}{1 + \omega_n^8 + \omega_n^2(b_1^2 + b_2^2 - 4) + \omega_n^4(6 - 2b_1^2 - 2b_2^2 + b_1^2b_2^2) + \omega_n^6(b_1^2 + b_2^2 - 4)} \quad (11\text{-}37)$$

To fit the desired Butterworth form, all terms other than ω_n^8 and the constant ($=1$) must be set equal to zero. The coefficients of the ω_n^2 and ω_n^6 terms are identical, namely

$$b_1^2 + b_2^2 - 4 = 0 \quad (11\text{-}38a)$$

The coefficient of the ω_n^4 term is

$$6 - 2(b_1^2 + b_2^2) + b_1^2b_2^2 = 0 \quad (11\text{-}38b)$$

Substituting (11-38a) into (11-38b) gives the relation in b_1 or b_2

$$b^4 - 4b^2 + 2 = 0 \quad (11\text{-}39a)$$

with solutions

$$b_1 = 1.8478 \quad \text{and} \quad b_2 = 0.7654 \quad (11\text{-}39b)$$

It should be noted that the desired constants, b_1 and b_2 [Eqs. (11-38), etc.], are the two positive roots of Eq. (11-39a). These fourth-order Butterworth filter constants are listed in Table 11-2.

The procedure for finding the values of b for higher-order Butterworth filters is similar. An expression equivalent to Eq. (11-37) is obtained and values of b chosen so that all terms in ω^{2n} are zero except the highest power (and constant term, -1). The resulting values of b for even order up to the tenth are listed in Table 11-2. The range of b is 0 to $+2$.

Odd-order Butterworth filters are practical and have properties similar to those of even order. Circuits other than the quadratic filter are required to implement them, however, and to save space are not discussed here.

Quadratic High-Pass Filter (Noninverting Configuration)

The transfer function of the quadratic high-pass filter of Fig. 11-18a is derived here. By analogy with that of the low-pass filter discussed above, the current sums at the v_2 and v_3 nodes are

$$(V_2 - V_i)sC + (V_2 - V_3)sC + \frac{V_2 - V_o}{R} = 0 \qquad (11\text{-}40a)$$

$$(V_3 - V_2)sC + \frac{V_3}{R} = 0 \qquad (11\text{-}40b)$$

$$V_o = (3 - b)V_3 = KV_3 \qquad (11\text{-}40c)$$

Variables V_2 and V_3 are eliminated to obtain

$$H(s) = \frac{V_o(s)}{V_i(s)} = \frac{3 - b}{1 + b\omega_x/s + (\omega_x/s)^2} \qquad (11\text{-}41)$$

which is similar to Eq.(11-29) except that s/ω_x is replaced by ω_x/s.

Quadratic High-Pass Filter (Inverting Configuration)

The transfer function of the quadratic high-pass filter of Fig. 11-18b is derived here. By analogy with the low-pass circuit discussed above [Eq. (11-31)] the current sums at the v_2 and v_- nodes are

$$(V_2 - V_i)sC + (V_2 - V_o)sC + \frac{3V_2}{bR} + V_2sC = 0 \qquad (11\text{-}42a)$$

$$-V_2sC - \frac{V_o b}{3R} = 0 \qquad (11\text{-}42b)$$

Eliminating V_2 from the two equations, we get

$$H(s) = \frac{-1}{1 + b\omega_x/s + (\omega_x/s)^2} \qquad (11\text{-}43)$$

which is similar to Eq. (11-32) except that s/ω_x is replaced by ω_x/s.

Normalized Frequency Equations

Several variations of the transfer functions are described in order to facilitate comparison of the equations here with those of other texts. Primarily these are normalizations which simplify the notation and facilitate derivation and analysis.

Useful normalized variables are

$$\tau = RC = \frac{1}{\omega_x} \qquad (11\text{-}44a)$$

$$s_n = \frac{s}{\omega_x} = s\tau \qquad (11\text{-}44b)$$

$$\omega_n = \frac{\omega}{\omega_x} = \omega\tau \qquad (11\text{-}44c)$$

The normalization is equivalent to choosing $\tau = RC = 1$. Sometimes the normalization is expressed as $R = C = 1$, which implies not only frequency normalization ($\omega_x = 1$) but also that the value of R or C can be chosen arbitrarily. In any case, the quadratic low-pass transfer functions for $a = 1$ (Butterworth filters) become

$$H(s) = \frac{H_{dc}}{1 + bs_n + s_n^2} \qquad (11\text{-}45a)$$

$$A(\omega) = \frac{H_{dc}}{1 + jb\omega_n - \omega_n^2} \qquad (11\text{-}45b)$$

The more general transfer function ($a \neq 1$) is

$$H(s) = \frac{H_0}{a + b's_n' + (s_n')^2} \qquad (11\text{-}46)$$

where
$$H_0 = aH_{dc} \qquad (11\text{-}47a)$$

$$s_n' = \sqrt{a}\, s_n = \sqrt{a}\, \tau s = \sqrt{a}\, RCs = \frac{\sqrt{a}\, s}{\omega_x} \qquad (11\text{-}47b)$$

$$\omega_n' = \sqrt{a}\, \omega_n = \sqrt{a}\, \tau\omega = \sqrt{a}\, RC = \frac{\sqrt{a}\, \omega}{\omega_x} \qquad (11\text{-}47c)$$

$$b' = \sqrt{a}\, b \qquad (11\text{-}48)$$

For $a \neq 1$ the break frequency ω_x' of each quadratic filter differs by a factor of a from the break frequency ω_x of the composite filter, which consists of several quadratic filters in cascade. Since the value of a differs for each quadratic filter, the value of $\tau = RC$ is different for each.

Calculation of actual component values from the normalization parameters can be done as follows. The parameters H_{dc} and b [Eq. (11-45b)] or H_0, b', and a [Eq. (11-46)] do not depend on the specific op-amp circuit. It is assumed that the crossover frequency $f_x = \omega_x/2\pi$ (-3-dB point) is given. From this the RC product is calculated

$$RC = \tau = \frac{1}{\omega_x} = \frac{1}{2\pi f_x} \qquad (11\text{-}49)$$

Knowing ω_x (or τ), we can express the unnormalized variables (s or ω) in terms of the normalized variables (s_n or ω_n) through Eqs. (11-44).

Next specific values of R and C are calculated. Of course the relations $RC = 1$ or $R = C = 1$ do not hold for the unnormalized variables. Either R or C can be chosen arbitrarily, and then the other is calculated by Eq. (11-49). Practical considerations limit the choice of R and C. Values of capacitors are readily available in rather large steps (0.010, 0.02, 0.05 μF, for example), and close-tolerance (under 2 percent) or variable capacitors are uncommon. Another problem is that higher-value capacitors (over about 0.5 μF) are available in electrolytic form, which is rather unstable and must be polarized. By contrast, stable resistors in 1 percent increments and variable (trim) resistors of nearly any range are generally available. However, most op-amp circuits work best if the resistance level is between about 1 and 100 kΩ. The usual design procedure therefore is to select a capacitor from the limited set of values available and then to calculate the resistor value required (within the desired range)

$$R = \frac{1}{\omega_x C} \tag{11-50}$$

If high precision is required, the resistor can be made variable over a limited range (± 10 percent) to compensate for imprecise capacitor values.

Gyrator

The response of the gyrator op-amp of Fig. 11-21 is derived here. More specifically the current i_1 at port 1 due to an applied voltage of v_1 and a load of Z_2 across port 2 is calculated. Amplifier IC2 is a noninverting amplifier of gain 2, so that $v_3 = 2v_1$. When we use the infinite-gain approximation, we see that the two inputs of IC1 are equal, so that $v_- = v_1$. Analysis shows that the currents through R_1 and R_2, that is, i_1 and i_2, are

$$i_1 = \frac{v_1 - (v_1 + v_2)}{R_1} = -v_2 G \tag{11-51a}$$

$$i_2 = \frac{v_3 - v_1}{R_2} = v_1 G \tag{11-51b}$$

$$i_2 = \frac{-v_2}{Z_2} \tag{11-51c}$$

where the relations $v_3 = 2v_1$ and $R_1 = R_2 = 1/G$ have been utilized. Eliminating v_2 and i_2 between these three equations gives

$$i_1 = G^2 Z_2 v_1 \tag{11-52}$$

and thus the effective input impedance Z_1 is $1/G^2 Z_2$.

Chapter Twelve

Oscillators and Signal Sources

In this chapter various oscillators and waveform generators are discussed. For the most part the circuits selected are relatively simple and are intended for applications where the versatility of commercial waveform generators, in particular wide tuning range, is not required. Aside from tuning range, the most critical waveform-generator characteristics are amplitude control, frequency stability, and low distortion. For most applications many of these characteristics are unimportant, and therefore the relatively uncomplicated waveform generators given here may prove quite adequate.

12-1 Wien-Bridge Oscillators

An excellent sine-wave generator at low frequencies is the Wien-bridge oscillator. Two simplified circuits are shown in Fig. 12-1. As with other sine-wave oscillators, the conditions for sustained oscillation are that (1) the phase is 0° at the operating frequency (positive feedback) and (2) the closed-loop gain (product of stage gain and feedback network gain) is exactly unity. The zero phase condition is met by the feedback network, where a phase lead of approximately 45° by the series network (R,C) is matched by an equal phase lag by the parallel network ($10C$, $R/10$). In order to meet the gain condition a nonlinear element is provided which reduces the gain at high signal amplitude. The stability control (amplifier gain) is adjusted so that the closed-loop gain is slightly greater than unity (divergent oscillations) for small signals. The oscillation amplitude

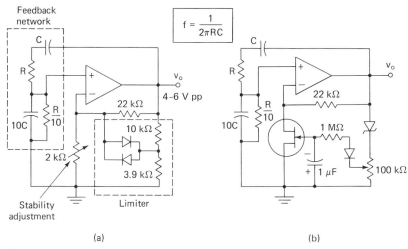

$$f = \frac{1}{2\pi RC}$$

(a)   (b)

Fig. 12-1 Simplified Wien-bridge oscillators.

increases until it self-limits (another type of limiter is shown in Fig. 12-3). Some degree of distortion is inherent. Without the limiter the waveform approaches a square wave, and the frequency shifts to a lower value. It can be minimized by adjusting the gain to the verge of oscillation, but at this point the stability is also minimal and the oscillation might stop with only a slight change of the feedback network with temperature or age. More detailed analysis of the Wien bridge can be found in Sec. 11-10.

In the second version of the Wien-bridge oscillator an FET type of automatic gain control is provided. The ac output is rectified, filtered, and applied to the FET gate. As the gate voltage (negative for an n channel) rises, the FET resistance also rises and the stage gain drops, thus lowering the ac amplitude. The amplitude depends on the FET characteristics, and the regulation is modest.

More precise amplitude control and lower distortion are provided by the circuit of Fig. 12-2. The amplifier (IC1) and feedback network are the same as in the circuits above. Rectified current proportional to the signal voltage flows through D1, which is matched under constant amplitude by a current of opposite sign flowing through R_2. An unbalanced current causes the integrator output (IC2) to change and thus cause a compensating change in the FET resistance and amplifier gain. Diode D2 compensates for the voltage drop across D1, and the filter at the noninverting input of IC2 compensates for any dc bias on the output v_o.

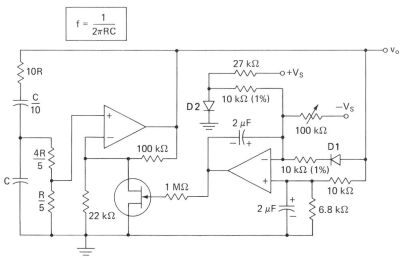

Fig. 12-2 Amplitude-controlled Wien-bridge oscillator.

12-2 Phase-Shift Oscillators

Phase-shift oscillators are convenient sine-wave generators at low frequency. A common version of the phase-shift oscillator is shown in Fig. 12-3. Its principle of operation is similar to that of the Wien-bridge

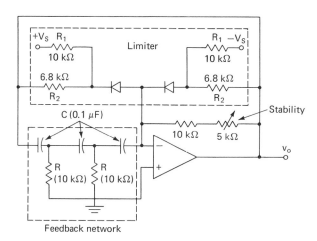

Fig. 12-3 Phase-shift oscillator.

oscillator except that here the amplifier is inverting and the feedback (phase-shift) network shifts the phase by 180° (approximately 60° per section) to meet the 0° closed-loop phase-shift requirement. The unity-closed-loop-gain requirement is met by adjusting the small-signal gain to slightly greater than unity (divergent condition) and limiting (clipping) the large-signal amplitude by a nonlinear element. The diodes conduct and therefore reduce the amplifier gain when forward-biased. The point at which limiting begins is determined by the ratio of R_1 to R_2. Alternately, the limiter of Fig. 12-1a may be used.

12-3 Colpitts Sinusoidal Oscillator

At high frequencies the Colpitts oscillator (Fig. 12-4) is usually the most satisfactory. It is easily tuned by an untapped variable inductor, which at frequencies over perhaps 30 kHz is of reasonably small physical size and oscillates over a wide frequency range. An emitter-follower buffer or op-amp unity-gain amplifier may be used for output isolation. Replacement of Q1 with a BJT works well, but the base should be connected to a resistor voltage divider of 2 to 3 V dc with a bypass capacitor to provide an ac ground for the base.

Analysis of the Colpitts oscillator is not as simple as it may appear

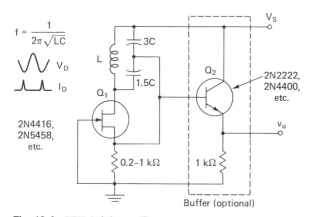

Fig. 12-4 FET Colpitts oscillator.

because it is basically a nonlinear circuit. Over most of the cycle Q1 is cut off. On negative (or less positive) peaks Q1 conducts, and the resonance circuit receives an impulse which compensates for resistive losses. Distortion is minimal for high-Q circuits. An IC version of the circuit based on the amplifier described in Chap. 13, is shown in Fig. 12-5.

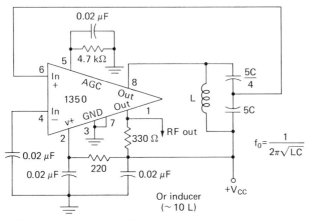

Fig. 12-5 IC Colpitts oscillator.

12-4 Pulse Generators

One of the most convenient pulse generators is the astable configuration (Fig. 12-6) of the IC timer. Pulse width and frequency f are controlled by the indicated RC products by adjusting the ratio of R_A to R_B. For narrow

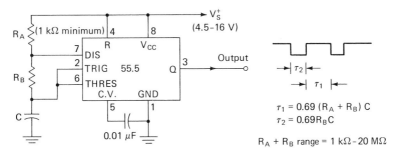

Fig. 12-6 Pulse generator based on the astable IC timer.

pulses, $R_B \ll R_A$ and $f \approx 1.1/R_A C$. Since R_A cannot be zero, this circuit will not produce a completely symmetric square wave. Operation of IC timers is discussed in Chap. 17.

Another version of the pulse generator is based on digital ICs. The inverters (Fig. 12-7) alternately charge the capacitors, which force the output of the following inverter to the high state. Actually R' and the diode can be omitted, in which case the output waveform is approximately square.

A simple pulse generator based on the unijunction transistor (UJT) is the circuit of Fig. 12-8. As the capacitor charges to a threshold voltage, the transistor conducts and partially discharges the capacitor. The conduction point is determined by the transistor characteristic (intrinsic standoff ratio) and is typically 40 to 70 percent of the supply voltage. As pointed out in Chap. 2, a PUT is functionally equivalent except that the threshold voltage V_T is adjustable by choosing the ratio of external resistors.

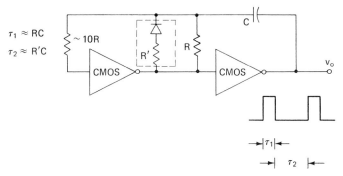

Fig. 12-7   Pulse generator based on two inverters.

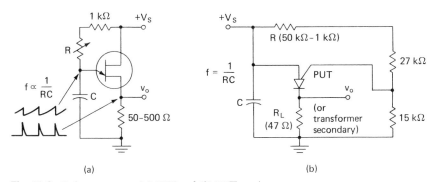

Fig. 12-8   Pulse generator: (a) UJT and (b) PUT versions.

Triggered pulse generators are identical in function to the monostable or one-shot multivibrators and to timer circuits discussed elsewhere. An op-amp type of pulse generator has been discussed previously (Fig. 4-22).

12-5   Square-Wave Generators

Several versions of a square-wave generator have been discussed. Both the op-amp version (Fig. 4-22) and digital IC (Figs. 12-6 and 12-7) of the pulse generators have adjustable width and can be converted into an approximately symmetric square wave with careful adjustments.

Other versions of the square-wave generator are discussed as an aspect of the triangular-wave generator (Fig. 12-11) and of the IC function generator (Fig. 12-17).

If high symmetry is required, a suitable circuit consists of a frequency divider (single-stage binary counter) driven by a stable pulse generator at twice the output frequency (Fig. 12-9).

A satisfactory square-wave generator consists of a stable sine-wave oscillator followed by a fast (digital-style) comparator, as illustrated in Fig.

Fig. 12-9   High-symmetry square-wave generator.

12-10. Symmetry can be trimmed by adjusting the offset voltage or by returning one input of the comparator to a variable voltage source instead of to ground (see Fig. 4-19).

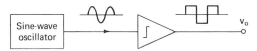

Fig. 12-10   Square-wave generator.

12-6 Triangular-Wave Generators

Nearly all waveforms which involve a linear rise or ramp utilize an integrator connected to a fixed or switched voltage source. A simple ramp generator is shown in Fig. 12-11. Here IC2 is an integrator, and IC1 is a hysteresis-type comparator with a threshold determined by the ratio of R_3 to R_4. If the output of IC1 is at negative saturation, the input to the inte-

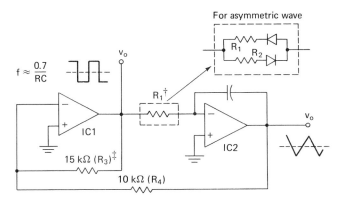

† Frequency control, 2–200 kΩ
‡ Amplitude control, 15–300 kΩ

Fig. 12-11   Triangular-wave generator.

grator will be constant and the output of IC2 will be a positive ramp. As the ramp crosses the threshold, the output of IC1 switches to positive saturation and the output of IC2 is a ramp with negative slope.

Note that a square-wave output as well as a triangular-wave output is available. If R_3 is increased, the threshold will decrease, causing the amplitude to decrease and the frequency to increase (note that $R_3 > R_4$). If R_1 is decreased, the frequency will increase. An asymmetric triangular wave can be produced by the diode circuit, which allows the positive and negative slopes to differ. A sawtooth wave is produced if R_2 is small.

For some applications, a single ramp initiated by a trigger or step input is desired. The start of the ramp in the circuit of Fig. 12-12 is controlled by the FET. It is held on by a positive voltage (usually positive saturation voltage of an op-amp) and is turned off by a negative voltage (usually negative saturation). The remaining part of the circuit is an integrator with a constant negative input, which results in a positive ramp.

Another version of the triangular-wave generator is discussed as an aspect of the function generator (Fig. 12-17).

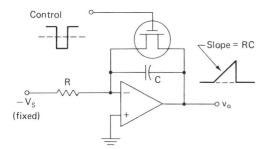

Fig. 12-12   Ramp generator (step control).

12-7   Crystal Oscillators

Crystal oscillators are often employed as accurate frequency or time references. Crystals are commonly available between 100 kHz and 10 MHz. The Pierce oscillator (Fig. 12-13) is a convenient circuit which works under a wide variety of conditions. Actually this oscillator is basically similar to the Colpitts oscillator (Fig. 12-4) except for the position of the ground. The similarity becomes apparent once it is realized that the crystal very near resonance is electrically equivalent to an inductor. The capacitors form a parallel IC resonance circuit with the crystal. Because the crystal inductance varies over a wide range very close to resonance, the values of the capacitors are not critical but must be made smaller at higher frequencies.

A circuit which is particularly convenient for digital applications is given in Fig. 12-14. The resistors bias the circuits into a quasi-linear region (see Chap. 13), thus allowing the oscillation to build up. Positive feedback is provided by the two inverters in cascade. The output approximates a square wave and is suitable for driving counters and frequency dividers.

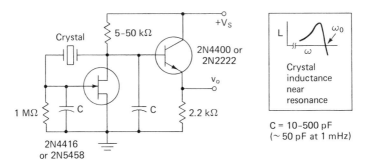

Fig. 12-13 Pierce crystal oscillator.

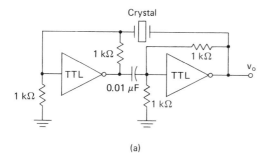

(a)

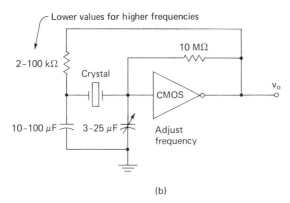

(b)

Fig. 12-14 Crystal oscillators, digital versions: (*a*) TTL and (*b*) CMOS.

12-8 DC-Signal Sources

There is often a need for an accurately regulated dc signal source in addition to the usual regulated power supplies (Chap. 21). Two circuits are given in Fig. 12-15. The first is simply a variable potentiometer followed by a unity-gain amplifier to eliminate loading effects. Of course, the output regulation is no better than that of the main supply ($+V_S$, $-V_S$).

Improved stability can be achieved with a high-quality voltage-reference Zener diode. Most of these diodes regulate between 6 and 9 V, a region where the temperature coefficient is small for Zener diodes. Regulation to 0.01 percent or better is possible if the load across the diode is constant, as it is here, and the main supply is also regulated. High resolution of the voltage control can be achieved with a 10-turn potentiometer, preferably with a calibrated dial. Calibration of the dial in terms of voltage requires

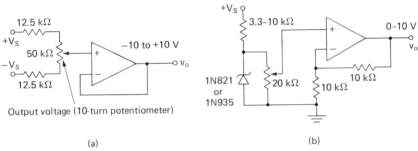

(a)   (b)

Fig. 12-15 Controlled dc sources (*a*) supply-reference and (*b*) Zener diode, semiprecision.

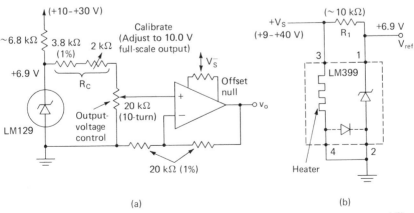

(a)   (b)

Fig. 12-16 Precision voltage-reference sources: (*a*) variable and (*b*) temperature-stabilized.

that the voltage across the 10-turn potentiometer be exactly 5 V, accomplished by a calibration trimming potentiometer.

Precision voltage-reference ICs are available which have excellent temperature stability (0.01 to 0.001 percent per degree Celsius) and long-term stability (3 to 20 parts per million). Most have two terminals and are used as a normal Zener diode (Fig. 12-16). Dynamic impedance R_z is very low (<1 Ω). While the change in output voltage with time or temperature is very small, the initial voltage tolerance is appreciable (±2 to ±5 percent) and some means of calibration or adjustment of the output to a standard voltage is usually necessary. In the circuit of Fig. 12-16a, a variable series resistor allows adjustment of the voltage across the 10-turn potentiometer to 5.0 V and thus the circuit output v_o to +10.000 V full scale. Previous nulling of the amplifier offset voltage to within ±1 mV is necessary. Since 10-turn potentiometers are usually linear to within ±0.1 percent, the output voltage should be accurate to ±10 mV or better.

Higher-precision regulators require temperature control of the internal Zener diode. Fortunately IC regulators with built-in heaters are available, as indicated in Fig. 12-16b. The heater-section supply can be connected to any 9- or 40-V supply of adequate current capacity (200 mA peak). Long-term stability of about 10 parts per million is achievable.

12-9 IC Function Generator

Because the need to generate a triangular or square wave or related waveform occurs often, a specialized IC has been designed to meet this need. The circuit is functionally similar to the triangular-wave generator of Fig. 12-11 but is contained within one package. Typical external connections are shown in Fig. 12-17. Features of this circuit are simplicity, accurate linear outputs, high stability, and ease of frequency adjustment. A further

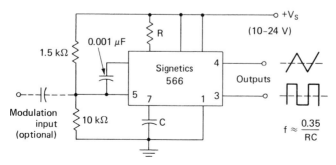

Fig. 12-17 Function generator.

feature is the ability to frequency-modulate the output frequency with high linearity by the application of a modulating voltage to the threshold control.

Voltage-controlled IC sine-wave generators currently available produce a waveform of fair quality over a moderately wide frequency range. The waveform generator of Fig. 12-18, for example, produces a reasonably

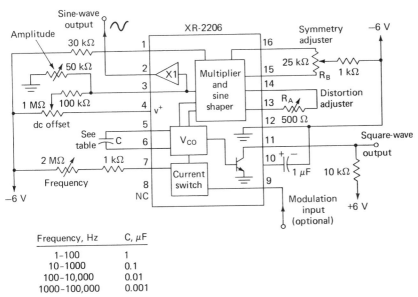

Frequency, Hz	C, μF
1–100	1
10–1000	0.1
100–10,000	0.01
1000–100,000	0.001

Fig. 12-18   IC sine-wave generator.

undistorted output over a frequency range of 100:1 by adjustment of R_F. These devices synthesize or approximate the sine wave by a multiplier, a circuit which is frequency-insensitive. Some filtering, however, may be necessary to smooth the waveform. Resistor (R_A, R_B) trimming reduces distortion to under 1 percent.

Stepwise or digital synthesis of a sine wave is possible with the circuit of Fig. 12-19. It is basically a five-section serial register (see Chap. 18) with its outputs connected as a simple D/A converter (Chap. 14). Weighting of the resistors to the analog summer is chosen to approximate a sine wave.

Fourier analysis indicates that the lowest harmonic is 9 times the fundamental f_o and is easily filtered. Low-pass Butterworth filters are especially effective.

The five-stage register is connected as a *ring* or *Johnson counter*. Except for the initial cycle (following clear on power up), each stage has the same pattern; **1** for 5 clock cycles followed by **0** for 5 cycles. The pattern is delayed by 1 clock cycle between stages. Zero crossing on the sine wave corresponds to turn-on of only the R and $1.62R$ resistors (in addition to the output zero resistor) into the summing amplifier.

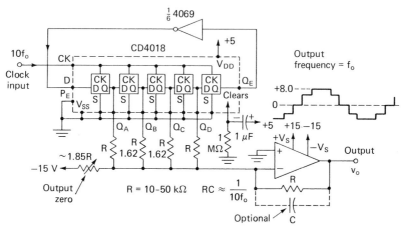

Fig. 12-19 Digital sine-wave generator.

Expansion to N stages with a correspondingly better approximation to a sine wave is easily achieved by adding more stages. Resistors for each stage are given by

$$R_N = \frac{R \sin [\pi(N - 1)/N]}{\sin (\pi/N)}$$

The lowest harmonic is $(2N - 1)f_o$ with an amplitude signal equal to $1/(2N - 1)$ times the fundamental.

Chapter Thirteen

High-Frequency Amplifiers

For most purposes the division between the low- and high-frequency regions of integrated circuits is within an order of magnitude (or two) of 1 MHz. Circuit configurations which work well at low frequencies may fail if the signal frequency is high. Often the cause is the limited frequency response of the device, a problem which can usually be cured by selection of a higher-frequency type. The main drawback of high-frequency devices, aside from higher cost, is the necessity of careful component placement, wiring, and ac bypassing. The frustration of eliminating high-frequency oscillations with no apparent cause and trying to achieve the advertised performance can dampen enthusiasm for high-frequency devices. It is wise to select a device with a frequency response only moderately beyond that needed to accomplish the required task.

13-1 Higher-Frequency Op-Amps

The most popular internally compensated op-amps, such as the 741 or 1458, have gain-bandwidth products of 1 to 4 MHz and slew rates of about 1 V/μs. For most applications a maximum operating frequency of

10 to 100 kHz results from these limitations. High-speed units are available which extend the frequency limitation by a factor of 10 or a 100. Some types are internally compensated. The 318, for example, which is pin-compatible with the 741, has a bandwidth of 15 MHz and a slew rate of 50 V/μs. However, an external-frequency-compensation-network capacitor tailored to the particular configuration, gain, and IC type is usually necessary to obtain optimum performance (Fig. 13-1). Manufacturers

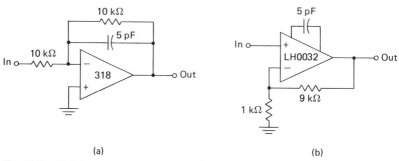

(a)   (b)

Fig. 13-1   High-frequency op-amp examples: (a) fast inverter; (b) gain 10 dc 5-MHz buffer amplifier.

usually suggest values for the compensation networks but for critical applications a mathematical analysis may be required to select the proper values.† Amplifiers with gain-bandwidth products in the gigahertz region are available (Teledyne 1435). Circuit layout is important, as discussed in the next section.

13-2 Wideband (Video) Amplifiers

In many applications a fixed-gain wideband amplifier is more satisfactory than the op-amp at higher frequencies. The input to these amplifiers is often single-ended (not differential like op-amps) and cannot utilize standard op-amp configurations. Various units are available, one of which is shown in Fig. 13-2. Its gain is 300 with a frequency response (-3 dB) from 30 Hz to 30 MHz with useful gain beyond 100 MHz. Extension of the low-frequency response is achieved by increasing C_1. An automatic gain control (AGC) is available. Gain is sharply reduced by applying 5 to 7 V dc to the AGC terminal.

Care must be taken in the layout and mounting in any wideband circuit. Capacitor and other component leads are kept short to reduce inductance and stray coupling. ICs are often soldered directly to the circuit board to

†See R. Roberge, "Operational Amplifiers," Wiley, New York, 1975, for a complete discussion of compensation methods.

reduce lead length. Metal-clad boards are preferred with blank or unused areas of the circuit board made of metal, which acts as a ground plane. Power-supply leads are bypassed (ac-grounded) by a ceramic or other low-series-inductance capacitor close to the IC itself. In critical applications, where good isolation between stages is needed, mounting the IC and associated components on a small separate circuit board inside a metal box as a shield is suggested. Input and output leads are coaxial cable.

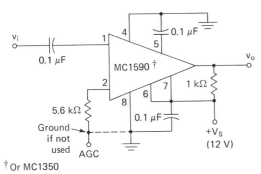

† Or MC1350

Fig. 13-2   A wideband (30 Hz to 30 MHz) amplifier.

13-3   Tuned RF Amplifiers

Often high-frequency amplification over a narrow band is desired or acceptable. In a radio receiver, for example, the RF and IF sections are sharply tuned to pass only the desired signals. The tuned circuit is an external IC network, which is chosen to resonate at the desired frequency. Generally speaking, tuned amplifiers are more trouble-free than wideband amplifiers, in part because the capacitance to ground is tuned out by the inductance. A typical RF amplifier is shown in Fig. 13-3.

Compared with that of most amplifiers, the input impedance is quite low (2 kΩ at low frequencies, dropping to about 0.3 kΩ at 60 MHz).

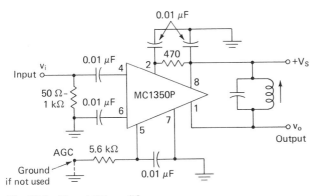

Fig. 13-3   Tuned RF amplifier.

Actually in most RF applications a lower impedance input, typically 50 Ω, is often desired to match the input coaxial cable. This is easily accomplished by adding a 50-Ω resistor to ground near the IC input.

13-4 Limiting IF Amplifiers

Limiting amplifiers provide a constant-amplitude output provided the input amplitude is above a certain minimum or threshold. They are used where the frequency, and not amplitude, of a signal is of interest, as in an FM receiver or frequency-counter applications. Basically they are simply high-gain (60-dB) amplifiers operated so that the output is saturated or clipped. In other respects they are similar to the tuned or video amplifiers discussed above. An example of such an amplifier is given in Fig. 13-4.

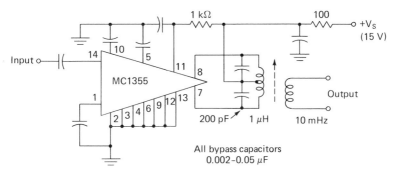

Fig. 13-4 Limiting IF amplifier.

13-5 Inside the Open-Collector RF-IF Amplifier

Some understanding of the IC arrangement (Fig. 13-5) is helpful, especially in connection with the differential inputs and open-collector outputs. A pair of input transistors (Q1, Q2) is operated as a differential amplifier. Constant-bias voltage is provided by an internal circuit (with an external bypass capacitor), so that the inputs have a dc potential several volts above ground. Neither input can have a dc ground return, although an ac ground at one input through a bypass capacitor is often used. Outputs are open-collector and require an external load, generally a tuned *LC* circuit. Note that a dc return to a power supply is required for both collector outputs (Q3, Q4). A tuned transformer which is electrically equivalent to the *LC* circuit is often employed. A center-tapped primary may be used, in which case the center tap is connected to the power supply and the remaining two terminals to the two amplifier outputs in a push-pull arrangement (see Fig. 13-4). A resistive load is permissible if a tuned

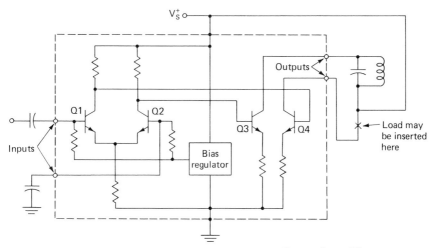

Fig. 13-5 Simplified circuit for a differential-input open-collector IC amplifier.

output is not wanted. When only one output is loaded, the other output (collector) should be connected directly to the supply.

13-6 CMOS and TTL Inverters as Linear Amplifiers

Although intended as a digital device, the digital inverters are quite satisfactory (and inexpensive) video (ac) amplifiers for some applications. Biasing of CMOS devices (Fig. 13-6) consists only of a resistor to feed the dc output to the input. The dc level of the inverter input v_i is equal to that of the output although the ac voltages differ. With this configuration the input and output dc potential is approximately half the supply voltage, which is intermediate between the **1** and **0** voltage levels. The exact voltage varies slightly with the particular unit and can be calculated (graphically) if the input-output characteristics (Fig. 5-29) are known. As pointed out in Chap. 5, appreciable supply current can be drawn under these conditions, and operation with a supply above 10 V could damage the device. The low-frequency response is determined by the input time constant RC. Actually the CMOS inverters do not have especially good high-frequency gain. The midfrequency gain of a single stage of a standard device (A series) is about 20, and the high-frequency breakpoint is about 100 kHz. Double-buffered devices (mostly B series) have an extra internal stage of amplification and have a better ac gain ($\sim$100). Gates such as NOR or NAND are equally satisfactory although, of course, devices with Schmitt-trigger inputs will not work. Distortion is rather high, especially at higher signal levels, and therefore this circuit should not be used where this is a consideration. High input impedance is one of the advantages of this circuit.

Several variations of the quasi-linear amplifiers are shown in Fig. 13-6. Control of the high- and low-frequency cutoff points as well as midfrequency gain are accomplished by the circuit of Fig. 13-6b. Higher gain is

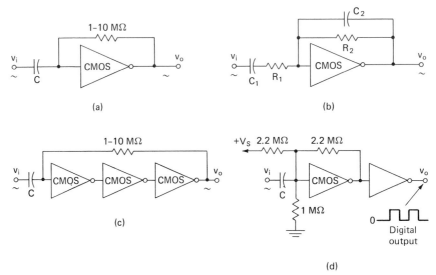

(a)

(b)

(c)

(d)

Fig. 13-6 Linearly biased CMOS ac amplifiers: (a) basic amplifier; (b) controlled-frequency-response amplifier; (c) high-gain amplifier; (d) digital level output stage.

achieved by cascading three inverters (Fig. 13-6c) so that the resultant feedback is negative.

Often the objective of the linearity-biased digital amplifiers is to bring the signal level to logic level, as, for example, in frequency-counter

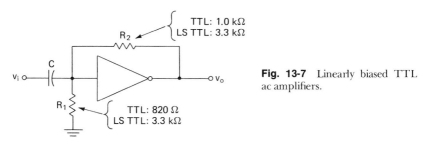

Fig. 13-7 Linearly biased TTL ac amplifiers.

applications. While it is possible simply to overdrive the linear stage, the output stage of Fig. 13-6d will provide a cleaner output. The input of the first stage is biased slightly away from the linear region, so that the output v_o is at **0** rather than an intermediate logic level when no ac signal is present. The last stage (optionally a Schmitt trigger) shapes the signal and acts as a buffer-driver. A fairly high-level ac input signal (peak-to-peak amplitude of at least 30 percent of the supply voltage) is required such as supplied by a previous quasi-linear stage.

Linear-mode TTL amplifier (Fig. 13-7) design is much more restrictive than for CMOS because the dc resistance to ground must be chosen to develop the required bias, taking into account the current flow out of the input terminal (typically 0.25 mA for LS TTL and 1 mA for TTL). Some dc feedback R_2 is needed to correct for variation in the operating point of individual units (alternatively R_1 can be made variable). The main advantage of TTL-type quasi-linear amplifiers is good high-frequency response. A drawback is the low input impedance.

Chapter Fourteen

Analog-to-Digital Conversion

Considered here is the conversion of an analog signal or voltage which can vary continuously within a given range into a digital signal registered by a set of flip-flops. In general, the cost and complexity of the A/D converter increases with the speed and accuracy required. Conversion speed is often very important in computer applications (such as the interface between an instrument and a minicomputer control) but may be of little concern in other applications, e.g., the digital voltmeter. Emphasis here will be on simpler types of converters. Many of the circuits are described in terms of block diagrams because detailed circuits for displays and counters are described elsewhere.

14-1 D/A Converter Specifications

A digital-to-analog (D/A) converter provides an analog output v_A proportional to a digital number on the input, usually presented in parallel binary form. Output changes when the input number changes. A/D converters do the reverse except that they require some time to convert following an input command (start) pulse. Basic converter specifications can be illustrated by the 3-bit D/A converter of Fig. 14-1. Most converters are

8-, 12-, or even 16-bit, with correspondingly higher resolution. For illustration purposes, a periodically incremented 3-bit counter is the input, and the output is therefore a staircase waveform. Ideally each step is the size of the least significant bit (LSB). In practice errors equal to $\pm\frac{1}{4}$ LSB or perhaps $\pm\frac{1}{2}$ LSB are acceptable. Errors over ±1 LSB are unacceptable

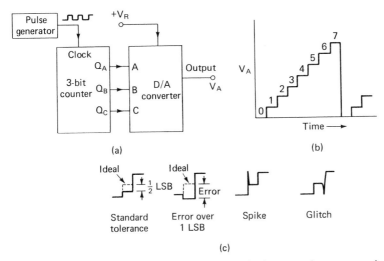

Fig. 14-1   Block diagram of a D/A converter. The input to the converter is periodically incremented by an up counter.

since the analog output-voltage change might have the wrong sign (Fig. 14-1c) which in control-feedback applications is likely to result in instability. Other deviations from clean steps are *spikes* and *glitches*, most troublesome when the steps are small and are changed rapidly. A spike is usually due to stray-capacitance feedthrough from the changing digital onput. A glitch is usually due to a delay between switching of larger and smaller bits. As an example, consider the change from **0111** to **1000**, a net step of 1 LSB, which requires the most significant bit (MSB) to switch. A glitch occurs if the lower bits turn off before the higher bit turns on. A low-pass filter or sample-and-hold circuit is the remedy.

The speed of a D/A is expressed in terms of *settling time*, which is usually defined as the time required for a large step to change in voltage or settle to within $\frac{1}{2}$ LSB of the final value. Higher-resolution units are slower, with typical values in the range of 0.1 to 50 μs.

A reference voltage and op-amp are necessary for all D/A and A/D converters but are often internal. Many units have dual-polarity outputs, for example, -5 to $+5$ V. Binary codes for polarity are discussed in later sections.

The speed of an A/D is expressed as a *conversion time*, which is the time between the start command and time finished. A *busy* or *done* output is always provided. Typical speeds are 1 to 100 μs. Cost rises rapidly with speed. Most A/D converters provide serial as well as parallel output.

14-2 Voltage-to-Frequency Converters

A voltage-to-frequency converter (VFC) has an output frequency which is proportional to the input voltage. Output amplitude or waveform is generally unimportant and usually is a pulse train. A voltage-controlled oscillator (VFO), as discussed in Chap. 12, performs a similar function except that its output waveform is sinusoidal or a symmetric square wave. Because a VFC can be used as an A/D converter, it is considered in this chapter.

Most VFC circuits involve an integrator which automatically resets itself when the output reaches a certain threshold. A suitable circuit is shown in Fig. 14-2.

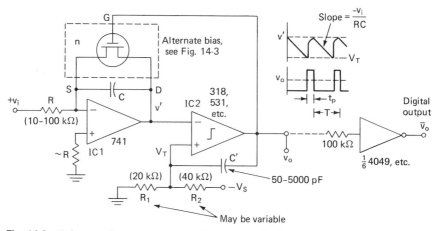

Fig. 14-2 Voltage-to-frequency converter (op-amp version).

Application of a positive dc signal v_i to the input of the integrator (IC1) results in a negative-going ramp at the integrator output v' with a slope proportional to v_i. A comparator (IC2) connected to the integrator has an output v_o which is in negative saturation if v' is less negative than the threshold voltage V_T but rapidly switches to positive saturation at the point which the ramp v' becomes more negative than V_T. As v_o becomes sufficiently positive, the FET is switched on (resistance $> R_{on}$), causing C to discharge. Capacitor C' provides positive feedback and determines the output pulse length, which must be long enough to allow C to fully discharge ($C'R \approx 5CR$). Depending on the type of comparator used, which in turn depends on the FET, an interface driver like that shown may be needed for digital compatibility.

The FET in Fig. 14-2 is an *n*-channel enhancement type (MOSFET) presumed to switch on by the application of a positive voltage, i.e., when the comparator goes high, and off when the comparator output goes low. It is desirable that the FET resistance when switched on be low (<50 Ω) so that the capacitor discharges rapidly. Actually nearly any FET can be used in this application *if biased properly*. The main points to be recognized are that the JFET requires a gate resistor and that the voltage ranges of the comparator and FET must match.

In Fig. 14-3 alternate methods of biasing the FET as a switch are

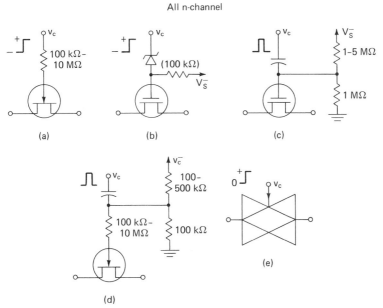

Fig. 14-3 Alternate *n*-channel FET-switch bias methods: (*a*) JFET with limiting resistance; (*b*) MOSFET with Zener bias, (*c*) MOSFET biased off, pulsed on momentarily; (*d*) JFET biased off, pulsed on momentarily; (*e*) transmission gate.

presented. Gate voltages sufficient to switch the particular FET on and off (cutoff) should be known. Nearly all available FETs will turn full on and off when driven by an op-amp which switches from positive to negative saturation, assuming a ±15-V supply. Although *n*-channel JFETs turn on for a slightly positive voltage, they may require −2 to −8 V for cutoff. However, because the gate of a JFET (*n*-channel) cannot be biased more than 0.6 V positive without damage and/or breakthrough of the switching signal, it is necessary to limit the gate current to a low level. A high-value resistor (Fig. 14-3*a*) in series with the gate provides current limit and allows FET turn-on with a positive voltage. The resistor does not interfere with the application of a negative or cutoff voltage since little gate current is then drawn.

Digital comparators which have only a positive output cannot turn off a JFET (and many MOSFETs as well) if driven directly. Bias networks (Fig. 14-3) are required. If, as required for the VFC described above, the turn-on pulse is short, one method is to bias the FET just to cutoff and to couple the turn-on pulse by a capacitor. If the turn-on or turn-off pulses are of arbitrary length, as may be required in other applications, the dc-coupled or Zener-diode method should be used. Another approach (Fig. 14-3*d*) is to substitute a transmission gate for the FET.

Substitution of *p*-channel FETs simplifies the match to a digital-comparator output, but the polarity at the comparator input must be reversed.

A narrow output pulse t_p is required for a reasonably high conversion linearity. If the ramp were ideally linear, so that the time t_i between the start and stop of the ramp would be strictly inversely proportional to voltage, the frequency would be strictly proportional to voltage. Unfortunately the net period T is not t_i but $t_i + t_p$. This produces a fractional deviation from linearity equal to t_p/t_i, which becomes an important nonlinearity at higher frequencies. If operation at higher frequency (>1 kHz) is desired, it is necessary to make the time constants CR_{on} and $C'R_1$ as small as possible. Furthermore, IC2 must have a high slew rate, and for this reason a high-speed op-amp or digital comparator is desirable. Care must be taken to ensure that the voltage-output swing is sufficient to drive the FET. As with any integrator, a quality capacitor, e.g., Mylar, is required. Accuracies somewhat better than 1 percent are possible. The conversion factor β is defined as

$$f_v = \beta v_i \qquad (14\text{-}1)$$

where f_v is the frequency of the output train. The value of β is determined by the ramp slope v_i/RC and the threshold voltage V_T. It is easy to show that $\beta = V_T/RC$ assuming that t_p is negligible.

14-3   IC Voltage-to-Frequency Converters

An IC version of the VFC (Fig. 14-4) is based on a comparator-triggered one-shot. The standard version has two timing RC circuits, one charged from precision current source i_o and the other associated with a one-shot multivibrator. The current source is switchable and is on when Q (one-shot output) is high.

Triggering of the one-shot (width $= \tau$) occurs when the comparator output is high, i.e., when $v_i > v_B$, and retriggering occurs if it remains high. In normal operation, after a starting transient, the one-shot delivers a standard charge $i_o\tau$, which charges the capacitor to the value $v_B = v_i + \Delta v_B$, where

$$\Delta v_B \approx \frac{i_o\tau}{C_B} \qquad (14\text{-}2)$$

Next v_B decays (through R_B) relatively slowly until it reaches the comparator threshold value v_i, when the one-shot fires again. Since the time T

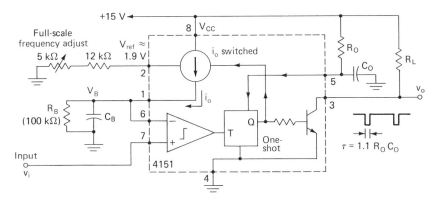

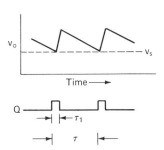

Fig. 14-4 IC voltage-to-frequency converter, standard version.

required to decay by an amount Δv_B is inversely proportional to v_B and thus v_i, the output frequency f_o is approximately proportional to input voltage, as seen from

$$v_B = v_i - \Delta v_B = v_i e^{-T/R_B C_B} \approx v_i\left(1 - \frac{T}{R_B C_B}\right) \qquad (14\text{-}3)$$

In the last term, the binomial expansion for the exponential has been employed. Combining Eqs. (14-2) and (14-3) gives an expression for the frequency $f_o = 1/(T + \tau)$

$$\frac{1}{f_o} = \tau \left(1 + \frac{i_o R_B}{v_i} \right) \tag{14-4}$$

or if $\tau \ll T$,

$$f_o \approx \frac{v_i}{i_o R_B \tau} = \frac{R_s v_i}{R_B(1.89)(1.1 R_o C_o)} \tag{14-5}$$

where the relation $\tau = 1.1 R_o C_o$ and the inverse dependence of i_o on R_s, that is, $i_o = V_{ref}/R_s$, have been substituted in the latter term. Note that the value of C_B drops out.

Because v_B and v_i must remain positive, the frequency should not drop below about 5 to 10 percent of the maximum. While the standard version of the circuit has the advantages of simplicity and operation from a single supply, the conversion accuracy is no better than 1 percent. Another disadvantage is a delay before the output frequency follows a step change in input voltage. Range is 1 Hz to 100 kHz.

An improved VFC circuit (Fig. 14-5) uses an op-amp to linearize the

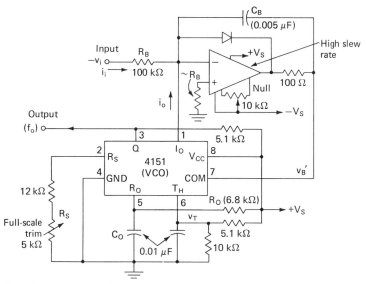

Fig. 14-5 Precision voltage-to-frequency converter.

charging of C_B. The input voltage v_i must be negative. Here the (negative) input current i_i and current source i_o when on, are summed and integrated to provide a ramp $v_i/R_B C_B$ which is compared with the threshold voltage V_T arbitrarily set at $\frac{2}{3}V_S$. Derivation of the conversion gain parallels that given above, with v_B' replacing v_B. The result is identical to Eq. (14-5). Conversion accuracy can exceed 0.1 percent, and response is within one f_o cycle.

14-4   VFCs as A/D Converters

Utilization of a VFC as an A/D converter requires the addition of a time-interval generator and a pulse counter, as indicated in Fig. 14-6. The VFC produces a continuous train of pulses, and the number of pulses which

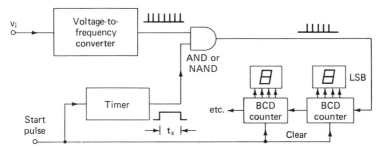

Fig. 14-6   A/D converter (VFC version).

pass through the NAND gate to the counter is determined by the timer pulse length t_{on}. A more detailed discussion of pulse counters is given in Chap. 17 and the design example sections. The number of pulses counted is

$$N = v_i \beta t_s \tag{14-6}$$

where β is the VFC conversion factor. In nearly all cases, the product βt_s is chosen as an even power of 10 so that N is numerically equal to the input voltage with an appropriate placement of the decimal point. For example, if $t_s = 0.500$ s and $\beta = 200$, a 1.00-V input would result in $N = 100$ displayed on the digital readout. The display decimal point between the second and third digit would be turned on. Usually t_s is made slightly adjustable as a calibration control.

14-5   Ramp Converters

A simplified single-ramp converter is shown in Fig. 14-7. Before the start of the conversion cycle the integrator IC1 or ramp output v_B is at positive saturation. To initiate conversion, a positive pulse v_t is supplied at the "convert" or trigger input which causes the transmission gate to conduct and thus the capacitor to discharge so that v_B drops to zero. Simultaneously v_t clears the counters. When v_t returns to zero, the transmission gate is turned off and the ramp begins. While the input v_i is more positive than the ramp input v_B the comparator output v_C is at **1**. When the NAND-gate input is high, pulses from the clock applied to the other input of the NAND gate pass through to the counter. After a time interval t_g the ramp voltage v_B will exceed the input v_i, and the counting will cease. It can be seen that the time interval will be proportional to the input voltage, provided, of course, that the ramp is linear, i.e.,

$$t_g = \frac{v_i RC}{v_R} \tag{14-7}$$

where v_R is a negative fixed reference voltage, which may be the supply voltage in less critical applications. During this time the number of pulses counted N is

$$N = t_g f_c \tag{14-8}$$

where f_c is the clock frequency, usually an even power of 10. With the proper choice of R, C, v_R, and f_c it is possible to make N numerically

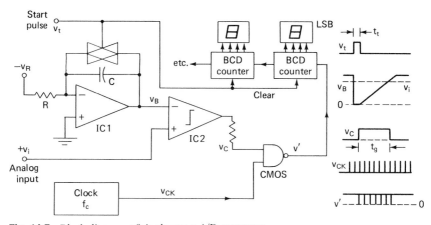

Fig. 14-7 Block diagram of single-ramp A/D converter.

equal to v_i with an appropriate placement of the decimal point. For example, if we choose $R = 100$ kΩ, $C = 0.50$ μF, $v_B = 5$ V, and $f_c = 10$ kHz, then t_g would be 10.0 ms for $v_i = 1.00$ V and N would be 100. Placing the decimal between the second and third digit on the output gives a readout directly in volts.

As the discussion on the VFC emphasized, the pulse applied to the FET transmission gate (comparator pulse in this case) must be long enough to discharge the capacitor. Conversion accuracies of better than 0.2 percent are achievable with a stable clock frequency and a quality capacitor (polystyrene or Mylar, not electrolytic or ceramic).

Dual-ramp (dual-slope) converters represent an improved version of the single-ramp converter. Conversion takes place in two phases. During the first phase the integrator input is connected to the analog voltage v_i, as indicated in Fig. 14-8.

During this phase the AND gate is held off, and a convert pulse resets the integrator (shorts the capacitor) and clears the counter. After a predetermined time t_r, when the integrator output ramp $v_B = -V_M = -v_i t_r/RC$, the second phase begins by switching the integrator to V_R, which has an opposite polarity to the input v_i. Since the slope is now equal to V_R/RC, the time required for the integrator output to return to zero t' will correspond to $v_B = V_R t'/RC$, and thus t' is given by

$$\frac{v_i t_r}{V_R} = t' \tag{14-9}$$

Note that the factor RC cancels, so that the capacitor need not be precision but should still be of high quality to avoid problems with

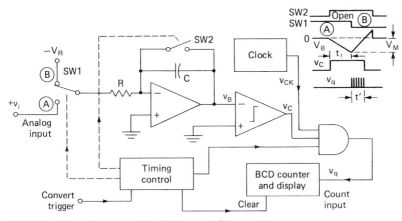

Fig. 14-8 Block diagram of the dual-ramp A/D converter.

hysteresis. The comparator holds the gate on for the time interval of t' so that the number of pulses counted is

$$N = \frac{t_r f_c}{V_R} v_i \tag{14-10}$$

As before, N is proportional to v_i, and setting the constant of proportionality equal to an even multiple of 10 gives a readout directly in voltage.

If, as usual, t_r is derived from the clock frequency f_c by frequency division so that the count number $N_s = t_r f_c$ is known exactly, and N depends only on V_R, or

$$N = \frac{N_s}{V_R} v_i \tag{14-11}$$

The advantage of the dual-slope converter is that the accuracy depends only on the reference voltage V_R, which can be precisely controlled without difficulty. The input voltage is averaged over the time interval t_r so that the measurement is relatively insensitive to noise (with the single-ramp converter noise pulses near the end of the cycle can prematurely trigger the comparator). Many commercial digital voltmeters or panel meters utilize the dual-slope method and are available at relatively low cost. They may include circuits for polarity reversal as well.

14-6   D/A Converters

Converting a digital signal to analog is quite simple, at least in principle, as shown in Fig. 14-9. The CMOS buffers act as a switch (see Fig. 5-9) which

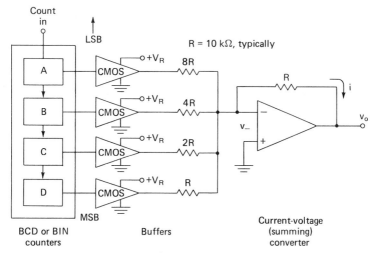

Fig. 14-9   Simple 4-bit CMOS D/A converter.

connect the output resistors (R, etc.) to the reference voltage V_R when the input is at **1** or to ground if at **0**.

Since the op-amp input v_- is at virtual ground, the current through the resistors $i = V_R/R$ is simply summed (see Secs. 4-4 and 4-6), and the output is

$$v_o = -V_R \left(M_d + \frac{M_C}{2} + \frac{M_B}{4} + \frac{M_A}{8} \right) \qquad (14\text{-}12)$$

where M_A, etc., are **1** or **0** depending on the logic state of each bit. Further bits are weighted as $R/10$, $R/20$, etc., for a BCD counter or as $R/16$, $R/32$, etc., for a binary counter.

Accuracy depends not only on the number of bits but on the precision of the reference voltage V_R and the resistors. Included in the total resistance must be the (on) resistance of the CMOS FET driver (typically 50 to 1000 Ω). With an 8-bit BCD counter and 1 percent resistors ($R \approx 10$ kΩ) an overall accuracy of 2 percent is achievable.

To avoid a large number of different-valued resistors, a ladder network like that in Fig. 14-10*a* is often employed. The CMOS buffers are represented by a SPDT switch with V_R as the supply voltage. Circuit analysis is expedited by recognizing that each section can in turn be represented by the Thevenin equivalent of Fig. 14-10*b*. The result for this voltage-mode converter is

$$v_o = \frac{V_R}{2}\left(M_D + \frac{M_C}{2} + \frac{M_B}{4} + \frac{M_A}{8}\right) \tag{14-13}$$

Extension to 8 or more bits is straightforward. Commercial IC packages of 8, 10, and 12 bits are available, most with internal amplifiers, but an exter-

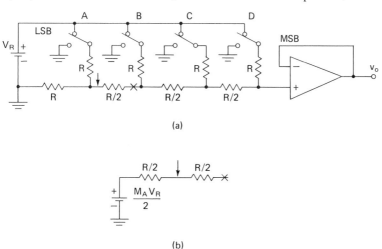

(a)

(b)

Fig. 14-10 Ladder network (*R*-2*R* type): (*a*) in a 4-bit D/A converter; (*b*) circuit equivalent.

nal reference V_R is often required. Some devices, termed *multiplying converters,* work with any polarity or magnitude of voltage reference, including alternating currents, and more properly can be thought of as a digitally switched attenuator.

Polarity switching (bipolar output) can be implemented by either of the circuits of Fig. 14-11. The circuit must match the particular bipolar binary code, which is illustrated in Table 14-1 for 3 bits. With the binary-offset method, a fixed negative voltage equal to the weight of the most significant bit (MSB) is summed with the ladder-network output. When the MSB switches on, it just balances the negative offset and the output is zero, as can be seen from Table 14-1.

With the twos-complement method, an extra or sign bit (now MSB) is added to the code, which is equal to **1** for negative numbers. Many computers utilize this code because of the ease of manipulating negative numbers.

A third method of implementing polarity is to use the sign bit to reverse the sign of the output voltage. This can be implemented by a switchable inverter.

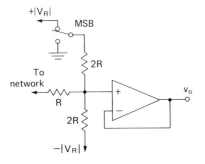

Fig. 14-11   Polarity-conversion circuits for offset binary. Twos-complement code is implemented by the same circuit except that the MSB is complemented.

TABLE 14-1   Bipolar Binary Codes

Decimal number	Offset binary	Magnitude + sign	Twos complement	Relative voltage v_o
+3	111	011	011	$+\frac{3}{4}$
+2	110	010	010	$+\frac{1}{2}$
+1	101	001	001	$+\frac{1}{4}$
+0	100	000	000	0
−0	—	100	—	0
−1	011	101	111	$-\frac{1}{4}$
−2	010	110	110	$-\frac{1}{2}$
−3	001	111	101	$-\frac{3}{4}$
−4	000	—	100	−1

Many higher-speed D/A converters use switchable BJT current sources. The currents are added by an analog summer (inverting amplifier), as indicated in Fig. 14-12.

14-7   Counter and Servo A/D Converters

A counter type of A/D converter is illustrated in Fig. 14-13. It uses a D/A converter, which allows comparison of the converter analog output v_A with the analog input v_i, assumed positive. Initially the counter is cleared by the convert pulse, so that $v_A = 0$ and the NAND gate allows clock pulses to pass. As time progresses, v_A increases in steps. At the point where v_A

exceeds v_i the comparator output switches negative and the pulses to the counter cease. At this point the digital output is essentially equal to the equivalent analog input.

Clock frequency determines the rate of conversion but has no effect on accuracy. While the illustration shows only 4 bits, this can be increased to

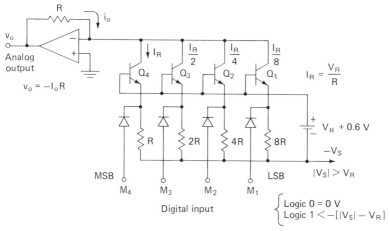

Fig. 14-12   Binary-weighted BJT current sources as a D/A; $i_o = M_1 I_R/8 + M_2 I_R/4 + M_3 I_R/2 + M_4 I_R$.

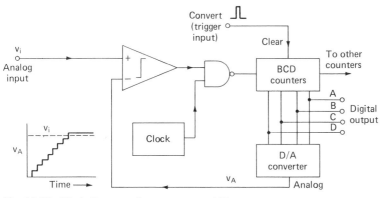

Fig. 14-13   Block diagram of a counter-type A/D converter.

any number although the time required grows rapidly with the number of bits, a significant drawback for many applications.

A related A/D converter (Fig. 14-14) is the *servo* type, so named because it follows or tracks the analog voltage like a servomotor with feedback. It is based on the up-down counter. Instead of turning off the clock pulses the comparator reverses the counter direction so that it tracks the analog voltage. If the input voltage is constant, the counter will hunt or dither ±1 LSB about the proper value. Tracking is best for slowly changing input signal voltages and worse for large steps. Sometimes the clock and comparator output is transmitted to a remote counter, which follows the main counter. This method of operation is known as *delta modulation*.

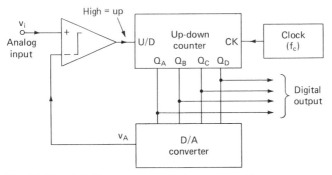

Fig. 14-14 Block diagram of a servo-counter-type A/D converter.

14-8 Successive-Approximation A/D Converters

Perhaps the most popular converter is the successive-approximation type (Fig. 14-15). It is similar to the counter type except that the bits are tested

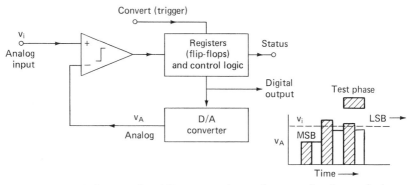

Fig. 14-15 Block diagram of an A/D converter (successive-approximation version).

in succession, i.e., incremented in large steps starting with the most significant bit (MSB), rather than incrementing by the smallest step, as done when counting pulses. To do this, the bit is switched on during a test phase. If the bit causes v_A to exceed v_i as detected by the comparator, it is switched off and kept off; otherwise it is left on. Sequentially the control logic tests the next most important bit until the conversion is completed.

Clearly the process is much faster than counting pulses, especially if the number of bits is large. The main disadvantage is the rather involved control logic required.†

Since successive-approximation converters are normally purchased as complete units (Fig. 14-16) or implemented by microprocessor software,

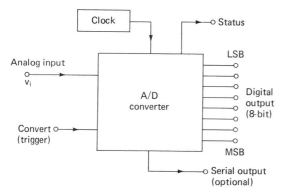

Fig. 14-16   Block diagram of an A/D converter module.

the circuit details will not be given here. Some converters require an external clock, which may be synchronized with other digital circuits. More often the clock is internal. The conversion process begins immediately after a convert or start pulse is applied. During the time the conversion process (turning on bits and testing analog levels, etc.) takes place, a *status* or *busy* output goes to **1**. Digital output is not read until the status output returns to **0**. In some models the output does not change (is latched) until the conversion is complete. Serial output is often provided as an option.

†For a more complete discussion of converters see "Analog-Digital Conversion Handbook," Analog Devices, Norwood, Mass., 1972.

Chapter Fifteen
Modulation and Demodulation

Often an analog signal of interest is proportional to the amplitude or frequency of a sine wave or other wave train. The process of extracting the signal from the ac wave is *demodulation,* and the reverse process is *modulation.* Included in demodulation is the conversion of an ac into a dc signal. Modulated signals may arise from transducers or transducer readout circuits. The purpose of demodulation then is to extract the desired signal. In other applications the signal is modulated to facilitate transmission to a remote site, where it is then restored by demodulation. Several signal-recording techniques also involve modulation and demodulation.

It is helpful to keep in mind that modulation and demodulation are inherently nonlinear and that the frequency content of the signal is transformed in the process.

15-1 Diode Rectifiers

Undoubtedly one of the simplest demodulation circuits is the diode rectifier. Transformer-coupled circuits, which have a dc return, require only the diode and filter network (Fig. 15-1). The principle is similar to that employed in power supplies (Chap. 21). It is necessary that the input voltage be greater than the diode drop (0.6 V for silicon diodes) for output current to flow. Charging the input capacitor C_i to the peak voltage ($v_i - 0.6$) each cycle is followed by a slight discharge with a time constant $\tau = R_D C_i$, which is made much larger than the period of the higher-fre-

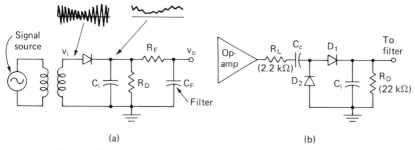

(a) (b)

Fig. 15-1 Diode AM demodulators: (a) transformer-coupled, (b) capacitor-coupled.

quency carrier ($1/f_c$). Often the ripple is removed by a subsequent low-pass filter (Chap. 11). Of course τ must not be too long, and the filter break frequency ($f_x \approx 1/2\pi R_F C_F$) not too low or the higher-frequency components of the demodulation signal will be attenuated along with the carrier. If the carrier and modulating frequencies f_c and f_m are not widely separated, a sharp cutoff filter, e.g., the Butterworth filter, may be required.

Capacitor coupling is usually preferred because the often bulky transformer is eliminated. A second diode (or a low-value resistor) is required to provide a dc ground return. Without D2 the coupling capacitor C_c would charge to the peak signal voltage after an initial transient and no dc voltage would be developed across R_D since, of course, C_c does not pass direct current. In effect the diode D2 limits the negative signal excursion to -0.6 V, a process referred to as *diode clamping*. An input-current-limiting resistance R_L prevents overload of the driving amplifier on signal peaks where the capacitor acts as an ac short circuit.

If $R_L' \ll R_D$, the capacitor C_i will charge to the *peak* signal voltage, where R_L' is the sum of R_L and the output (Thevenin) resistances of the driving stage and the current drawn by the output filter as well as the diode drop, assumed negligible. If instead $R_L' \gg R_D$, C_i will charge to the *average* rectified signal voltage. For nonsinusoidal or noisy signals this distinction can be important.

Diode demodulators have the advantage of simplicity but can be quite nonlinear, especially at low signal levels. They are suitable for noncritical low-accuracy applications or at very high frequencies where other methods fail.

15-2 AC-to-DC Converters

A substantial improvement in performance over the diode rectifier is provided by the ac-to-dc converter. Where the output is a meter, specifically a microammeter calibrated in voltage, the circuit of Fig. 15-2 is ideal. It is

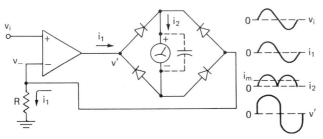

Fig. 15-2   An ac-to-dc converter (meter version).

a combination of the voltage-to-current converter (Chap. 4) and a bridge rectifier. Current i_2 from the op-amp output is identical to that flowing through the meter and R. Further the magnitude of i_1 is equal to the current flowing through the meter; that is, $i_2 = |i_1|$ is the absolute value of i_1 or the rectified signal. The meter responds to the average value I_m of the current through it, which for a sinusoidal signal is

$$I_m = \frac{2}{\pi} i_2 = 0.6366\,|i_1| \tag{15-1}$$

For a sinusoidal input, $v_i = v_a \sin \omega t$, the meter current will therefore be

$$I_m = \frac{0.6366}{R} v_a \tag{15-2a}$$

or if v_i is calibrated in rms, where $v_{\text{rms}} = v_a/\sqrt{2}$,

$$I_m = \frac{0.9003}{R} v_{\text{rms}} \tag{15-2b}$$

It must be stressed that these factors are strictly true only for a pure sinusoidal signal.

As an example, suppose that the meter sensitivity is 100 μA full scale and a calibration of 5.00 V rms full scale is desired; in this case R should be 45.0 kΩ. Often R is made adjustable over a ± 5 percent range to allow compensation for a slight difference in meter sensitivity.

A capacitor in parallel with the meter helps averaging at the signal frequency. It is also helpful or even essential at high frequencies as a bypass of the meter inductance across which an excessively high voltage may develop.

A major advantage of this circuit is that the diode drop (~0.6 V) is unimportant, and good linearity is achieved even at low signal levels, down to the millivolt range. Diode or meter drop has a negligible effect because, as discussed above, current feedback is employed and the meter current is forced by the op-amp to equal $v_i = v = Ri$, provided the infinite-gain approximation holds. The op-amp output voltage v' is not important provided saturation does not occur. Actually it must make up the diode drop and switches polarity, or jumps, as the signal passes through zero, resulting in a waveform which approximates the sum of the sine wave and a 1.2-V peak-to-peak square wave. A disadvantage of this circuit is that the output is floating and cannot drive a subsequent stage (without a differential-input amplifier).

Often an ac-to-dc converter with a single-ended (dc) voltage output is required. For example, an ac-to-dc converter can be connected to a dc digital voltmeter so that the combination acts as an ac voltmeter. This circuit is also equivalent to an AM demodulator. An excellent circuit for this purpose is shown in Fig. 15-3.

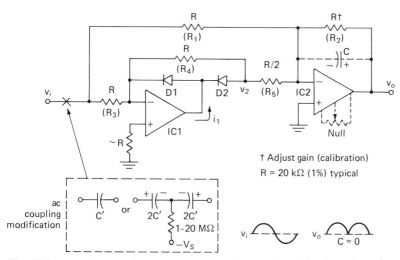

Fig. 15-3 Precision ac-to-dc converter; without the capacitor C the circuit is an absolute-value converter.

Because its principle of operation is not immediately obvious, its function can better be explained by considering the circuit with the capacitor C removed; the circuit then acts as a precision full-wave rectifier or absolute-value converter. For positive inputs, the current i_1 from IC1 passes through D2 (not D1), and thus IC1 acts as a unity-gain inverter with $v_2 = -v_i$. IC2 acts as a summing amplifier with two inputs (v_i, v_2). Since $R_5 = R_1/2$, the gain with respect to the v_2 input is twice that of the v_i input; that is, $v_o = -(v_i + 2v_2) = +v_i$. The net gain for positive inputs is therefore unity (with $R_2 = R$), and the output is, of course, positive. If the input v_i is negative, i_1 passes through D1 (not D2), so that v_2 is at ground potential. Again summing the two inputs to IC2, we see that the output v_o is $-(v_i + 0) = -v_i$. The net gain for negative inputs is unity but inverting, and the output is again positive. Good frequency response is required of IC1 at higher input frequencies.

It is clear the circuit acts as a full-wave rectifier or absolute-value converter. The term precision is applied because, like the circuit of Fig. 15-2, it is linear even for small-signal inputs (smaller than the diode drop) as a consequence of the current feedback through the diodes.

Addition of a capacitor across R_2 causes IC2 to act as an active low-pass network or averager (Chap. 11). The output may be considered to be direct current but is more accurately thought of as a demodulated signal with frequency components lower than the break frequency f_x, where $f_x = 1/2\pi C R_2$.

Although for the purposes of discussion, it was assumed that $R_2 = R$, it is unessential to the circuit operation. Often R_2 is adjusted to provide an output calibration. In particular the dc output can be adjusted to correspond to either the peak or rms value of the ac input. By analogy with Eqs. (15-1) and (15-2) for a sinusoidal input, $v_i = v_a \sin \omega t$, the output v_o will be

$$v_o = \frac{0.6366 R_2}{R} v_a \tag{15-3}$$

Values of R_2 for peak and rms calibrations are

$$R_2 = \begin{cases} 1.57R & \text{peak} & (15\text{-}4a) \\ 1.11R & \text{rms} & (15\text{-}4b) \end{cases}$$

Sometimes R_2 is made adjustable over a narrow range (± 5 percent) to allow calibration against a standard.

Without an input capacitor C' the circuit will respond to dc inputs, an advantage when measuring very low frequency signals but a disadvantage if the ac input signal has a dc component which is to be ignored. If a strictly ac reading is desired, an input capacitor C' is required. At high frequencies a (nonpolar) low-value ceramic or paper capacitor will do, but at low frequencies a polarized high-value electrolytic capacitor may be needed. Since the input direct current may be of either polarity, two back-to-back capacitors (each equal to $2C'$) are required to prevent reverse polarity. The junction is returned to a high negative potential through a high-value resistor to maintain the proper bias. This technique can be used whenever a large dc blocking capacitor is required.

15-3 Phase-sensitive Detectors

A phase-sensitive detector (PSD), lock-in amplifier, or synchronous demodulator is similar in function to the ac-to-dc converter except that the output is determined by the phase of the input signal (with respect to a reference) as well as its amplitude. The principle is illustrated in Fig. 15-4,

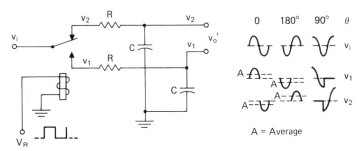

Fig. 15-4 Principle of the phase-sensitive detector.

in which a relay or chopper driven by a reference square wave acts as a switch. The output is here assumed to have the same frequency as the input signal but differs in phase by an angle θ. A consideration of the waveforms at the various points in the circuit shows that the average voltage on capacitor $\bar{v}_1$ will be positive for $\theta = 0$, negative for $\theta = 180°$, and zero for $\theta = \pm 90°$. The voltage across the other capacitor $\bar{v}_2$ will be just the negative of $\bar{v}$, and the output, which is the difference, $v_o = \bar{v}_1 - \bar{v}_2$, will be twice $\bar{v}_1$. In other words, the sign of the output reverses as the phase reverses by 180°.

A sinusoidal input with a frequency ω equal to the reference frequency ω_0 is assumed

$$v_i(t) = v_a \sin (\omega_0 t + \theta) \tag{15-5}$$

A low-pass filter is present which removes the reference ω_0 and higher-frequency components. It can be shown by Fourier analysis that for an in-phase reference ω_0 the output response v_o is

$$v_o = \frac{2}{\pi} v_a \cos \theta \qquad \omega = \omega_0 \text{ (in-phase reference)} \tag{15-6a}$$

By shifting the reference by 90°, the quadrature phase-sensitive detector response is

$$v_o = \frac{2}{\pi} v_a \sin \theta \qquad \omega = \omega_0 \text{ (quadrature reference)} \tag{15-6b}$$

A simple but excellent circuit at low to moderate frequencies is shown in Fig. 15-5. The FETs (one an n-channel, which switches on when the comparator output is positive, and the other a p-channel, which switches on during the negative cycle) act as a single-pole double-throw switch. Gate-current-limiting resistors are needed for JFETS. Biasing methods for the FET as a switch driven by a comparator are discussed in Chap. 14. Transmission gates are substitutes.

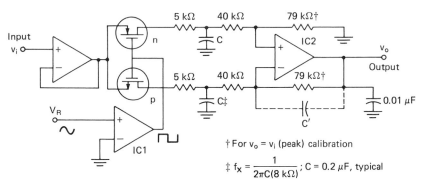

Fig. 15-5 Phase-sensitive detector (FET-switch version).

Output response time is determined by the capacitors in the low-pass filter. Because of the FET in series with the resistors, which is on only half the time, their effective resistances is $2R$. A differential-input amplifier IC2 provides a voltage output with respect to ground. Sometimes a second capacitor C' is added so that IC2 acts as a second low-pass filter.

At high frequencies the limited speed of the FET as well as switching transients reduce the performance of the FET type of PSD. A high-frequency version is shown in Fig. 15-6. A difference-input-gated amplifier (see Chap. 13) is switched between inputs A and B at the reference frequency f_0. Since both inputs are connected to v_i but with different polarities, the switched output v_o is the same as the other phase-sensitive detectors [Eq. (15-6)] except for an additional gain $A \approx 8$.

As with other demodulators, the "dc" output actually varies with time and is more appropriately thought of as a demodulated ac signal processed by a low-pass filter. Instead of deriving Eqs. (15-6) again for a modulated signal or signal with a frequency ω not equal to the reference frequency ω_0, it is convenient to consider the phase to be time-varying, i.e.,

$$\theta(t) = \omega_m t \tag{15-7}$$

where $\omega_m = \omega - \omega_0$ and in this case the input signal is expressed as

$$v_i(t) = v_a \sin \omega t = v_a \sin (\omega_0 t + \omega_m t) \tag{15-8}$$

Since Eq. (15-8) is the same form as Eq. (15-5) with θ given by Eq. (15-7), the result is Eq. (15-6) in the form

$$v_o(t) = \frac{2v_a A_L \cos(\omega_m t + \phi)}{\pi} \qquad \text{where } A_L = \frac{1}{\sqrt{1 + (\omega_m RC/2\pi)^2}} \qquad (15\text{-}9)$$

Here A_L is the gain of the low-pass filter (Chap. 11), which is a part of the

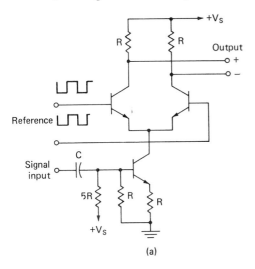

(a)

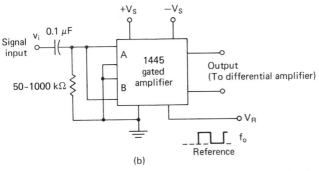

(b)

Fig. 15-6 Phase-sensitive detectors (high-frequency versions): (a) gated collector; (b) gated dual amplifier.

PSD. The filter should highly attenuate the carrier ω_0 while passing the beat or modulating frequency ω_m. The PSD may be thought of as a narrow-band filter with a width which can be made arbitrarily small by the choice of the time constant RC. Alternatively it is equivalent to a mixer with ω_m as the intermediate, or beat, frequency.

When the input amplitude v_a is held constant, usually as a result of clipping the input signal by passing through a limiting amplifier, the output v_d of the PSD responds only to phase difference θ between the input signal and the reference signal. In this case the detector is termed a *phase comparator* with the following response

$$v_d(t) \approx K_d \sin \theta \qquad\qquad (15\text{-}10a)$$

$$v_d = K_d\theta \qquad \text{for } \frac{-\pi}{2} < \theta < \frac{\pi}{2} \qquad (15\text{-}10b)$$

where K_d is a constant for the device which depends on the clipping level. The response curve is actually a triangular wave (as in Fig. 15-7b) but the

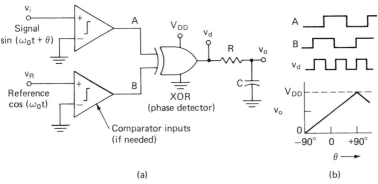

(a)   (b)

Fig. 15-7   Digital (XOR) phase detector: (*a*) circuit with comparator drivers and (*b*) response.

approximations of Eq. (15-10) are more useful in practice. By convention, the phase of the reference of a phase comparator corresponds to cos $\omega_0 t$, so that an in-phase response [Eq. (15-10)] corresponds to a quadrature PSD response [Eq. 15-6b)]. Also by convention, the low-pass filter is not considered a part of the phase detector, and any dc bias (here $+V_{DD}/2$) is ignored. If the reference phase is shifted by 90°, the response of a quadrature phase comparator becomes

$$v_d(t) \approx K_d \cos \theta \qquad\qquad (15\text{-}11)$$

A digital phase detector can be constructed from an XOR gate, as indicated in Fig. 15-7. From the XOR truth table (Fig. 5-6) it can be seen that when $\theta = 0$ (A and B displaced 90°), the gate output v_d is a **1** for half the time and therefore the averaged output v_o is half the **1** level. Phase shifts θ about this point result in a proportional change in output, and therefore the XOR phase detector is inherently a linear detector (between $\theta = \pm\pi/2$). Note that $v_o = V_{DD}/2$ for $\theta = 0$, resulting in a biased output response for $v_d = v_o - V_{DD}/2$.

15-4 Amplitude Modulation and Control

Modulation or control of the amplitude of an ac signal is ordinarily accomplished by a variable-gain or nonlinear amplifier. When the ac signal is a high-frequency, usually sinusoidal, carrier ($v_a \sin \omega_c t$) with an amplitude change proportional to the instantaneous value of a low-frequency modulating signal v_m, the composite signal v_s is considered to be amplitude-modulated:

$$v_s(t) = v_a A_m(t) \sin \omega_c t \tag{15-12a}$$
$$v_s(t) = v_a[1 + m(t)] \sin \omega_c t \tag{15-12b}$$

where $A_m(t) =$ instantaneous gain of modulator
$\quad m(t) = K v_m(t) =$ instantaneous modulation index $[0 \leq m(t) \leq 1]$
$\quad K =$ modulation sensitivity

Discussions of amplitude modulation as found in texts on communications demonstrate that the modulation process generates two sideband frequencies, $\omega_c + \omega_m$ and $\omega_c - \omega_m$, which are present in addition to ω_c and ω_m. An important practical consideration is the linearity of the modulation assumed by Eq. (15-12).

For amplitude-control applications, A_m is adjusted so that the magnitude of the high-frequency signal is a particular value, and in this case A_m is not thought of as a modulation. Linearity here is usually of secondary importance. The FET gain controls of modulators shown in Fig. 15-8

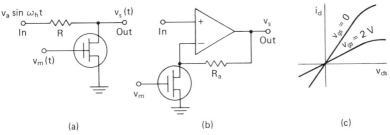

(a) (b) (c)

Fig. 15-8 Amplitude gain control (FET versions): (a) attenuator, (b) variable gain, and (c) typical FET characteristics.

depend on the variation of drain-source resistance R_{ds} with gate voltage V_{gs}. It should be noted that the FET acts as a pure resistance at low signal levels for signals of either polarity. Biasing techniques are discussed in Chap. 14 (Fig. 14-3).

Although the FET provides a simple, high-speed method of varying gain, it can be quite nonlinear in two respects. The first type of nonlinearity is a curvature in the current-voltage characteristics (I vs. V_{sd} at higher voltages (typically over 0.5 to 2 V) which tends to clip the signal at higher (drain-source) voltage levels. This type of distortion can be made negligible by keeping the signal across the FET small.

The second type of nonlinearity is the change of resistance with gate voltage. A sinusoidal voltage $v_m(t) = -|v_m| \sin \omega_m t$ applied to the gate will result in a distorted modulation. Such distortion can be avoided only if the modulation percentage is kept small or if corrective feedback is applied. A feedback scheme is suggested in Fig. 15-9. Either the ac-to-dc converter or the PSD is a suitable linear AM demodulator.

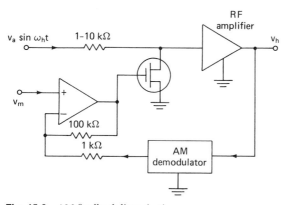

Fig. 15-9 AM feedback linearization.

A gated amplifier (see Chap. 13) is a particularly effective high-frequency modulator (Fig. 15-10a). Like the FET modulator, it suffers from some nonlinearity, which can be corrected by feedback similar to that of Fig. 15-9. A lower-frequency version (Fig. 15-10b) utilizes a variable transconductance (gain) op-amp (Chap. 3).

15-5 Frequency Modulation

Frequency modulation (FM) and voltage-to-frequency conversion (VFC) are basically the same process, but the distinction in practice is that FM involves a comparatively small frequency shift Δf about some center frequency f_0; that is,

$$f = f_0 + \Delta f(t) = f_0 + K'v_m(t) \tag{15-13}$$

where K' is the modulation sensitivity. Frequency deviations $\Delta f/f_0$ of 0.05 to 0.001 are common.

Many of the VFO circuits discussed in Chap. 14 can be used as FM modulators. Their input would be a dc bias, upon which is superimposed a relatively small ac modulating signal so that the output would be in the form of Eq. (15-13).

A high-frequency FM modulator based on a voltage-variable capacitor or varactor diode is comparatively simple to implement. Voltage-variable capacitors are reverse-biased diodes made to maximize the high-frequency capacitance change associated with the variation of the depletion region with voltage. Diodes act as capacitors best at high frequencies,

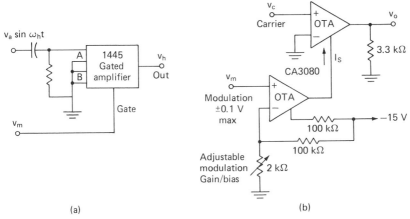

(a) (b)

Fig. 15-10 Amplitude modulators: (*a*) gated-amplifier version; (*b*) variable-transconductance version.

above 1 to 10 MHz. The capacitance change per unit voltage is highest at lower dc bias voltages, but the nonlinearity also increases in this region. Care must be taken to avoid higher modulation amplitudes, which might drive the diode into the forward conducting region. Typically diodes have nominal capacitances (e.g., at -4 V) in the 10 to 100 pF range and vary 50 percent about this value with factor-of-2 variations in voltage.

The low-frequency modulation voltage must be coupled to the diode without shorting or excessively loading the high-frequency resonance circuit. Conversely the low reactance of the inductor at low frequencies cannot be allowed to short out the modulation-voltage source. The coupling network shown (Fig. 15-11*a*) satisfies these requirements while also allowing an adjustment of dc reverse bias across the diode.

A suitable *LC* oscillator is the Colpitts circuit (Chap. 12) as implemented in the telemetry transmitter discussed in Chap. 19.

15-6 FM Detectors

A low-frequency FM detector is based on the time averaging of narrow pulses synchronized with the FM carrier (Fig. 15-12). A comparator (IC1) produces a square wave at the incoming frequency, which, after differentiation, triggers a monostable multivibrator (IC2). Since the output of

IC2 is a continuous train of pulses of fixed width and height, the time average of the pulses will be proportional to frequency. Averaging is accomplished by an RC network and amplifier connected as a low-pass filter. A zero control adjusts the dc output to zero when the input is at the center frequency f_0. The output of the detector v_0 is

$$v_0 = K' \, \Delta\omega = K(f - f_0) \tag{15-14}$$

The break frequency f_x of the filter is normally chosen to be at least an order of magnitude lower than the lowest frequency of the FM input.

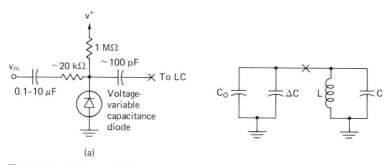

Fig. 15-11 Voltage-variable capacitor modulator: (a) coupling circuit; (b) electrical equivalent.

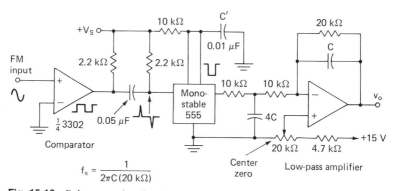

Fig. 15-12 Pulse-averaging (low-frequency) FM detector.

A common FM demodulator is the quadrature detector based on the detuning of a LC resonance network near resonance. An IC version which includes an IF amplifier is shown in Fig. 15-13. Linear output occurs only for a rather narrow frequency deviation (1 to 5 percent) around the resonance frequency of the LC network.

The phase-locked loop, an excellent FM detector, is discussed in the next section.

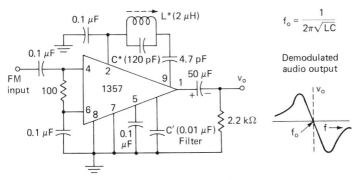

$$f_o = \frac{1}{2\pi\sqrt{LC}}$$

Fig. 15-13   FM quadrature detector (IC version).

15-7   Phase-Locked-Loop Theory

Many of modulation, detection, and frequency-synthesis methods have become practical with the introduction of the IC versions of the phase-locked loop (PLL). Figure 15-14 shows a block diagram of this device.

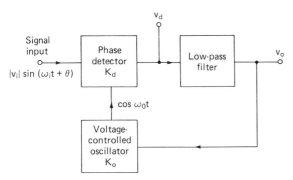

Fig. 15-14   Basic phase-locked loop.

The operation of the individual sections has been discussed previously. It consists of a phase detector (PD) connected to a voltage-controlled reference oscillator (VCO) which is automatically adjusted to equal the input frequency. By definition, lock occurs when the VCO frequency f_0 is equal to the input frequency f_i. The phase difference between the input and VCO ideally does not deviate far from zero when locked. Adjustment of the VCO to the input frequency f_i is accomplished by feedback from the PD output to the VCO input. The response time of the feedback circuit is determined by the low-pass-filter break frequency f_x and the system gain K_L. The demodulated output is proportional to the frequency deviation around the VCO center frequency f_0 since the VCO is designed so that its frequency shift is proportional to the control voltage. In other words, the PLL is an FM detector which follows the input frequency once lock is achieved.

The response of each section is summarized below. We adopt a common notation for PLL analysis; the PD output v_d in the linear region is

$$v_d(t) = K_d[\theta_i(t) - \theta_o(t)] = K_d\theta_e(t) \qquad (15\text{-}15a)$$

where θ_i = input-signal phase
θ_o = VCO phase
θ_e = difference or error phase
K_d = detector gain or sensitivity

Note that in general the phases are time-dependent, but when they are locked, θ must be small enough for the linear approximation to hold [Eq. (5-10)]. Outside the linear region the response of a PD is usually given approximately by

$$v_d(t) = K_d \sin \theta_e(t) \qquad (15\text{-}15b)$$

The VCO frequency f_o is proportional to the PLL output voltage v_o around a center frequency f_{oo}, or

$$f_o(t) = K'_o v_o(t) + f_{oo} \qquad (15\text{-}16)$$

When $v_o = 0$, or the loop unlocks, the VCO frequency reverts to f_{oo} (if ω_0 is substituted for f_o, then $K_o = 2\pi K'_o$ replaces K'_o).

As indicated in Chap. 11, the single-pole low-pass-filter response, expressed as a phasor, is

$$\frac{v_o}{v_d} = \frac{1}{1 + j(f_m/f_x)} \qquad (15\text{-}17)$$

where f_x is the low-pass-filter (or open-loop) break frequency and f_m is the modulator frequency related to θ_i by

$$\theta_e = \theta_i - \theta_o = 2\pi f_m t \qquad (15\text{-}18)$$

Of course, the phase-locked loop is a closed-loop system with negative feedback. Analysis of the closed loop is similar to that of feedback-amplifier analysis. It can be shown† that the stability of the loop depends on the total loop gain $K_L = K_d K_o$ and the response of the low-pass filter. If the filter and gain are improperly chosen, the loop will border on instability, as expressed, for example, as a ringing or overshoot of the output v_o if the input is rapidly changing. A simple RC low-pass filter often results in a system which is at the edge of instability and which tracks pulse FM

---

†A detailed discussion of phase-locked loops is given in F. Gardner "Phaselock Techniques," Wiley, New York, 1966.

modulation (step changes of f_i) poorly. By simply adding a resistor R_2, the filter of Fig. 15-15 is obtained, which can be shown to result in a substantially more stable system and reduced tracking error. Optimum filter values depend on many factors, e.g., input-frequency variation with time (step, ramp, sine) and noise level. There is no universal best choice which

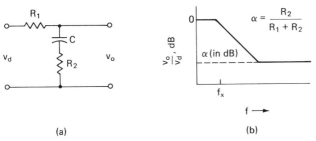

(a)   (b)

Fig. 15-15 Standard low-pass filter for PLL applications (single pole with zero): (a) circuit; (b) frequency response.

fits all situations, but the following design criteria can be used except for more exacting requirements.

1. Calculate (or obtain from the manufacturer's specification) the loop gain K_L

$$K_L = 2\pi K_o' K_d \qquad (15\text{-}19)$$

2. Set the closed-loop maximum frequency response f_n

$$f_n \approx \frac{1}{2\pi} \sqrt{\frac{K_L}{CR_1}} \qquad (15\text{-}20)$$

Demodulated frequency at PLL output v_o will be strongly attenuated above f_n, the closed-loop break (-3-dB) point. Often R_1 is set by the device manufacturer (perhaps an internal IC resistance, typically 2 to 20 kΩ) and then C must be calculated from Eq. (15-20).

3. Choose R_2 from the relation

$$R_2 \approx \frac{1}{\pi C f_n} \qquad \text{if } \frac{1}{K_L} < R_2 C \qquad (15\text{-}21)$$

It is assumed that $R_2 \ll R_1$. By setting the time constant ($\tau_2 = CR_2$) to this value a loop-damping factor of unity is chosen. A unity damping factor allows the most rapid response to a step input without overshoot. A more rapid response is obtained by decreasing R_2 (or τ_2) below the value specified by Eq. (15-21). The above relation holds for high loop gain; for low loop gain ($K_L < R_2 C$) the time constant τ_2 has little effect, and the loop is stable without τ_2 ($R_2 = 0$).

Two important characteristics of the PLL are *lock range* Δf_H (also called *hold-in range* or *tracking range*), and *capture range* Δf_C. By lock range is meant the range in input frequency f_i over which a PLL once in lock will remain in lock. The usual reason for the loop to drop out of lock is that the range of the phase detector is exceeded, in particular, that θ_e is greater than $\pm\pi/2$. When $\theta_e = \pm\pi/2$ (or $\pm90°$), according to Eq. (15-15), the PD output will be $v_d = \pm K_d$, so that the VCO frequency shift from center [Eq. (15-16)] will result in a maximum frequency deviation, or lock range, of

$$\Delta f_H = \pm K_o' K_d = \pm \frac{K_L}{2\pi} \qquad (15\text{-}22)$$

By capture range is meant the range in frequency an initially unlocked loop will automatically lock onto or acquire the input signal. Shifts of the VCO frequency to that of the input can take anywhere from less than 1 cycle to millions of cycles (seconds or minutes to pull in or acquire lock). It is not obvious that an initially unlocked loop will ever spontaneously lock, and indeed the analysis of the capture or acquisition process requires an involved nonlinear analysis. When unlocked, the output of the phase detector is a distorted beat or difference frequency between the input and VCO frequency. Analysis† shows that the distorted beat frequency has a small dc component which is always in the direction to pull the VCO frequency toward the input frequency. If the filter has a long time constant (small system bandwidth), the magnitude of the beat frequency (including the dc component) will be small and therefore the drift toward lock will be slow.

An estimate of capture range Δf_C is provided by a nonlinear analysis

$$\Delta f_C \approx \pm f_n \qquad (15\text{-}23)$$

An estimate of acquisition time τ_A is

$$\tau_A \approx \frac{f_i - f_{oo}}{4\pi f_n} \qquad (15\text{-}24)$$

where f_n is the closed-loop frequency response and f_i the initial input frequency.

In the preceding analysis it was assumed that the VCO frequency f_o and input frequency f_i are equal at lock. Actually the loop will respond to any odd harmonic of the input frequency (f_i, $3f_i$, $5f_i$, . . .) because the phase detector (or PSD) responds to odd harmonics (but with decreasing sensitivity). Often the higher harmonics are outside the capture and/or hold-in range of the PLL, so that harmonic response is not a problem.

15-8   PLL Integrated Circuits

A practical IC PLL is pictured in Fig. 15-16. It is intended for operation at frequencies up to 15 MHz and therefore is suitable as an FM detector at standard intermediate frequencies of 4.5 or 10.7 MHz. Operation in the

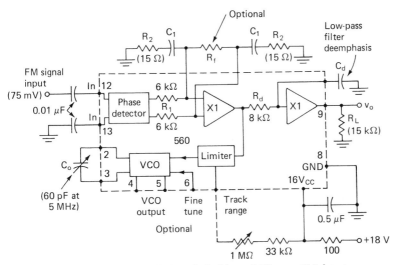

Fig. 15-16   An RF/IF integrated phase-locked loop (560) as an FM detector.

audio-frequency range is possible although the required capacitors are somewhat large and the loop gain, which is proportional to operating frequency, becomes small. Like other RF amplifiers (Chap. 13), the input amplifier has a differential input which must be capacitor-coupled. For input amplitudes in excess of 1 to 3 mV rms the amplifier limits, and therefore the PD output becomes independent of signal amplitude. The detector has a balanced output, like the PSDs described above, and therefore a dual low-pass filter ($\tau_1 = R_1 C_1$) is required (compare with Fig. 15-15). Internal loop feedback is provided through differential amplifier 1 and limited to the VCO. A capacitor C_0 controls the center frequency f_{00} of the VCO although an optional fine-tune control can also shift f_{00} over a modest range ($+40$ percent). The nominal value of C_0 is

$$C_0 = \frac{300}{f} \qquad f \text{ in MHz} \qquad (15\text{-}25)$$

The normal (open) tracking range Δf_H of 10 percent can be reduced to below 1 percent by connecting a resistor from the track-range input to

$+V_{CC}$ to control inject current into this input (open if not used). Current injection from a voltage source through a limiting resistor can also be used for the fine-tune control.

A second low-pass filter (break frequency is $1/2\pi R_D C_D'$), termed a *deemphasis filter,* is present in the output amplifiers. It is intended to provide a rolloff at higher audio frequencies to compensate for the high-frequency boost provided in the standard FM broadcasting. Note that it is not in the feedback loop but only (additionally) filters the output. The break frequency can be set slightly above the closed-loop break frequency f_n if no rolloff is needed. Because the output transistor amplifier has an open emitter, a load resistor to ground R_L is required.

A characteristic of the VCO is that the fractional change in frequency $\Delta f/f_o$ with control voltage is a constant, independent of C_o and thus f_{oo}. This implies that the VCO gain K_o and thus the loop gain K_L is proportional to frequency; specifically

$$K_L \approx 10 f_{oo} \qquad f_o \text{ in Hz} \qquad (15\text{-}26)$$

for the 560 series PLL, assuming that the input voltage is high enough to provide limiting (in which case K_d is about 1.5 V/rad). At higher frequencies (>5 MHz), the loop gain may be excessive and require a reduction by the addition of the resistor R_f to the low-pass filter, which in effect reduces the detector gain K_d. Since R_f (typically 1 to 10 kΩ) also affects the filter resistance, the filter time constant is affected and C_1 must be recalculated. Values of the low-pass filter were chosen for a closed-loop bandwidth f_n of 30 kHz (damping constant of unity) using the procedure outlined above.

Chapter Sixteen

Noise and
Noise Reduction

At low signal levels the background noise present in every amplifier may be comparable to or greater than the signal voltage. Noise-reduction techniques can be divided into two groups, the design of low-noise amplifiers and signal enhancement by signal averaging or bandwidth reduction. Emphasis here will be placed on the latter techniques. Discussion of amplifier noise is presented here primarily to help in the informed selection and use of low-noise amplifiers rather than their detailed design.

It should be pointed out that the noise analysis assumes random noise rather than unwanted signals of other kinds (hum pickup or oscillation). Many of the random-noise reduction techniques actually reduce nonrandom noise as well.

16-1 Noise Sources and Terminology

Noise generation is inherent in any electric conductor as a consequence of the discrete nature of the charge carriers (electrons). Consider a resistor consisting of a conducting solid containing randomly moving charge carriers. At any instant of time there is a certain statistical probability that the charge carriers will be concentrated more at one end than at the other, and as a consequence of this polarization a small voltage will develop across the terminals even in the absence of an external current or voltage source. This is referred to as *thermal* or *Johnson noise.* It is a basic statistical fact that the fractional deviation from the mean of a property decreases with an increase in the number of particles, so that the relative fluctuation of the charge carriers at either end will become smaller as the carrier concentration, or conductivity, increases. In other words, the voltage fluctuation decreases with conductivity. A more detailed statistical analysis shows that the fluctuation expressed in terms of a thermal noise power P_{n0} is given by

$$P_{n0} = 4KT\,\Delta f \qquad (16\text{-}1)$$

where Δf = bandwidth within which measurement is made, Hz
  K = Boltzmann's constant = 1.38×10^{-23} W·s/K
  T = absolute temperature, K

The proportionality to the bandwidth occurs because statistical deviations are more likely in a short time interval and conversely tend to average to zero after a long time. When the noise power per unit bandwidth is the same at all frequencies, it is referred to as *white noise.* Expressed in terms of rms voltage $\bar{v}_{n0}$, the terminal noise for an ideal resistor becomes

$$\bar{v}_{n0} = \sqrt{4KTR\,\Delta f} \qquad (16\text{-}2a)$$

$$\bar{v}_{n0} = 1.28 \times 10^{-4}\sqrt{R\,\Delta f} \qquad \mu\text{V rms at 25°C} \qquad (16\text{-}2b)$$

Since the bandwidth of any portional system is finite, the thermal-noise voltage is also finite and in fact small. For example a noise-free unity-gain amplifier with the extraordinary frequency response of dc to 1000 MHz ($\Delta f = 10^9$) when attached to an ideal resistor of 1000 Ω would have a noise of 128 μV. An equivalent circuit for a resistor consisting of a hypothetical noise-free resistor (as distinct from the ideal resistor) in series with a noise-voltage generator is shown in Fig. 16-1c.

Nonideal resistors—indeed all electrical components—exhibit a noise voltage in excess of that of the thermal noise of an ideal resistor. The ratio of the noise power of a real component P_n to that of the ideal P_{n0} is defined as the *noise factor* F_n of the component

$$F_n = \frac{P_n}{P_{n0}} = \frac{\bar{v}_n^2}{\bar{v}_{n0}^2} \qquad (16\text{-}3)$$

where P_{n0} and $\bar{v}_{n0}$ are the thermal-noise power and voltage given by Eqs. (16-1) and (16-2). When the noise factor is expressed in decibels, it is referred to as a *noise figure* (NF)

$$\text{NF} = 10\log F_n = 20\log\frac{\bar{v}_n}{\bar{v}_{n0}} \qquad (16\text{-}4)$$

Excess noise at all frequencies increases with the direct current through the resistor. A low-noise resistor (wire-wound or metal film) may have a noise figure below 0.1 dB even when a direct current is present, but some common carbon composition resistors have noise figures above 10 dB. Capacitor noise is uncommon.

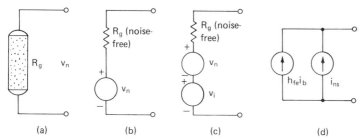

Fig. 16-1   Noise equivalent circuits: (*a*) resistor, (*b*) equivalent of a resistor, (*c*) signal voltage source with internal resistance, and (*d*) transistor signal current source with shot-noise generator.

16-2   Semiconductor Noise

Shot noise is a current noise associated with the flow of direct current across any junction. It is caused by the statistical fluctuation in the number of charge carriers passing the junction during a given time. For a reverse-biased diode, the shot-noise current i_{ns} is

$$\overline{i_{ns}} = \sqrt{2eI_{dc}\,\Delta f} \qquad (16\text{-}5a)$$

$$\overline{i_{ns}} = (5.6 \times 10^{-10})\sqrt{I_{dc}\,\Delta f} \qquad (16\text{-}5b)$$

where I_{dc} is the direct current flowing through the junction and e is the electronic charge. In the particular case of a BJT the diode generating the shot noise is the base-collector junction and I_{dc} is the quiescent collector current. The noise produced can be represented by a current generator or magnitude i_{ns} in parallel with the usual current generator $h_{fe}i_b$, as indicated in Fig. 16-1*d*. It can be seen that it is advantageous from the noise-reduction point of view to operate the transistor at as low a collector current as possible, taking into account the reduction of transistor-current gain h_{fe} and thus signal current as i_c is decreased. For many BJT input IC op-amps i_c is in fact quite low (in order to keep the input bias current low), and therefore the shot noise is not excessive. Shot noise is also present in vacuum tubes but is usually negligible in FETs.

At low frequencies the noise of transistors, or indeed all known amplifying devices, increases as the frequency decreases. This is referred to as $1/f$ (read "1 over *f*") noise since the noise power usually varies inversely with frequency. The low-frequency or $1/f$ noise P_{nl} is given by

$$P_{nl} = K \frac{\Delta f}{f} \tag{16-6}$$

where K is a constant for the device, Δf is the bandwidth as before, and it is assumed that $\Delta f \ll f$. In terms of voltage

$$\bar{v}_{nl} = \sqrt{\frac{K \, \Delta f}{f}} \tag{16-7}$$

Causes of low-frequency noise vary with the device, and there appears to be no fundamental reason why a device free of $1/f$ noise cannot be constructed. Techniques of reducing $1/f$ noise have been successful, but no device completely free of this noise has been made. It is even observed in light sources and nerve axons. More generally this low-frequency noise varies as $1/f^n$, where $0.5 < n < 2$. In the case of the transistor (BJT and FET) the noise is associated with the random generation and recombination of surface traps at interfaces (gate or pn junction) which slightly modulate the current passing through the device. In vacuum tubes it is referred to as the *flicker effect*. No cutoff below which the noise ceases to increase with decreasing frequency has been observed, but the noise voltage does not approach infinity even theoretically because Δf must approach zero as f goes to zero.

Voltage reference diodes can be extremely noisy, especially at higher voltages. While they are often referred to as Zener diodes, actually only low-voltage diodes (below 4 to 10 V) undergo breakdown by the Zener mechanism; higher-voltage diodes break down by the avalanche mechanism. The distinction is important to noise considerations because the Zener mechanism exhibits only shot noise, while the noise due to the avalanche mechanism is orders of magnitude higher, including a substantial amount of $1/f$ noise. Higher-voltage reference diodes should not be used to bias or regulate the voltage of low-noise amplifiers. Noise in reverse-breakdown diodes may occur if the diode is overloaded or abused and is a sign of early failure.

16-3 Amplifier Noise

In this section the noise characteristics of amplifiers are discussed to aid in the selection of low-noise amplifiers, especially ICs, with due consideration to such factors as source resistance and frequency response. Design of discrete low-noise amplifiers will not be presented, in part because of its complexity but primarily because the improvement in performance for most applications is marginal considering the effort required.†

---

†For a detailed discussion of low-noise-amplifier design, see C. Motchenbacker and F. Fitchen, "Low-Noise Electronic Design," Wiley, New York, 1973.

Amplifier noise is always referred to the input. For example, if the noise at the amplifier output is 50 μV and the gain is 10, the effective noise with respect to the input is 5 μV regardless of the stage within the amplifier at which the noise actually originates. Of course, in designing an amplifier it is necessary to know at what point the noise originates, and in any well-designed amplifier the noise of all stages other than the input stage is made negligible. An equivalent circuit for a high-input-impedance amplifier connected to a (low-level) signal source v_i with an internal (Thevenin) resistance R_g is shown in Fig. 16-2a. For a differential-input op-amp it is

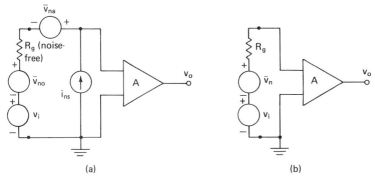

(a)   (b)

Fig. 16-2 Amplifier noise equivalent circuits (grounded input).

more accurate to apply the equivalent circuit shown to each input separately, but in practice the noise of one input is usually negligible because it is connected to a low-impedance source and/or can be referred (added) to the other input so that the simpler single-ended input model or equivalent is usually adequate. Here the $\overline{v_{no}}$ is the magnitude of the thermal noise associated with the source resistance R_o, $\overline{v_{na}}$ is the amplifier noise voltage (at a particular frequency) referred to the input, and $\overline{i_{ns}}$ is the noise-current generator due in part to shot noise coupled (directly or capacitively) to the input. The noise current is small but not zero for FET amplifiers.

Note that even in the simpler equivalent circuit, two noise-voltage sources and one noise-current source are present, and it is convenient to combine them into a single noise-voltage source (Fig. 16-2b) with a total voltage of

$$\overline{v_n} = \overline{v_{na}} + \overline{v_n} + \overline{i_{ns}} R_g \qquad (16\text{-}8)$$

The plot of noise voltage as a function of source resistance (Fig. 16-3a) illustrates the increase in noise with resistance. Alternatively the noise figure of the amplifier, obtained by inserting the noise voltage expressed by Eq. (16-8) into (16-4), can be plotted as a function of source resistance. It expresses the contribution of amplifier noise related to that of the unavoidable (once R_g is fixed) thermal noise of the source.

A more definite measure of amplifier noise is the signal-to-noise ratio S_n in decibels

$$S_n = 20 \log \frac{v_i}{v_n} \qquad (16\text{-}9)$$

It should be noted that a 3-dB noise figure achievable in a good but not extraordinary low-noise amplifier represents a noise voltage only 41

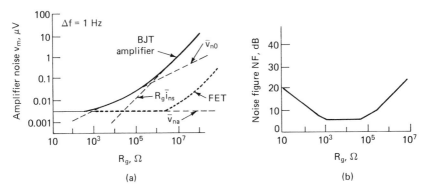

Fig. 16-3 (*a*) Typical amplifier noise and (*b*) noise figure as a function of source resistance.

percent above the irreducible thermal noise. Only infrequently would the effort or cost of a very low noise amplifier (0.1 to 1 dB) be justified.

It is important to construct the amplifier with proper shielding, grounding, and low-ripple power supplies if the injection of unwanted signals, often referred to as noise, is to be avoided. Placement of the preamplifier in a separately shielded box near the signal source and remote from hum-producing transformers is suggested (Fig. 16-4). Rigid mounting of components helps provide immunity to room vibrations. A separate ground return to the signal source avoids noise pickup on the ground line. Ripple and noise from standard regulated power supplies may be too high for low-noise applications. Batteries with bypass capacitors are good low-noise power sources. Alternatively the voltage regulator of Fig. 16-4 may be used. It is actually a low-pass network which removes hum and ripple and is not intended to improve on the dc regulation.

In summary the following consideration should be applied to the low-noise amplifiers:

1. Choose an IC, amplifier, or transistor with a low noise figure, below 5 dB and preferably below 2 dB. If the source impedance is high ($R_g > 10 \text{ k}\Omega$), an FET amplifier usually has a lower noise. If the signal frequency is low, a low $1/f$ noise unit is needed.

2. Keep the source resistance R_g as low as possible if this is not a

specified (fixed) parameter. A plot of signal-to-noise ratio as a function of R_g aids in the selection of the optimum value.

3. Minimize external influences by proper shielding, grounding, and utilization of low-noise power supply.

4. The bandwidth should be limited, as discussed in more detail in the next section, to that just needed to pass the signal.

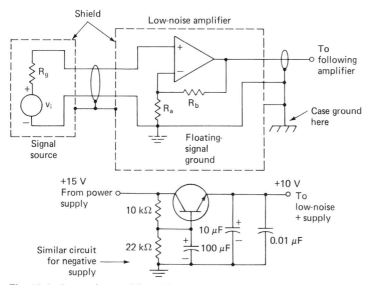

Fig. 16-4   Low-noise amplifier wiring.

16-4   Bandwidth Limitation

Once a low-noise amplifier and the optimum source resistance have been selected, in general, the only further means of reducing noise is to reduce the amplifier bandwidth since, as pointed out above, all random-noise voltage is proportional to the square root of bandwidth [Eqs. (16-2), (16-5), and (16-7)]. It is advantageous in particular to limit the low-frequency response in order to minimize $1/f$ noise. Commercial low-noise preamplifiers often have adjustable high- and low-pass filters to allow limitation of the low- and high-frequency response of the amplifier to that just required to pass the signal without distortion or attenuation. Excess bandwidth increases the noise without affecting the signal.

The filters which reduce the bandwidth need not be in the input amplifier. Often it is most convenient to limit the bandwidth in the second stage, where the signal level is much higher than any noise generated by the components making up the filter but not so high that the amplitude noise outside the desired bandwidth can drive the amplifier into a nonlinear region (on peaks, in particular) before being taken out by the filter. Suitable filters have been discussed in Chap. 11.

Under conditions where the signal frequency is nearly constant the bandwidth can be made very narrow and the noise greatly reduced correspondingly. Narrow-band amplifiers can be used to limit the bandwidth, but if the ac signal is to be demodulated, as is usually the case, the PSD (Chap. 15), which combines the bandwidth-limitation and detection functions in one circuit, is superior in terms of both performance and ease of design. Where the noise level is high, however, it may be advantageous to precede the PSD by a narrow-band amplifier to limit the noise peaks and thus reduce the linearity and dynamic-range requirements of the PSD.

As an example of a PSD as a narrow-band detection system, consider the circuit of Fig. 16-5, which is intended to measure very small changes

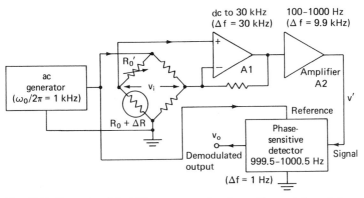

Fig. 16-5 Example of the PSD as a narrow-band amplifier and demodulator.

in resistance from the nominal value R_0. Many transducers or precision bridge measurements can use this or a similar circuit. As indicated in the discussion of the bridge amplifier, the output voltage is proportional to the bridge imbalance ΔR. Thus, the dc output voltage is also proportional to ΔR. The output frequency response of the PSD ($f_x = 1/2\pi RC$) is determined by the time constant of the output low-pass network, which can be made arbitrarily low by choosing sufficiently large capacitors (see Chap. 15). In other words, only input frequencies between $f_0 + f_x$ and $f_0 - f_x$ after demodulation are passed by the filter, resulting in a bandwidth $\Delta f = 2f_x$. If, for example, f_x is chosen to be 0.5 Hz (time constant approximately 0.3 s), the bandwidth will be 1 Hz. The noise can be predicted by Eq. (16-8) [and the signal-to-noise ratio by Eq. (16-9)] with $\Delta f = 1$ Hz. Note that the noise can be made very small by increasing the PSD time constant.

Any signal which does not bear a fixed phase relation to the reference wave, including hum or other undesirable signals, has a zero average and therefore will be rejected by the PSD. A narrow-band amplifier will also reject unwanted signals but has the disadvantage of being difficult to tune, especially if the bandwidth is very narrow. By contrast a diode rectifier or converter responds to noise and unwanted signals as well as the desired signal, and subsequent filtering is not effective in bringing out the signal.

16-5 Signal Averaging

An effective means of noise reduction for periodic or repetitive signals is signal averaging. Each repetition of the signal is equivalent to that seen as one sweep of an oscilloscope, and the averaging process can be thought of as taking the time average of many sweeps at each point along the time axis. The assumption here is that random noise or any unwanted signal not synchronized to, or correlated with, the desired signal will tend to average to zero over a long time while the desired signal will remain unchanged. An average $\bar{S}_n$ of repetitions of sweeps of the noisy signal-to-noise ratio S_n is given by

$$\bar{S}_n = \frac{S_n}{\sqrt{n}} \tag{16-10}$$

One way of interpreting this result is to think of the averaging process as increasing the effective period over which each point is taken, a procedure equivalent to reducing the bandwidth. Thus signal averaging in effect reduces noise by decreasing the system bandwidth. An average over $n = 10^4$ sweeps resulting in a signal-to-noise improvement of 100 times (40 dB) is not uncommon.

Signal averaging is often done by a computer after A/D conversion. However, the averaging often requires that a large number of bits per second over an extended time be processed, and the dedication of a substantial portion of the capacity of a general-purpose computer to this simple task may be economically unsound. Even when a computer is available, it may be wise to preprocess the data with a specialized signal averager or microprocessor before transfer to a larger computer for further processing.

All signal averagers break the sweep time into discrete intervals, or channels (Fig. 16-6). For high resolution the number of data points n should be high (perhaps 100, 400 or more), but for simplicity the principle of measurement will be illustrated here with a small number of points. In Fig. 16-7 a block diagram of the signal averager is given. The clock and counter which control the incrementing from one channel to the next are equivalent to a digital time sweep. Sweep time is determined by the clock frequency and may range from microseconds to seconds. It is important to note that an external trigger is required to start the sweep since the signal is usually too noisy to provide a reliable internal trigger. During read-in the counter-controlled data switch transfers the data into a digital

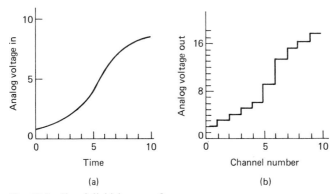

Fig. 16-6 Signal digitizing waveforms.

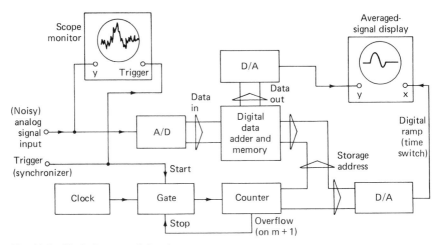

Fig. 16-7 Block diagram of signal averager.

storage element. The analog signal is first converted into digital form via an A/D converter before switching. During subsequent sweeps further data is added to each channel. After n sweeps the process is stopped and the data is held in memory. During the data readout process the data switch transfers the data stored in memory to the output in serial fashion. The analog data is connected to the y axis of the oscilloscope or recorder while the output of the counter (after D/A conversion), representing time, is connected to the x axis. This display is the averaged signal. It should be noted that the readout sweep rate can be quite different from the read-in sweep rate. For example, the input sweep interval could be 0 to 100 μs to record a fast transient while the readout could be slowed to 0 to 10 s so that a plot of this transient can be made on a standard xy or time-base recorder.

SIGNAL AMPLIFICATION AND PROCESSING DESIGN EXAMPLES

Example 6   Fourth-Order Butterworth Low-Pass Filter (4-kHz Cutoff)

DESCRIPTION: The filter attenuates signal frequencies above breakpoint sharply (80 dB/decade) while not appreciably attenuating lower frequencies.

SPECIFICATIONS

Voltage gain: 2.58

Breakpoint (-3 dB): 4.0 kHz†

DESIGN CONSIDERATIONS: A fourth-order filter is made by cascading two quadratic filters following the design procedure outlined in Sec. 11-7. For simplicity in obtaining capacitors, the equal-value-capacitor version of the op-amp configura-

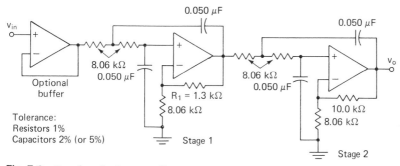

Fig. E-6   Fourth-order low-pass filter.

tion (Fig. 11-16a) was selected. A standard-value capacitor which has a reactance of the order of 10 kΩ (optimum impedance range for op-amps) was chosen. From the resonance condition the value of the resistors ($R = 1/2\pi C = 8.06$ kΩ) was then calculated. Butterworth filter constants b are 1.845 and 0.765, according to Table 11-2, so that the feedback resistors R_1 and R_2 are 0.16R and 1.24R, respectively. Gains of the two stages are 1.16 and 2.34, respectively, for a total gain of 2.58. A unity-gain input buffer is required if the filter is not driven by a low-impedance source, e.g., another op-amp stage.

Example 7   Amplifier with Digital Gain Control

DESCRIPTION: The analog amplifier has a gain which is determined by an 8-bit digital signal.

†Attenuation (relative to dc) at 3 and 6 kHz is 0.4 and 26 dB, respectively.

SPECIFICATIONS

Voltage gain: 0 to 1.00 in steps of $\frac{1}{255}$
Frequency response: dc to 1 MHz
Analog voltage range (input/output): ±10 V
Digital control: 8-bit TTL level (0 to 4 V)

DESIGN CONSIDERATIONS: A standard multiplying D/A converter based on a digitally switched ladder network (Fig. 14-10) acts as a variable resistor in an inverting-

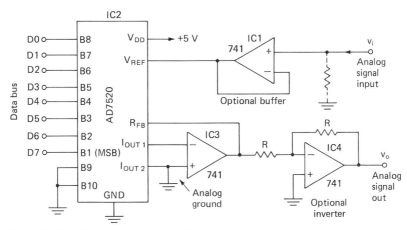

Fig. E-7 Digital gain-controlled amplifier.

amplifier configuration. The signal, which may have either polarity, is applied to the reference-voltage input. Internal (IC2) feedback resistors provide unity gain for the inverting amplifier (IC3). An output amplifier (IC4) reinverts the signal if an uninverted amplifier is required. The input amplifier (unity gain) is required only if the signal source does not have a low impedance.

Standard TTL-level signals constitute the 8-bit digital data bus, which can be connected directly to a microprocessor output port. All bits equal to **0** correspond to zero gain, and all **1** to maximum gain.

SIGNAL AMPLIFICATION AND PROCESSING DESIGN PROBLEMS

26 Design a pulse averager with an uninverted output equal to the average of a stream of irregular input pulses. The pulse frequency varies from 100 Hz to 100 kHz and the amplitude from -10 to $+10$ V.

27 Draw a circuit which will produce an output (inverted acceptable) equal to the slope of an input ramp signal v_i. The calibration should be such that a rise of 1 V/ms results in a 1-V output. Describe the output which occurs on the practically instantaneous return slope.

28 An ac amplifier of gain of 1.00×10^4 is required with a frequency response from 10 Hz to 10 kHz (-3-dB points). The dc offset voltage at the output should be under 0.1 V, and no potentiometer adjustments are allowable.

29 Design a notch filter to reject 60 Hz while passing all other frequencies from dc to 100 kHz. Specifications include a dc gain of 10.0, an input impedance over 100 kΩ, and an output impedance under 10 Ω. If desired, calculate bandwidth.

30 A bandpass filter is desired which will pass 2 kHz (−3-dB points at about 1.6 and 2.4 kHz) with unity gain.

31 Design a fourth-order low-pass Butterworth filter which will cut off at 1 kHz. A gain of either 1 or 10 is desired.

32 Design an EEG amplifier filter to pass frequencies between 4 and 16 Hz (−3-dB points) while sharply attenuating signals outside this band. In particular, the gain at 60 Hz should be at least 40 dB below the gain at 8 Hz (midband). A midband gain of 1000 is needed, and the dc offset voltage at the output should be below 0.1 V. *Hint:* Use high- and low-pass Butterworths.

33 Draw the circuit diagram of a sinusoidal oscillator with an output of 20 V peak to peak at a frequency of 400 Hz.

34 Draw the circuit diagram of a sinusoidal oscillator with an output frequency of 2 MHz (at 1 V peak to peak minimum).

35 Draw the circuit diagram of a square-wave generator with a 5-V peak-to-peak output at a frequency variable between 1 and 10 kHz.

36 Design a wideband amplifier with a gain of 10.0 from dc to 10 MHz.

37 Draw the diagram of a circuit which will produce an LS TTL-compatible output (0 to 3 V, pulse or square wave) with an input voltage of 1 mV peak to peak minimum with a frequency range of 20 kHz to 20 MHz.

38 Design a circuit for a voltage-to-frequency converter with a conversion factor of 100 Hz/V and a linearity of 1 percent at full scale (+10 V input). An output pulse of 0 to 5 V (CMOS-compatible) is desired.

39 Design an A/D converter (8-bit) with a range of 0 to 1 V dc based on either the single-ramp, counter, or servo method. A conversion time of 0.1 s is acceptable.

40 A digital sweep (sawtooth) generator with a 4-bit resolution is required. Combine a binary counter and a simple D/A to accomplish this.

41 Design an ac-to-dc converter-type demodulator for a strain-gage bridge excited by ac (1 kHz). Assume the output of a bridge amplifier at maximum strain ($\varepsilon = 0.01$) is 5 V peak to peak. A dc component may be present due to an uncompensated amplifier offset. An output of 1.0 V at full scale is required. Frequency components of the strain (modulation of the 1-kHz carrier) will not exceed 20 Hz. A PSD is suggested.

42 An FM oscillator (center frequency of 100 kHz) is frequency-modulated with a sine wave with a frequency between 50 and 500 Hz. Design an FM detector with an output of 10 V (peak) for a frequency deviation of 10 kHz maximum.

43 Design a PLL which will demodulate pulse-code modulation. The center frequency is 2.0 MHz and input amplitude 10 mV. The frequency deviation for a pulse (**1**) is 10 kHz, the pulse clock frequency is 200 Hz, and at least one **1** (and one **0**) pulse is transmitted every 10 clock cycles. The circuit output pulses must be CMOS-compatible (0 to 5 V). *Hint:* Use an ac amplifier to bring up the output level to the point where a comparator or Schmitt trigger will work well, possibly using a diode clamp to restore dc.

44 Design a constant-amplitude amplifier (automatic amplitude control) for a sine-wave source (100 Hz to 10 kHz) which has an output amplitude of 10.0 ± 0.5 V peak to peak at that frequency. The amplifier input is variable between 2.0 and 0.5 V peak to peak. Assume that the source amplitude changes slowly (under 5 percent per second).

45 Design a low-noise amplifier with a gain of 10^5 between 50 Hz and 1 kHz. Estimate the noise level for a source impedance of 10 kΩ.

46 Add a narrow-band amplifier with a center frequency of 400 Hz and gain of 10 to the low-noise amplifier in Prob. 45. Estimate the bandwidth and noise level.

Part Four

Data Switching, Control, and Readout

Chapter Seventeen

Pulse Timers and Counters

The various standard circuits for pulse timing, delay, counting, and decoding described here are major or minor subsections of numerous electronic instruments. Most are based on simpler digital devices described previously (especially in Chap. 5). The related pulse-generation circuits were discussed in Chap. 12.

17-1 IC Timers

IC timers like the popular 555 timer of Fig. 17-1 are versatile devices which can produce a single pulse of known duration or a continuous pulse stream. Operation can be understood by examining the timer connected as a triggered one-shot. Internally the timer is basically an RS-type flip-flop with comparators on the set (S) and reset (R) inputs. An extra open-collector output is provided for external capacitor discharge. The comparators have internal thresholds fixed at two-thirds and one-third the supply voltage V_{CC}, except for a few applications where an external voltage is injected into the control-voltage input. Note that the pin labeled "threshold" or THRES is somewhat misleading since usually the fixed-voltage input to a comparator is thus named. The discharge of an external capacitor by the transistor occurs when the flip-flop output Q is at **0**.

In the normal reset or rest state, before the arrival of the trigger pulse, the output Q is low and the discharge transistor holds the capacitor voltage at zero. Since the capacitor is connected to the THRES input, $\bar{R}$ is

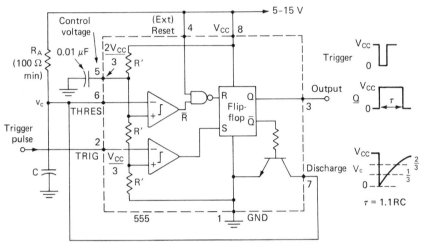

Fig. 17-1 The 555 timer as a one-shot (monostable) multivibrator.

high and R is low. It is necessary that the TRIG input be more positive than $\frac{2}{3}V_{CC}$ in this state because if it were not, S and therefore Q would be high.

Triggering of the timer requires that the input pulse drop from its positive rest value (above $\frac{2}{3}V_{CC}$) to a value less positive than $V_{CC}/3$. The required pulse is sometimes referred to as a *negative-going* or (inaccurately) a *negative-edge-triggered pulse*. When this occurs, S and therefore Q go high. Also at this time the discharge transistor turns off, and the capacitor C starts charging. As the capacitor voltage rises to the upper threshold value $\frac{2}{3}V_{CC}$, the flip-flop is reset (Q low), ending the timed period (provided the trigger pulse is no longer present). If the trigger pulse is less positive than the lower threshold $V_{CC}/3$, the flip-flop remains set (Q high).

The time τ required for the capacitor to charge to $\frac{2}{3}V_{CC}$ corresponds to $1.1R_AC$. Note that if R_A is too small, the current flow through the discharge transistor will be excessive. An external reset ($Q = \mathbf{0}$ for RESET $= \mathbf{0}$) is provided for special applications. External variation of timed period is accomplished by applying a signal to the control-voltage input. Capacitors across the power-supply terminal (0.1 to 5 μF) and perhaps from the output to ground (500 pF) may be required to eliminate multiple pulses (glitches).

Timing accuracy depends on the resistor and capacitor stability. With Mylar capacitors the accuracy will be better than 1 percent, and reproducibility from pulse to pulse is about 0.1 percent. Higher-precision timers, for example, the LM3905, are at least a factor of 10 more reproducible than the 555; however, accuracy still depends on capacitor and resistor stability, which are likely to be the limiting factors. If high timing stability (<0.2 percent drift) is required, a crystal-controlled timer is a better choice.

Connection of the 555 timer as an astable multivibrator or continuous-pulse generator is discussed in Chap. 12.

17-2 Digital One-Shot Multivibrators

One-shot multivibrators are also available in both TTL and CMOS digital lines. They feature high speed and direct compatibility with digital systems but have the disadvantages of a fixed threshold and hysteresis. Also they are impractical for long pulse widths (>1 s). Standard and retriggerable units are shown in Fig. 17-2. Both are triggered when the input trigger

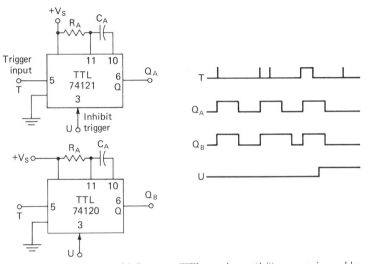

Fig. 17-2 One-shot multivibrators, TTL versions: (*left*) nonretriggerable; (*right*) retriggerable.

pulse reaches an upper threshold level and cannot be retriggered unless the trigger input drops below a lower threshold and again rises to the upper threshold. The retriggerable type differs only in that the output pulse width is measured from the time of the last trigger input, as indicated by the pulse-test sequence of Fig. 17-2. Output pulse width T is determined by the product of R_A and C_A. Variable widths are achieved by making R_A adjustable.

Often a variable trigger level (or hysteresis) is desired, and in this case a comparator can be used preceding the one-shot (Fig. 17-3). The threshold is adjusted by the voltage on the noninverting input. Hysteresis is adjusted by altering the positive-feedback resistors.

For applications where timing accuracy is not critical, the simple CMOS

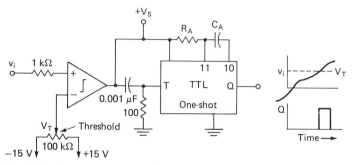

Fig. 17-3   Comparator-input-triggered single-pulse generator.

one-shots of Fig. 17-4 may suffice. They have the advantage of requiring only inexpensive standard gates, possibly unused sections of an existing package, rather than a special IC. The Schmitt-trigger type is based on the discharge of an *RC* network charged by the trigger pulse. Note that the

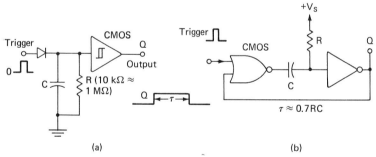

Fig. 17-4   One-shots built from CMOS gates: (*a*) Schmitt trigger, pulse-stretcher type; (*b*) edge-triggered.

output Q rises on the leading edge of the trigger pulse but the timing starts on the falling edge. The circuit therefore functions as a *pulse stretcher*. It is necessary for the input pulse to last long enough to charge the capacitor. Long output-pulse durations (at least 10 s) are practical. The Schmitt trigger can be replaced by the two-inverter equivalent (Fig. 5-36). Reversal of the diode polarity and return of R to $+V_{CC}$ instead of ground allows triggering by an inverted trigger pulse (negative logic).

A one-shot which is triggered by a positive edge is shown in Fig. 17-4*b*.

Positive feedback is employed to sharpen the transition. Timing is again determined by the RC product.

17-3 Triggered Sweep

A triggered sweep generates a ramp or linearly rising voltage with time following an input trigger pulse. An integrator with a constant-voltage input is employed to produce the sweep. The circuit of Fig. 17-5 has a flip-

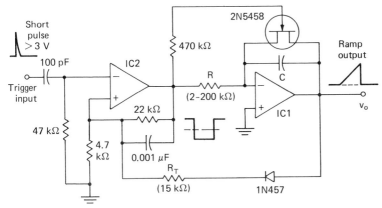

Fig. 17-5 Triggered sweep.

flop, set by the trigger signal. The output of the flip-flop v_o is integrated (IC1) to form the ramp. When the ramp reaches the threshold level, the comparator (IC2) output goes positive, resetting the flip-flop and the integrator (discharge of C through the FET). The ramp slope $S_R = dv_o / dt$ is given by

$$S_R = \frac{V_S}{RC} \qquad (17\text{-}1)$$

where V_S is the op-amp output voltage at saturation.

Termination time of the ramp can be made variable by adjusting the ramp threshold resistor R_T. In this case the pulse length T_A taken at the alternate output v_o' varies linearly with the threshold potentiometer setting, and the circuit acts as a one-shot multivibrator.

For some applications the simple sweep circuit shown in Fig. 17-6 is suitable. Current from a constant-current I_o diode charges the capacitor at a constant rate dv_o / dt. In this case the input pulse must last at least as long as the sweep desired and serves only to remove the FET short during this time. For less exacting applications or if the range of v_o is limited, the diode can be replaced by a resistor. The output must be connected to a high-impedance load such as the input of an op-amp or comparator.

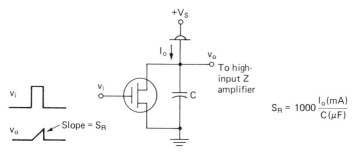

Fig. 17-6   Simplified sweep circuit utilizing a constant-current diode.

17-4   Pulse-Delay Circuits

Pulses must often be delayed in order to perform a series of operations in the proper sequence. In most cases the delayed-pulse width need not be the same as that of the input pulse, and in this case the circuit of Fig. 17-7

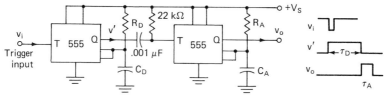

Fig. 17-7   Dual one-shot pulse-delay circuits.

would be suitable. The delay time τ_D is determined by the length of the first timer connected as a one-shot multivibrator. Because the 555 timer is triggered on a negative-going edge, the second timer will be triggered only at the end of the pulse v' produced by the first timer. Output pulse width τ_A is controlled by the time constants R_A and C_A of the second timer.

The more elaborate pulse-delaying circuit shown in Fig. 17-8 is especially useful if several delayed pulses v_1 are to be generated. It consists of a triggered sweep followed by one or more comparators. The output goes positive as the threshold (V_{T1}, V_{T2}, ...) is reached. Time delay is calculated from the relation

$$\tau_1 = \frac{V_{T1}}{S_R} \tag{17-2}$$

where V_{T1} is the magnitude of the threshold voltage and S_R is the ramp slope [see Eq. (17-1)]. Similar expressions hold for τ_2, etc. Note that the threshold voltages are linearly dependent on the potentiometer settings. For many purposes the simpler sweep of Fig. 17-6 will suffice.

Where the length of the delay is not critical, the CMOS circuit of Fig. 17-9a may be adequate. The output pulse is delayed by the RC network. Positive feedback is provided to reduce the CMOS power dissipation near the transition point (see Chap. 5).

Sometimes a dual-pulse output circuit is desired in which one of the

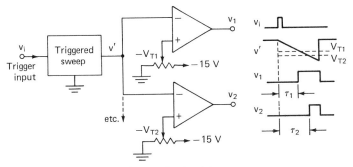

Fig. 17-8   Ramp-threshold type of pulse delay.

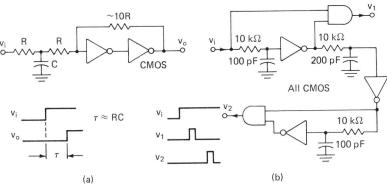

Fig. 17-9   Simple CMOS pulse-delay networks: (a) single delayed output; (b) dual output pulses, one delayed.

pulses is delayed slightly. One application is the latch-and-clear sequence needed when transferring digital data from a counter to a latch. The circuit of Fig. 17-9b is intended for this purpose. It provides a 1-μs pulse v_1 immediately following a step input and a second 1-μs pulse v_2 about 2 μs later.

17-5   Bounceless Switches

When a pulse sequence is initiated by a push-button switch or relay, it is usually necessary to add a special circuit to eliminate contact bounce or jitter, which can be interpreted by logic circuits as multiple pulse inputs.

Contact bounce generally lasts for a few milliseconds at most, and there-fore a circuit which has a time constant of 20 to 50 ms is effective in reducing the problem. The CMOS circuit (Fig. 17-10a) produces a pulse Q which lasts as long as the switch is depressed plus about 50 ms. Basically it consists of a capacitor which is rapidly discharged by the switch but is

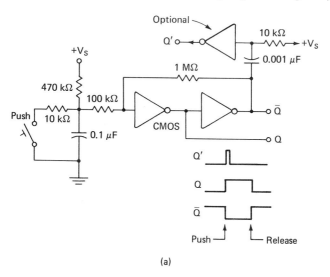

(a)

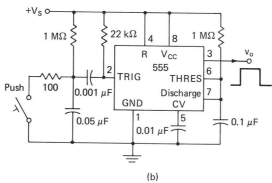

(b)

Fig. 17-10   Bounceless switch, timing versions: (a) CMOS version; (b) one-shot version.

charged relatively slowly through the resistor connected to the supply. A Schmitt trigger shapes the pulse. An optional output Q' produces a brief (1-μs) pulse when the switch is depressed and nothing when released. The second circuit employing the IC timer produces a 10-ms pulse when the switch is depressed.

The second type of circuit (Fig. 17-11), popular in microprocessor applications, consists of an SR flip-flop which is set by the first pulse from the single-pole double-throw push-button switch and reset by a stop pulse returned by the system after the desired series of operations has been completed. Note that bounce or multiple pulses to one input of the SR

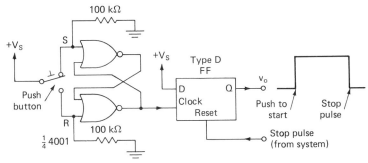

Fig. 17-11 Bounceless switch, flip-flop version.

flip-flop do not produce multiple output pulses of the SR since bounce is associated with only one switch contact at a time. If the switch is still depressed at the time the stop pulse occurs, the type-D flip-flop will reset but will not set until the switch is pushed again. In other words, it is practically foolproof.

17-6 Pulse Counters

Frequently the number of pulses occurring within a period must be known. The circuit of Fig. 17-12 is intended to count the number of pulses between an externally supplied start pulse S and a stop pulse R. An AND gate connected to the flip-flop output controls the flow of pulses into the counters. A NAND gate can replace the AND gate since the counter in either case requires one full clock cycle per count. A clear pulse is generated on the leading edge of the gate control pulse. For TTL the pulse-differentiator resistor R_p must be low enough to sink the currents flowing out of the clear inputs (see Chap. 5). Of course, if a gate control pulse with a width equal to the timing interval T_c is available, the flip-flop is not needed.

The counter and display sections are indicated in terms of a block diagram because circuit details have been described elsewhere (Chaps. 5 and 7). A specific design example was given in Example 3.

Often counters employ latches (described in Chap. 18) to hold the final count for display (Fig. 17-13). In this case, the counter can be cleared immediately after the data has been stored in the latch. Latching is

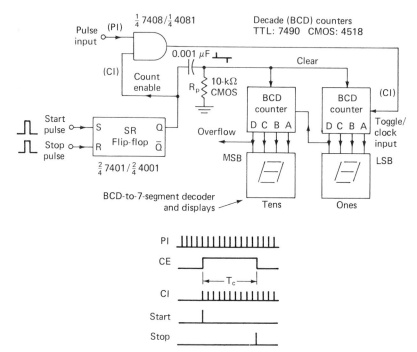

Fig. 17-12 Decade pulse counter.

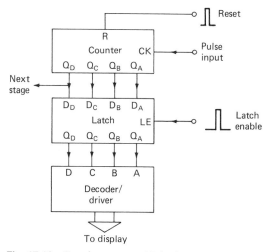

Fig. 17-13 Decade counter with latch.

controlled by the latch-enable pulse, which usually occurs immediately following turnoff of the counter. The clear pulse must of course follow the latch pulse. The delay may be brief, as generated by a pulse-delay circuit (Fig. 17-9), but often the clear pulse occurs instead on the rising edge of the counter-enable pulse.

Sometimes counters based on a scale other than 2 (binary) or 10 (decade) are desired, and in this case external logic can be used to provide the reset. As examples the scale of 3 and 6 are suggested by Fig. 17-14. Such counters find application in frequency dividers (see Sec. 17-7).

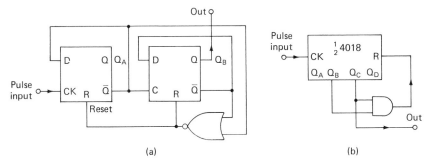

Fig. 17-14 Counters: (a) base 3 and (b) base 6.

17-7 Frequency Dividers and Multipliers

A number of simpler frequency-divider circuits have been discussed previously. These include divide by 2 (Fig. 5-20), divide by 10 (Fig. 5-19), divide by 3 and divide by 6 (Fig. 17-14), and divide by $2N$ (Fig. 5-17). Division by multiples of these numbers, for example, 100, 20, 60, can be accomplished by cascading several stages. For applications where the division is by a large number which is not a multiple of a standard divider or is variable, it is best to use a general-purpose divide-by-N counter. Either a counter specifically designed as a divide-by-N counter, for example, the 4522, or a presettable down counter (Fig. 17-15) is suitable.

The cycle of a divide-by-N counter starts when the individual counters are simultaneously preset by the application of a pulse to the preset-enable (PE) inputs. Each counter stage corresponds to one (decimal) digit (n_1, n_2, etc.) of the number N. For example, if $N = 93$, counters 1 and 2 are preset to $n_1 = 3$ and $n_2 = 9$. Note that n_1, n_2, . . . are actually 4-bit binary numbers (here BCD), which are supplied by thumbwheel encoders or, in more complex circuits, by a set of registers or latches. The first time counter 1 counts down, it will require n_1 clock pulses before it reaches zero count, at which point the carry output $\bar{C}_0$ will go to **0**. Subsequent counter 1 cycles will require 10 counts to produce a carry $\bar{C}_o = \mathbf{0}$. The carry out $\bar{C}_o$ of one stage is connected to the carry in $\bar{C}_i$ of the next. Actually the $\bar{C}_i$ acts as a counter enable, and the stage will count only if the $\bar{C}_i$ of that stage is at $\mathbf{0}$. Countdown from N is completed when the $\bar{C}_o$ of all stages are at $\mathbf{0}$. At this point the gate (negative-logic NAND) output pro-

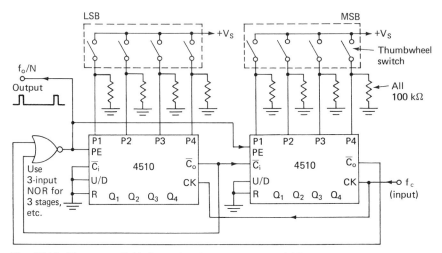

Fig. 17-15   Two-stage divide-by-N decade counter (expandable).

duces a $\mathbf{1}$ (for one clock cycle). This is the circuit output ($f_o = f_c/N$) and also the preset pulse which restarts the cycle. Note that the output is not a symmetric square wave. If symmetry is required, the input frequency should be made $2f_c$ and the circuit followed by a standard divide-by-2 counter. Expansion to three or more stages requires only that the $\bar{C}_o$ of all stages be connected to the gate (three or more inputs).

Digital frequency multiplication is more difficult than frequency division. A ×2 frequency multiplier can be made by differentiating the input wave and then counting both positive- and negative-going edges. In Fig. 17-16a the NOR gate receives a positive input pulse from one input or the other each time the input makes a transition. Thus the number of pulses out is twice that of the input. The output, of course, is not a square wave. Because sufficient time must be allowed for the RC decay, the circuit speed is more limited than the gate speed.

A more generally applicable but more elaborate method of frequency multiplication utilizes a PLL with a frequency divider (divide by N) inserted between the phase detector and VCO, as indicated in Fig. 17-16b. In many IC versions of the PLL, for example, the 565, the output of the VCO is a square wave directly compatible with digital (CMOS) counters, and the connection between the VCO and phase detector is external and therefore separable. This allows a digital frequency divider (divide by N)

to be added easily. The VCO center frequency is adjusted to approximately Nf_c, where f_c is the input (clock) frequency, assumed symmetric. The output is also symmetric at N times the frequency. A disadvantage of this method is the limited tracking range of the VCO, which limits the output frequency range (unless the VCO center frequency is readjusted). As with any PLL, stability can be a problem.

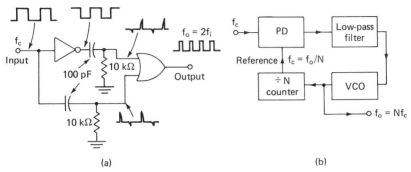

(a)   (b)

Fig. 17-16   Frequency multipliers: (*a*) edge counter (×2); (*b*) block diagram of PLL method.

17-8   Digital Interval Timers

An interval timer measures the time interval between a start and stop pulse with the result displayed on a digital readout (Fig. 17-17). It consists of a digital counter plus a clock (often a crystal-controlled oscillator).

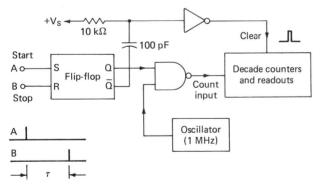

Fig. 17-17   Interval-timer block diagram.

Decade counters and crystal oscillators (Fig. 12-15) are described elsewhere.

A 1-MHz oscillator will result in a count equal to the time interval in microseconds. To convert into milliseconds or seconds only requires a shift of decimal point on the counters. If high resolution is not required, the displays can be omitted on the least significant digits. An alternative approach is to divide the oscillator output by some power of 10. The factor can be selected by a switch which acts as a range control. Often the decimal point is selected by the same switch. If the input-pulse rise times are slow, comparators (Fig. 17-3) can be added to the inputs.

17-9 Frequency Counters

A frequency counter records the number of cycles during a fixed time interval, commonly 1 s. As indicated in Fig. 17-18, it is a variation of the

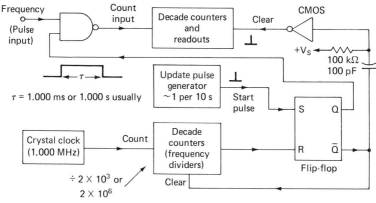

Fig. 17-18 Frequency-counter block diagram.

decade counter and interval timer (Fig. 17-17). The NAND (or AND) gate allows pulses to pass to the counter during the desired time interval τ_1, which is made precisely 1 s or an even multiple of 10. A flip-flop controls the NAND gate and is turned on by a start-pulse generator. Simultaneously the time-base counters connected to an oscillator ($f_o = 1$ MHz) are started. When the time base counters reach N_0 pulses (10^6 in this example), the carry output turns off the flip-flop at a time τ equal to N_0/f_o, for example, 1 s. During this interval the input pulses (frequency) are counted so that the number of pulses N_f registered is equal to τf. A brief clear pulse is generated as the gates are switched on. Usually the start-pulse generator rate is chosen to be variable in the 1- to 10-s range. Note that the clock frequency dividers are inhibited (held in clear) before the start pulse. Frequencies up to at least 1 MHz with CMOS, 30 MHz with LS TTL, and 200 MHz with ECL devices are achievable.

17-10 Decoders and Encoders

The decoders described here have N output lines, only one of which is at **1** at any time, as determined by an input binary (or BCD) number. A 2-to-

4-line decoder has four output lines controlled by two input lines, as indicated by Fig. 17-19a. In this example the decoder is built from individual gates although integrated units are available. As seen from the truth table, an output goes to **1** if the corresponding binary number appears at the inputs; e.g., output 2 will turn on if the input is **01**. The

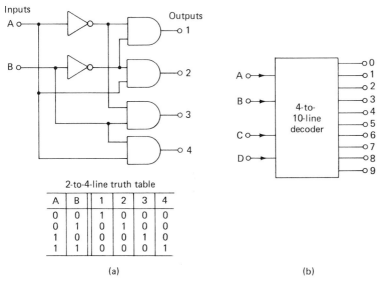

2-to-4-line truth table

A	B	1	2	3	4
0	0	1	0	0	0
0	1	0	1	0	0
1	0	0	0	1	0
1	1	0	0	0	1

(a)   (b)

Fig. 17-19   Decoders: (*a*) 2-to-4-line; (*b*) 4-to-10-line.

extension to 3-to-8-line, 4-to-16-line, 5-to-32-line, etc., decoders is straightforward. A 4-to-10-line decoder (for BCD data) is the same as a 4-to-16-line decoder except that the upper six outputs are missing (Fig. 17-19b). Many other types of decoders are also available.

Some decoders have an enable input capable of turning off all the outputs, regardless of the data input. Note that the D input of the 4-to-10-line decoder acts as an (inverted) enable input if only eight output lines are used (3-to-8-line decoder).

One application of decoders is to minimize the number of lines or wires needed to control or select data; e.g., 4 input lines (bits) can control 16 output lines. Another application is the control or scan of output lines by counters, as described below.

A common kind of encoder is the thumbwheel, a special panel-mounted switch used to enter digital data into an instrument. Perhaps the most popular is the 10-position BCD-coded switch described in Fig. 17-20. There are four output lines corresponding to the BCD code and an input line corresponding to the **0** or **1** voltages. Logic **1** is connected to the common line in Fig. 17-20, and external resistors to ground provide **0** when the switch is open.

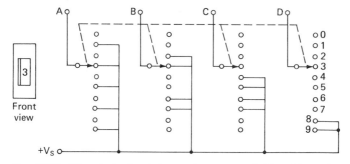

Fig. 17-20 Thumbwheel switch encoder; a ganged 10-position BCD-coded switch is shown.

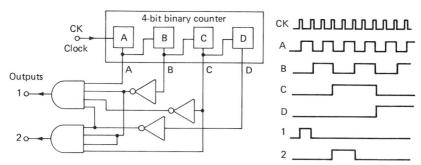

Fig. 17-21 Example of a synchronous pulse-sequencing unit.

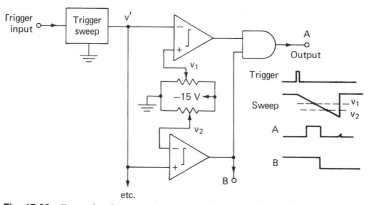

Fig. 17-22 Example of an asynchronous pulse-sequencing unit.

17-11   Pulse Sequencing

Often there is a need to generate a series of pulses in a particular time sequence. Techniques can be divided into synchronous and asynchronous methods. Synchronous techiques require a clock control, and all output pulses transitions occur as the clock pulse makes a transition. Generally some type of decoder is used to generate the output pulse, as illustrated in Fig. 17-21. In this example, two output pulses are produced in sequence, one twice the width of the other. Complex sequences can be obtained by proper decoder design, but the output pulse widths are limited to multiples of the clock period.

A cyclic or repetitive sequence of pulses is produced by the circuit of Fig. 17-21. If a single series of pulse is required, a NAND gate controlled by an *SR* flip-flop is added, as illustrated by the circuit of Fig. 17-17. The flip-flop is switched on by a start pulse and turned off by an output pulse from the counter, e.g., the *D* output in Fig. 17-21.

Asynchronous pulse-sequencing units are based on one-shot multivibrators and/or triggered sweeps. An example of the use of multivibrators for a two-pulse sequence is given in Fig. 17-7, and the extension to multiple pulses is straightforward. An example of the triggered-sweep method is illustrated in Fig. 17-22 and is a variation of that shown in Fig. 17-7. Two comparators are used for each output pulse. Both the beginning and the end of the pulse are adjusted by the threshold controls.

Chapter Eighteen
Multiplexing

Multiplexing is a technique of transferring several independent signals through a common line or circuit. In time multiplexing, considered here, the several input signals are sampled, transmitted sequentially through a common circuit, and separated at the output or receiver. Presumably the cyclic scanning or sampling rate is fast compared with the rate of change of the signals, so that transmission of sampled portions of the signals is nearly equivalent to transmission of the entire continuous signal. Both digital and analog signals can be multiplexed.

There are several reasons for multiplexing signals or data. Transmission of two or more signals to a remote site along a single telephone line or radio-telemetry link may be required. In digital systems, especially, data may be multiplexed to avoid duplication of decoders or more complicated devices. Often data is multiplexed simply to reduce the number of signal lines or pins which must be passed in or out of an integrated or printed circuit. The reduction in costs or wiring time can be significant and more than offsets the additional complexity of the multiplexing and demultiplexing circuits at the input and output.

18-1   Analog Switches

Analog (and some digital) multiplexing requires an electronic switch. A convenient switch is the CMOS transmission gate (Fig. 18-1). The basic operation of the gate has already been described in Fig. 5-9, where it was

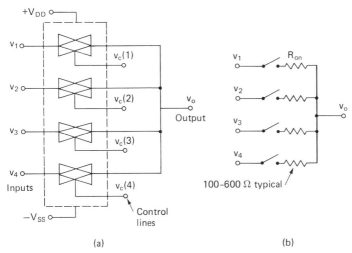

(a)   (b)

Fig. 18-1   Transmission gates: (*a*) quad CMOS gates; (*b*) equivalent circuits.

pointed out that the signal voltage (input and output) must lie between V_{DD} and V_{SS} with the 4016 gate. If V_{SS} is grounded, the signals must always be positive. This is normally the case for digital signals.

If analog signals of both polarities are to be switched, a dual power supply is needed, for example, $V_{DD} = +5$, $V_{SS} = -5$ V. The control signals must likewise span V_{DD} and V_{SS} for the 4016, an inconvenience since usually the digital signals are positive (0 to V_{DD}). In this case separate interface drivers are required (Fig. 18-2). However, many other analog switches (e.g., AH0014) do not require this driver.

18-2   Analog Multiplexers and Demultiplexer

An analog multiplexer consists of a set of analog switches which are turned on in sequence by a counter and decoder (Fig. 18-3). The composite signal is transmitted along the lines (Fig. 18-4). Signal amplifiers and line drivers may be required. Analog multiplexers are frequently employed at the inputs to A/D converters so that a single converter can serve multiple analog inputs.

A demultiplexer at the receiver end is identical to the multiplexer. Capacitors hold the voltage for the time between samples in the version

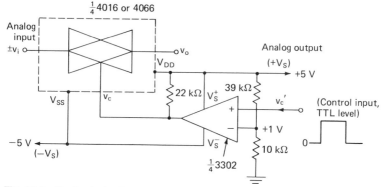

Fig. 18-2 Dual-polarity CMOS 4016 switch with logic-level driver.

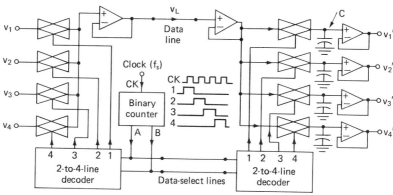

Fig. 18-3 Four-channel multiplexer and demultiplexer.

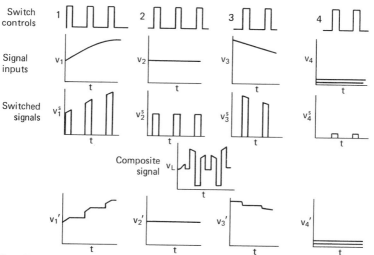

Fig. 18-4 Four-channel multiplexer signal (examples). Normally the signal changes more slowly than shown.

shown. This method is similar to that employed in the track-and-hold circuit. The output follows the respective input signals in a stepwise form (Fig. 18-4).

Presumably the sampling rate is rapid compared with the rate of change of the signals. According to the Nyquist sampling theorem, the sampling frequency f_s must be at least twice the highest signal frequency. It need only be slightly higher than the signal frequency provided a sharp-cutoff filter, e.g., high-order Butterworth, is used at the output. With the simple capacitor filter shown, however, f_c must be one or two orders of magnitude higher.

Synchronization of the multiplexing counters and decoders can be done most simply by driving the respective decoders (data selectors) by the same counter, as shown in Fig. 18-3. Extra lines can be avoided by utilizing separate clocks with almost identical frequencies (f_c and f'_c) at the transmitter and receiver ends. Some means of synchronization is necessary. In this example, however, there is little point in multiplexing because the number of lines running from the multiplexing side (transmitter) to the demultiplexer side (receiver) is only reduced from 4 to 3. With a larger number of channels there is a greater savings in lines required, but it can be seen that it would be desirable to reduce the data-selection lines. In Fig. 18-5 only one line (synchronizer line) in addition to the data line is

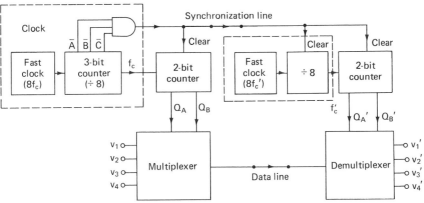

Fig. 18-5 Data-scanning clock and synchronization method. Expansion from 4 to 16 or more channels is straightforward.

required. A timing diagram is shown in Fig. 18-6. The frequencies of the two independent clocks are approximately but not exactly the same without synchronization. Reset of the demultiplexer or receiver clock when the multiplexer clock reaches **0** is accomplished by the synchronization pulse. To avoid extreme shortening (or lengthening) of the demultiplexer clock pulse, which would occur if the synchronizer line or reset pulse happens to occur in the wrong phase of the clock cycle, a fast clock ($8f_c$) with a divide-by-8 counter is employed. Reset occurs when the 3-bit counters reach **0** and the uncertainty is only one-eighth clock cycle (or one fast clock cycle). Without the extra divide-by-8 counter, the uncertainty in reset time is nearly one clock cycle and thus the demultiplexer-counter Q_A

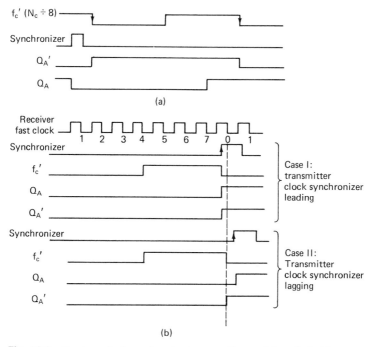

Fig. 18-6 Receiver clock-synchronization waveforms: (a) undesirable waveforms if fast clock not utilized; (b) waveforms with leading and lagging synchronizer pulse.

output on count zero could range from a tiny fraction of a clock cycle to nearly two clock cycles long, an unacceptable situation. The extra divide-by-8 counter may be thought of as providing a means of resetting the *phase* of the receiver clock close to zero. Subcycle timing, if needed, can be controlled by the divide-by-8 counter through another decoder. This synchronization method is widely used in serial data transmission, especially when it involves UARTs (see Sec. 18-6). Often the fast clock is $16f_c$, rather than $8f_c$ as discussed here, so that the synchronization is even more tightly controlled. It is desirable that the transmitter and receiver clocks (f_c and f'_c) be matched so that the accumulated difference is under 1 cycle between synchronizer pulses. For the divide-by-8 counter and 2-bit counter (four channels) the frequency tolerance should be 1 part in 64.

18-3   Latches

A latch is a set of flip-flops, usually four to eight to a package, used to store binary data temporarily. Often they are connected to counter (BIN or BCD) outputs to hold data for display while the counters are cleared and recycled. Both RS and type-D flip-flops are employed.

When the latch enable of the SR-type flip-flops (Fig. 18-7a) is high, the

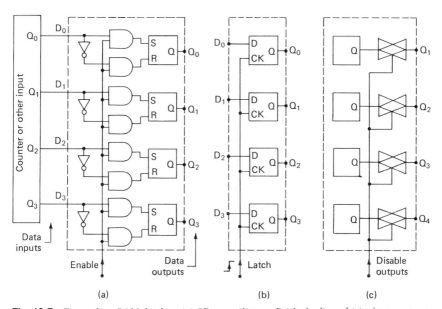

(a)   (b)   (c)

Fig. 18-7   Examples of 4-bit latches: (a) SR type, (b) type D (clocked), and (c) tristate output.

latch flip-flops follow the inputs and therefore the input and outputs are identical. When the enable goes low, the flip-flops hold the data and therefore the outputs are equal to the input data when the enable is switched.

Latches built from type-D flip-flops (Fig. 18-7b) transfer data from the input to the output on the rising edge of the clock (usually) or latch-enable pulse. These differ from the RS type in that the output does not change if the input data changes while the latch enable remains high.

For multiplexing a tristate latch is desirable. It consists of a standard latch with a transmission gate connected to the output. The name tristate implies that a third high-impedance (or disabled or disconnected) state exists in addition to **1** and **0**. As discussed later in this chapter (Sec. 18-5), the ability to disconnect allows the outputs of two or more latches to be connected in parallel. Also it is convenient where a single line is used for data input and output. In operation care must be taken to see that the outputs of all but one of the parallel latches are disabled to avoid shorting outputs.

18-4 Digital Switches

Digital data may be switched by transmission gates or by combinations of gates (decoders). The four-channel data switch shown in Fig. 18-8a is

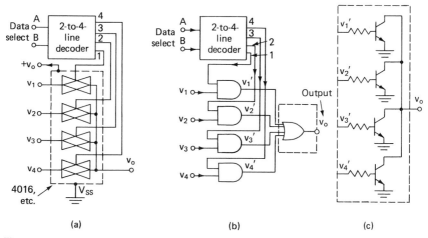

(a)   (b)   (c)

Fig. 18-8 Digital data switches: (a) CMOS transmission-gate type, (b) AND-OR-gate type, (c) open-collector (wired-OR) option.

essentially the same as the analog switch described above (but with V_{SS} grounded). If the decoder has an enable E input, all the gates can be turned off. In this case the output is tristate. In Fig. 18-8b the appropriate data line is switched on by an AND gate, controlled by a decoder. Expansion to more than four channels is straightforward.

The OR gate can be replaced by an open-collector or wired-OR gate. When all lines are turned off (all AND gate outputs at **0**), the output is in the off state. Other similar devices can be tied to the same output and turned on when required. In this sense the wired-OR acts as a tristate output (Fig. 18-8c).

18-5 Display Multiplexers

Multiplexing of seven-segment LED displays with a number of digits is done primarily to reduce the number of decoders and drivers required. Each digit is turned on or scanned in sequence at a fast rate (>100 Hz), so that the flicker cannot be discerned by the eye. The scanning technique is similar to that for the analog multiplexer except that the four BCD data lines are switched simultaneously. Only one BCD-to-seven-segment decoder is required. All the seven-segment LED (Fig. 18-9) common-anode displays are connected in parallel, but only when the anode of a particular display is connected to the positive supply by the driver transis-

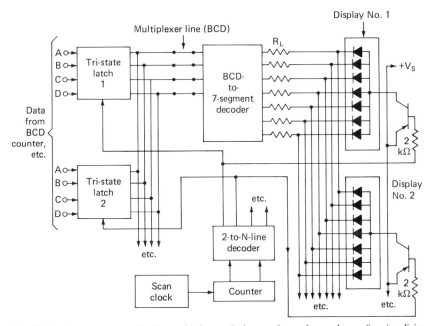

Fig. 18-9 Seven-segment display multiplexer. Only two channels are shown for simplicity.

tor will that display light. Calculation of the current-limiting resistance R_L is the same as for a one-display segment except that it is chosen such that the *average* rather than the peak current is equal to the dc value per segment. Switching of the anode driver and latch output is synchronized by the decoder–data selector driven in turn by the scan clock and counter. Typically the scan frequency, which is not critical and which need not be sychronized with any other operation, is 0.1 to 10 kHz.

18-6   Serial and Parallel Digital-Data Conversion

Digital data is normally presented in parallel form, i.e., as the states (**0** or **1**) of a set of flip-flops or a set of lines equal to the number of bits of data. All the data bits are available simultaneously. In serial form the data is transmitted sequentially, 1 bit at a time along a single line. One data bit is transmitted per clock pulse.

Conversion to, and transfer of, serial data usually is implemented with *D* flip-flops, as illustrated in Fig. 18-10 for 2 bits (more bits simply require additional flip-flops). According to the truth table for the *D* flip-flop (Chap. 5), the state present at the *D* input *before* the clock transition will appear at the output *Q after* the clock transition. Since the output of one

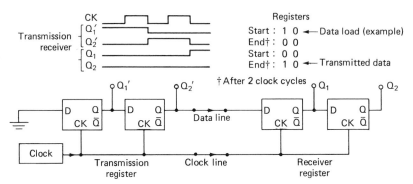

Fig. 18-10   Serial-output *D* flip-flops (2 bits). For simplicity loading and synchronization techniques are not shown.

flip-flop is the input of the next, a particular state assumed preloaded (say 1) will be transmitted serially from one flip-flop to the next. To move the bit by *n* flip-flops, *n* clock cycles are required. Transfer of serial data out of a standard 8-bit register (set of flip-flops) therefore requires exactly 8 clock cycles (additional clock cycles would transmit zeros or meaningless data).

Conversion of serial data back into parallel form is done by a similar set of *D* flip-flops except that the data are transmitted into the inputs of the first *D* flip-flop, as shown. After 8 clock cycles for an 8-bit register, the data originally loaded into the transmitting register will have been transmitted to the receiving register, where it is available in parallel form (Fig. 18-11).

Loading of the original data (parallel form) into the transmitting register is done before the first clock cycle. Since some means of synchronization must be provided, the data is taken from the receiving register only after the required number of clock cycles following loading of data into the transmitting register. Obviously, if the data is taken out even 1 cycle too soon or too late, the data is entirely erroneous.

Elimination of the clock line is possible by transmission of synchronizing pulses along the data line. Sometimes a higher-voltage synchronization pulse (if allowable) is sent and detected at the other end by a comparator set to reject the standard-level signals. The pulse synchronizes a clock at the receiver end, as described in connection with the analog multiplexing technique above. A more common synchronization technique is to add synchronization (start and stop) pulses to the data stream. Separation of the data and synchronizing pulses requires a substantial amount of additional logic or a special-purpose IC. The complete serial transmitter-receiver unit, which includes clocks and synchronization logic, is referred to as a UART and is described in the next chapter.

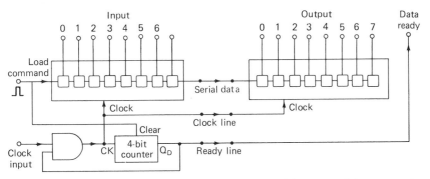

Fig. 18-11   An 8-bit parallel-input–serial-output and serial-input–parallel-output register connection.

In Fig. 18-12 CMOS series-parallel registers are utilized for serial data transmission. Parallel input (to the 4014) occurs on the rising edge of the clock pulse if the parallel-series control is high. Data is clocked out (serially transmitted) on the rising edge of the clock pulses if parallel-series control

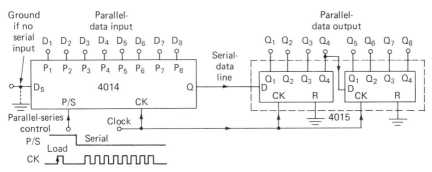

Fig. 18-12   CMOS series-parallel register.

is low. After eight clock pulses the data is positioned in the receiver register (4015). A counter (3-bit, not shown) must keep track of the clock pulses and thus control the transmission process.

18-7   Track-and-Hold Circuit

A track-and-hold circuit is a switched analog amplifier with an output which is equal to the input when switched to the track mode but keeps the output constant at its current value when switched to the hold mode (Fig. 18-13). It consists of an FET switch or transmission gate followed by a capacitor. When the gate is on, the capacitor tracks the input. A low-output-impedance (unity-gain) amplifier supplies enough current to charge and discharge the capacitor rapidly ($\tau \approx CR_{on}$). A high-input-

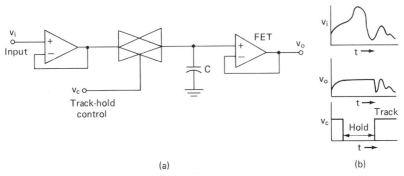

(a)   (b)

Fig. 18-13   Track and hold: (a) circuit; (b) typical response.

impedance (FET) op-amp (also unity-gain) minimizes capacitor discharging in the hold anode.

Analog multiplexers often employ track-and-hold circuits on the inputs, especially when connected to A/D converters. The voltage is held during conversion to eliminate errors which might result from rapidly changing signals.

18-8 Charge-Coupled Devices

Charge-coupled devices (CCD) and the closely related bucket-brigade devices (BBD) are special cases of ICs termed *charge-transfer devices* (CTD). They are capable of storing and transferring charge between capacitance-storage elements under clock control. Input signals are subdivided into a series of time segments for storage, as in time multiplexing. Both analog and digital signals can be handled.

Basically these devices are a series of linearly connected FETs on a single substrate (Fig. 18-14). The gate-channel capacitance stores the charge. Transfer from one FET to another requires a sequence of voltages to be applied to the gates as synchronized by a multiphase clock. Initially the charge is injected into the input capacitance in proportion to the input signal. The ac signal is superimposed on a positive dc bias. Next a gate-voltage bias is applied such that the input gate (gate 0) is cut off, thus trapping the charge on the first FET stage. Gate 1 is closed, so that at this time no charge flows into section 2, but during the next phase gate 1 is opened by the application of an appropriate control voltage (clock ϕ_1).

The charge is then drawn out by lowering this potential of the channel of FET section 2 (clock ϕ_2). Each complete clock cycle results in transfer of charge from one FET storage element to the next. At the end of N elements (perhaps 64 or 256), the charge is monitored by an output FET, which acts as a unity-gain voltage amplifier. A total of N complete cycles, each of period T_c, is required to transfer the charge from the input to the output. Thus the total delay T_D is NT_c.

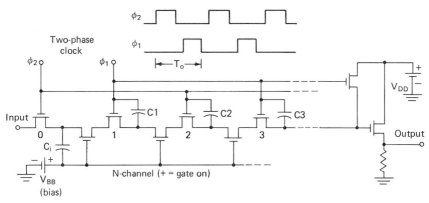

Fig. 18-14 Basic bucket-brigade-device organization.

A limitation of the CTD is charge loss during storage and transfer. Typically one-ten-thousandth of the charge is lost per transfer and one-ten-thousandth of the charge will leak per millisecond. These losses limit the total time delay. Often the output is amplified to make up the charge-transfer (voltage) loss. Recirculation of the analog signal from the output to the input is possible, but since the voltage transfer even after amplification is not exactly unity, an analog signal cannot be stored for long. Digital signals can be stored indefinitely, however, since they can be restored to standard digital level before being transferred back to the input.

Digital charge-transfer devices are also utilized as serial memories. Because of the complexity of the read, write, and data-recirculation logic, these circuits will not be described here. It should be pointed out that serial registers (Sec. 18-11) are functionally similar, and the main advantage of the CTD is low cost when a large number of bits are to be stored.

Chapter Nineteen

Data Transmission and Recording

The discussion of data recording and transmission devices here is rather general, the intent being to present standard techniques rather than circuit details. In some cases this is because ICs are available to accomplish the task. In other cases only slight adaptations of widely available instruments are required. In still others, the circuit complexity is beyond the scope of this text.

19-1 Analog Telemetry Techniques

Transmission of a signal to a remote site, usually along a single line or radio link, is referred to as *telemetry*. Most common communication transmission and recording systems (radio, telephone, tape recorders) are, of course, intended for speech or music. The audio-frequency range is typically 0.3 to 3.3 kHz for a standard telephone line and 20 Hz to 20 kHz for a quality music-reproduction system. Some communication systems can transmit the full audio range but few can transmit direct current directly. Generally also the signal amplitude (transmission-link gain) is poorly controlled because the ear is rather insensitive to variations in absolute sound levels.

It will be assumed here that the "analog" signal to be transmitted over the audio-frequency link has a dc component and cannot tolerate a variable gain (or loss), so that some type of modulation or coding is required. Frequency and pulse-width modulation are the most common analog methods. Digital techniques, which can also be employed to transmit analog signals after conversion, are discussed in the next section.

With frequency modulation (FM, Fig. 19-1) a difference in frequency Δf with respect to a center or carrier frequency f_0 is proportional to the signal or modulation voltage v_m; that is

$$\Delta f = f - f_0 = K v_m \qquad (19\text{-}1)$$

where K is a modulation index. The carrier frequency is set at the center of the transmission band; for example $f_0 \approx 1.8$ kHz for a frequency of 0.3

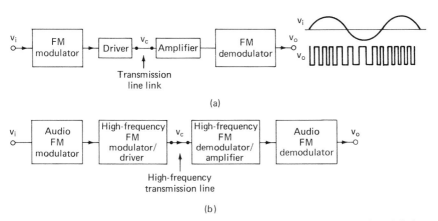

Fig. 19-1   Analog FM telemetry block diagrams: (*a*) single modulation; (*b*) dual modulation.

to 3.3 kHz. Modulation and demodulation techniques are discussed in Chap. 15. The maximum modulation frequency of a sinusoidal signal component $v_m(t) = |v_m| \sin 2\pi f_m t$ permissible without distortion is about 10 to 20 percent of the carrier frequency. There is no lower frequency limit; that is, dc can be transmitted (dc corresponds to a constant Δf). Carrier amplitude variation due to a variation in transmission-link gain does not affect demodulated signal amplitude because signal amplitude depends on Δf not carrier amplitude. Note that the price paid for FM encoding and its associated advantages is a reduction in bandwidth to 5 to 20 percent of that possible with direct (ac or audio-frequency) transmission. Also circuit complexity is substantially greater.

Another technique for transmitting analog signals is pulse-width modulation (Fig. 19-2). The carrier frequency and amplitude are fixed, but the width of the pulse is proportional to the signal amplitude. As in FM, the highest-frequency components of the modulating signal are at least an order of magnitude lower than the carrier frequency. Here carrier frequency (500 Hz) is determined by the pulse generator. The width of the pulse is controlled by a following monostable multivibrator. The pulse width (0.4 to 1.6 ms) of the timer is proportional to the control-voltage

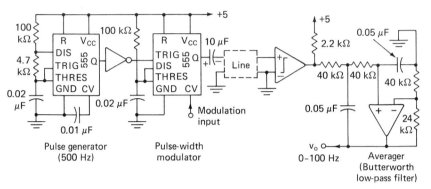

Fig. 19-2 Simplified pulse-width-modulation telemetry system.

input, to which is applied the modulating voltage, assumed positive. The received pulse is restored to a standard level (5 V peak) by a comparator, and the average of the pulse is extracted. A second-order Butterworth filter provides better removal of the carrier than a standard integrator-averager (Chap. 11). Linearity is mediocre because of the modulation limitation and probable distortion of the pulse due to transmission-line frequency response.

Delta modulation (Fig. 19-3), which might be classified as a combination of digital and analog techniques, is a method of transmitting changes of an analog signal. It is a variation of the servo method of A/D conversion (Chap. 14) in which a second up-down counter at the receiver end follows the transmitter counter. The transmitted signal consists of a positive pulse for an up count and a negative pulse for a down count. Because the maximum rate of change is limited to one least significant bit per clock period, the response time (slew rate) is limited. It is best suited to analog signals which require high transmission accuracy but do not change rapidly with time.

In practice some drift in the transmitter or receiver center frequency can occur, so that the dc output zero drifts similarly. High-frequency telemetry systems are the most susceptible to drift because the frequency deviation is comparatively narrow and because available high-frequency (over 1 MHz) FM detector circuits have a dc output bias which is not constant. These problems are much less severe at lower frequencies (below 10 to 100 kHz), and therefore in this range low-drift telemetry systems with a dc response can be made without difficulty. For applications where a dc signal must be transmitted through a high-frequency link (and a digital technique is unsuitable) a dual modulation (FM-FM) is often employed (Fig. 19-1b). In this case an FM signal with a low center frequency, for example, $f_0 = 2.0$ kHz with a ± 0.4 kHz deviation, modu-

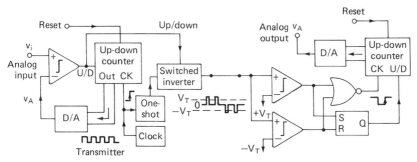

Fig. 19-3 Block diagram of delta modulation method.

lates a high-frequency signal, for example, $f = 100$ MHz with a deviation of ± 0.2 MHz. The output of the high-frequency FM detector (center frequency f_0) is an ac signal at the audio frequency (1.6 to 2.4 kHz in this example); the dc level is unimportant. This signal is then sent to a low-frequency detector (center frequency $f_0 = 2.0$ kHz) for the required dc output. Although seemingly complex, this technique is actually much easier to implement with such standard devices as PLLs than a single FM detector which has a drift that is difficult to control. The description of FM telemetry systems can therefore be divided into two parts, the high-frequency FM transmitter and receiver for ac signals and a low-frequency FM modulator and demodulator for dc (and low-frequency ac) signals.

A simple high-frequency telemetry system for short-distance (10 to 100 m) communication is easily assembled from widely available components (Fig. 19-4). It utilizes the Colpitts oscillator (Chap. 12) with a voltage-variable-capacitor modulator (Chap. 15) tuned to the standard FM broadcast.† A standard FM radio receiver demodulates the high-frequency signal into an audio signal. If conversion into an analog signal is required, a second demodulation (FM or pulse-width) can be used, as suggested above (Fig. 19-1).

---

†RF emissions are controlled by law, which in the United States is administered by the Federal Communications Commission. Experimenters are unlikely to run afoul of the law if the dc power into the oscillator (or last RF stage) is under 100 mW and the transmission is in the FM band (88 to 100 MHz) at a frequency which does not interfere with local radio broadcasts.

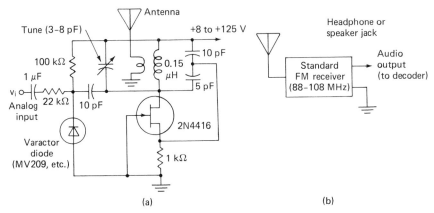

Fig. 19-4   Simple 100-MHz system: (*a*) Colpitts transmitter and (*b*) FM receiver.

19-2   Digital Telemetry Techniques

Techniques for transmitting digital data are similar to those employed in transmitting analog data except that only two voltage levels (**0** and **1**) are required. Typically only one line of serial data is transmitted. Often dc response is needed since the data might consist of a long string of **1**s or **0**s. Data in parallel-data form must be converted into serial form, as discussed in Chap. 18. Usually this is done by a UART (see below), which also adds synchronization logic.

Two general modulation methods are tone modulation, or frequency shift keying (FSK), and pulse-width (or position-phase) modulation. Tone modulation offers better noise immunity although at the expense of speed. A standard tone decoder (Fig. 19-5) actually utilizes two phase detectors, one in-phase (I) and one quadrature (II). When a tone of the proper frequency is present and the loop has locked, the phase difference θ and therefore the output of dectector I will be at a minimum [Eq. (15-10)] and that of detector II [Eq. (15-11)] a maximum. If no tone is present, both will be close to zero. Detector filter time constants are set so that detector I responds more rapidly than detector II. This assures establishment of lock before the comparator connected to detector II responds. Response time is typically longer than 10 to 100 carrier cycles and as such is most suited for lower-speed data links.

Often two tones are transmitted (Fig. 19-5*b*), as is done in touch-tone telephone dialing, to reduce the chance of noise or spurious tones (speech) being interpreted as a data pulse. Only when both tones, chosen to be harmonically unrelated, are present is the data considered valid, and a pulse appears at the output. With other schemes **1** is transmitted on one tone and **0** on the other, so that no tone indicates disconnect. Various standard frequencies and data-transmission rates have been established for digital-data communication systems. The combination transmitting-receiving circuits for telephone links (MODEMs) are discussed in the next section. Some digital tape recorders also utilize tone modulation, as discussed in subsequent sections.

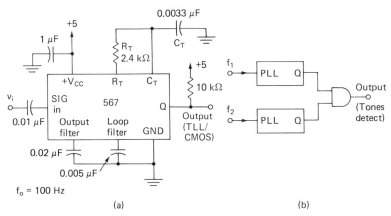

Fig. 19-5 A PLL tone decoder: (*a*) single tone; (*b*) block diagram of dual method.

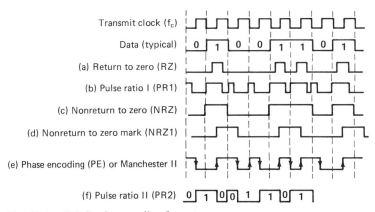

Fig. 19-6 Digital pulse-encoding formats.

Popular pulse-modulation encoding formats are presented in Fig. 19-6. They are suitable for asynchronous operation and are often employed in connection with UARTs (see below). If a second (clock) channel is available, the method is by definition synchronous instead. With several formats (phase-encoding and pulse ratio I) the clock pulse can be derived from the single line of transmitted data. For all but the last methods (pulse ratio II) the receiver-clock frequency must roughly match that of the transmitter clock (or, equivalently the pulse widths must match).

The return-to-zero method (line *a*) simply produces a pulse for each **1** data pulse. It is a satisfactory method for remote, aperiodic pulse-counting applications (perhaps the output of a voltage-to-frequency converter), but since it is not efficient or easily synchronized, it is little used in general data transmission. A better method is the pulse-ratio I code (line *b*), which transmits a narrow pulse, say one-fourth clock period, for **0** and a wide pulse, say three-fourths clock period, for **1**. Encoding and decoding is rather simply implemented, as indicated in Fig. 19-7. The one-shot trig-

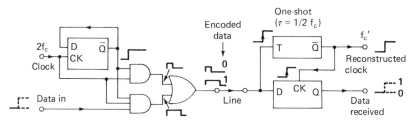

Fig. 19-7 Pulse-ratio I encoding-decoding circuits (fixed clock rate).

gers on the rising edge of the data pulse, and after a time delay τ equal to half the clock period latches (clocks) the type-*D* flip-flop. Since the encoded-data logic level at this time is the same as the input data, the flip-flop output Q will equal the input data.

Transmitted data with the nonreturn-to-zero (NRZ) method are identical to the input data. Decoding consists simply of a type-*D* flip-flop connected to the clock transmitted in a second channel (see discussion of digital tape recorders below). A related format is nonreturn-to-zero-mark (NRZ1), in which the transmitted data inverts (toggles by a divide-by-2 counter) each clock period if the input data is at **1**. This format is often employed to encode each channel of the seven- or nine-track digital tape recorders found in larger computer installations. It is an efficient code in the sense that a nearly maximum number of bits per unit time can be recorded with a given recorder high-frequency limit, but it is not self-clocking.

Phase encoding (Manchester II) utilizes the transition direction of the transmitted pulse at the *center* of the clock period (rising edge, or *data time*) to indicate the logic sense. Logic **0** and **1** correspond to the falling and rising transitions, respectively. Data transitions at the *beginning* of the clock period (phase time) are ignored during decoding and are necessary whenever the serial data does not change sign (**0** followed by **0** or **1** by **1**). A one-shot multivibrator ($\tau = 1/f_c$) triggered by both transmitted-data pulse (at data time) edges is used to disable an edge-sensitive latch during the phase time. Advantages of phase encoding are modest bandwidth requirements, high bit rate, and the self-clocking feature. The main disadvantage is the more complex circuitry required. A related format, Manchester I, is similar except that the transition times are shifted by a half clock cycle.

Examples of self-clocking codes which can be received at an arbitrary rate are the pulse-ratio formats. It is necessary only that the rate not change appreciably within 1 clock cycle. A more complex decoding circuit than that of Fig. 19-7 is required, however, perhaps utilizing PLL or microprocessor. These formats or variations are often employed by optical bar-code readers, which convert black and white bars printed on paper into digital data. The clock period for the pulse-ratio I format is reconstructed from the transmitted signal since it consists of the time between rising edges of the pulses.

Pulse-ratio II code does not have a definite clock frequency in the sense that **1** and **0** data pulses do not require the same time to transmit. A transition is made after each data bit, and the width of the pulse distinguishes **1** from **0** (**1** is twice as wide). It is a compact code but cannot be decoded with a UART.

19-3 MODEMs

A MODEM (*modulator-dem*odulator or *data set*) is an instrument which converts tones on a telephone line into digital-level data and the reverse. It implements the frequency-shift-keying modulation method. Many computer (teletype or CRT) terminals require a MODEM as an interface unit. In the transmit mode, serial input data is clocked out at a predetermined rate in the form of audio tones. The tone pulses are connected to the line by a transformer (hard-wired) or through an acoustical coupler, consisting of a microphone and small loudspeaker mounted in a cradle shaped to fit a standard telephone headset. The level of the ac signal at the coupler terminals (input and output on the same line) is roughly the same level as on the telephone line (0.05 V on receive, 0.3 V on transmit).

A MODEM which can transmit and receive at both ends is termed *duplex*. If one unit can only send and the other only receive, the operation is termed *simplex*. If both units do not transmit data at the same time, the operation is *half duplex*, but if both transmit and receive simultaneously on the same line, the operation is *full duplex*. Under full duplex the transmitter and receiver operate at separate frequencies. Normally in full-duplex operation the received signal is retransmitted, or echoed, back to the originating MODEM as a check on transmission accuracy. Even in half-duplex operation the reverse transmission of a steady tone is standard so that the data-transmitting station has a check that the receiver station remains connected. Most MODEMs can operate in either half- or full-duplex modes without modification; the mode is controlled by the connected terminal.

The American (Bell Telephone) frequencies are shown in Table 19-1. Often the old teletype terms *mark* and *space* are used to designate **1** (high)

TABLE 19-1   Standard MODEM Frequencies

	110 and 300 b/s		1200 b/s (half
Logic	Transmit, Hz†	Receive, Hz†	duplex), Hz
1 (mark)	1270	2225	1200
0 (space)	1070	2025	2200

Frequency tolerance ±5%
†For originate MODEM.

and **0** (low), respectively. The MODEM which initiates the call, termed an *originate MODEM,* transmits at the lower-frequency band and receives at the higher. The *answer MODEM* responds by transmitting on the higher-frequency band and receiving on the lower. Usually the remote terminal originates the call and the computer answers. Various intervals defining maximum and minimum contact and disconnect times have been established.

A block diagram of an originate MODEM is given in Fig. 19-8. After

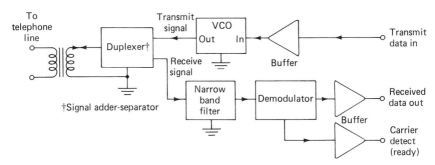

Fig. 19-8 Block diagram of an originate MODEM.

dialup by a standard telephone handset, the line is connected to the MODEM. The duplexer, a type of ac bridge circuit, separates the transmitted and received signals. The main purpose is to keep the transmitting signal (especially second harmonics) from interfering with the received signal. Separation is not complete, and typically 10 to 20 percent of the transmitted signal is transferred into the receive section. A notch filter keeps the transmitting frequency and harmonics from the receive section.

The tone demodulator may be PLL (low baud rates) or a digital detector based on tone frequency or period. The receiver bandpass filter must be carefully designed to allow sharp discrimination against interfering signals without producing excessive ringing or phase differences between the two received frequencies. During transmission the voltage-controlled oscillator switches between the two transmitting frequencies.

If the standard RS-232 interface is required, a special inverting buffer amplifier intended for this purpose should be added. Logic levels for this interface are defined as $+5$ to $+15$ V for **0** and -5 to -15 V for logic **1** (-3 to $+3$ V is no signal). Output impedance ranges are also specified. Alternative outputs are TTL drivers or a 20-mA current loop. With the current loop, current normally flows (mark) unless momentarily broken by either transmitter or receiver producing a "space."

19-4 UARTs

A universal asynchronous receiver and transmitter (UART) is a relatively complex IC which can encode parallel input data for serial-data transmission along a single channel. It also performs the inverse function. One UART transmits while the other is in the receive mode. Basically the transmitter is a serial-output parallel-input register (Chap. 18), but additional synchronization information (start and stop bits) is added to the data system so that additional (clock) channels are not needed for this purpose. Operation is asynchronous in the sense that the transmitter and receiver clocks need not be exactly the same. Also the attached system and UART clocks need not be synchronized.

The clocks of the transmitter and receiver are nominally but not exactly the same. Rates of data transmission are chosen through adjustment of the clock rate to be compatible with the frequency response of the line. Standard data transmission rates, among others, are 110 and 300 bits per second (b/s, or baud).

The synchronization method involving an internal code requires further discussion. A common code (used at 110 b/s) as illustrated in Fig. 19-9 has 11 bits per word, 8 of which are data bits. There is one *start bit,* which is **0**, and two *stop bits,* which are **1**. Usually only one stop bit is used at 300 b/s (10 bits per word), and UARTs allow this option. The first pulse phase synchronizes the receiver clock by clearing on the leading edge, a divide-by-16 counter connected to the fast clock, as discussed in Chap. 18. Only 8 bits appear at the data output bus since the start and stop bits are only used internally. The start bit can appear immediately after the stop bit or after a longer period of time, as would occur with a normal break and resumption of transmission. Only the middle section of each pulse is checked for valid logic levels since the leading and trailing edges of each bit normally jitter somewhat.

Commonly a UART and MODEM are employed in tandem. The frequency- or tone-modulated information on a telephone line is converted into serial digital form by the MODEM and then into parallel form

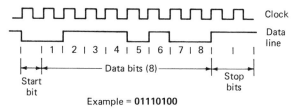

Example = **01110100**

Fig. 19-9 Standard serial-data-transmission synchronization code.

by the UART. One byte (8 bits) is transmitted at a time. When the byte has been received, a ready (receive-buffer-full) signal is generated (Fig. 19-10). While the byte may represent any binary number, it usually represents one alphanumeric character in ASCII code (see Sec. 19-5). A UART

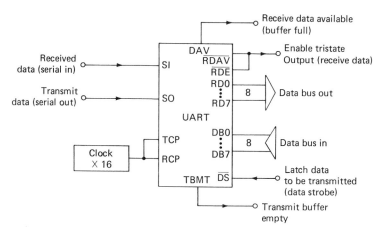

Fig. 19-10 Input/output connections to an IC UART.

can be thought of as a device which allows transmission of a keyboard character to a remote location.

19-5 Keyboard and Character Generators

A minimal computer terminal consists of a typewriter-style keyboard and some means of recording alphanumeric information. The keys of a keyboard are often organized electrically as a matrix (Fig. 19-11). When a key is depressed, current will flow out of a specific X line into a specific Y line. The encoder produces the corresponding 4-bit number for each key. Although a 16-key array is shown here, a similar IC is capable of handling 64 or more keys. A signal is provided when a key is pressed, which can be used to latch the encoder output. Switch debouncing is included.

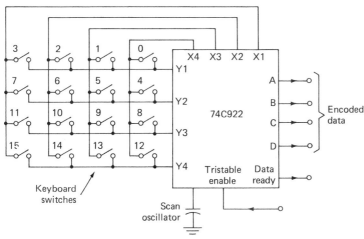

Fig. 19-11 Keyboard encoder.

The keyboard output (6- or 7-bit) is usually ASCII encoded (American Standard Code for Information Interchange). As Table 19-2 indicates, only 6 bits are needed to encode the standard 64-character (uppercase) keyboard. If lowercase is added, 7 bits are required. An eighth bit is a parity bit, but it is often ignored. For even-parity convention, the sum of all 8 bits is even for error-free transmission. Note that the BCD code for numerics is embedded in the code (add 3 for the higher half byte). A number of code words are utilized for control, e.g., carriage return.

Alphanumeric display in smaller systems is usually done by LED 5×7 dot displays (Chap. 7) or a cathode-ray tube (CRT). Both utilize an IC character generator to produce the desired character. The process is most easily demonstrated by the LED alphanumeric display (Fig. 19-12). Basically the character generator is a read-only memory (ROM, see Chap. 20) in which the dot pattern of all characters to be generated is stored; 1 bit represents one displayed dot. To display all 64 uppercase ASCII characters on a 5×7 dot display requires a 2240-bit (minimum) ROM. The data output word required is 7 bits, and the address is 9 bits, composed of the 6-bit ASCII-encoded input word and the 3-bit column address. A clock with counter and decoder scans the columns repeatedly. Only one column of the display is lit at any instant, but because of the rapid scan (minimum 30 Hz, usually much faster) the eye perceives the complete character. Driver and scan circuitry is similar to that for seven-segment display multiplexing (Chap. 11) except for the larger number of LEDs which must be scanned. If, as usual, multiple characters are to be scanned, the

TABLE 19-2   ASCII Code Hexadecimal Equivalent†

			X_U					
	Control		Numeric		Uppercase		Lowercase	
X_L	0	1	2	3	4	5	6	7
0	NUL	DLE	Space	0	@	P	`	p
1	SOH	DC1	!	1	A	Q	a	q
2	STX	DC2	"	2	B	R	b	r
3	ETX	DC3	#	3	C	S	c	s
4	EOT	DC4	$	4	D	T	d	t
5	ENQ	NAK	%	5	E	U	e	u
6	ACK	SYN	&	6	F	V	f	v
7	BEL	ETB	'	7	G	W	g	w
8	BS	CAN	(	8	H	X	h	x
9	HT	EM	)	9	I	Y	i	y
A	LF	SUB	*	:	J	Z	j	z
B	VT	ESC	+	;	K	[	k	{
C	FF	FS	,	<	L	\	l	|
D	CR	GS	–	=	M	]	m	}
E	SO	RS	.	>	N	^	n	~
F	SI	US	/	?	O	—	o	DEL

† X_U is upper half byte, and X_L is lower half byte. Most significant bit (D7) is **0** (parity ignored). Control characters are not printed. Example: code for T is $X_U X_L$ = 54 (= **01010100**).

multiplexing is expanded to allow selection of displays, in effect extending the horizontal scan to $5N$, where N is the number of display characters.

Character generation for CRT displays employs basically the same method except that rows rather than columns are scanned. Because of the large number of characters and lines which must ordinarily be displayed, the circuitry is complex. This is especially true of refresh type (nonstorage) CRT terminals, which must recall the stored characters (data) from memory (RAM) during each high-speed scan of the screen while keeping track of the column and row number of each character.

19-6 Tape Recorders

Magnetic tape recorders, including the home-entertainment type, are based on amplitude modulation of the tape at an audio frequency. As

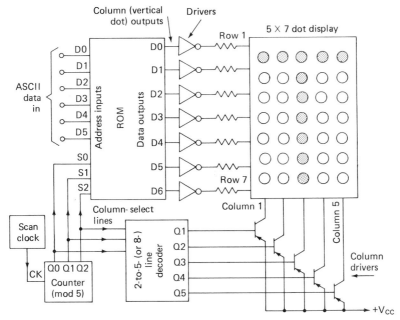

Fig. 19-12 Single LED alphanumeric display with column-scanned character generator.

discussed above, any device which can record an audio signal can be used for recording analog or digital signals by proper modulation-demodulation methods. It is instructive to discuss the recording and playback principles in order to understand the frequency and dynamic-range limitations.

A magnetic tape has small magnetic particles embedded in the plastic matrix. The particles can be magnetized by an external field produced by the recording head (Fig. 19-13). Current-drive requirements for the recording-head coil are similar to those of loudspeakers. The head has a narrow gap filled with a nonmagnetic material across which and close to which the magnetic field is concentrated. The field drops off very rapidly away from the gap and can be neglected elsewhere. All the particles within the gap at a given instant have approximately the same magnetization.

Assuming that the head-drive current is sinusoidal with time, the magneti-zation M along the tape will be sinusoidal with distance x, with $x = ut$, where u is the tape velocity. Only a fraction of a wavelength, at most a half wave, can be magnetized at any instant. Since the length of the half wave on the tape cannot be shorter than the gap distance D, the highest

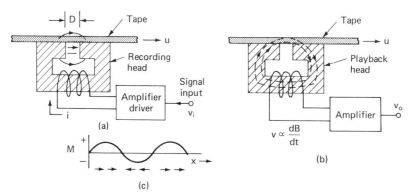

Fig. 19-13 Principle of tape recording: (*a*) record; (*b*) playback; (*c*) tape magnetization.

frequency that can be recorded corresponds to a wavelength of $2u/D$. High-frequency recording therefore requires a high tape speed or a small recording-head gap (1 to 4 μm). Small cassette recorders typically have a frequency limitation of about 4 to 10 kHz, while a reel-to-reel recorder (15 in/s) may respond above 40 kHz.

The playback head has the same general construction as the recording head. During playback the magnetized particles near the gap produce a small magnetic field B in the head. A voltage proportional to dB/dt is induced in the coil. For a sinusoidal wave ($M_0 \sin \omega t$), the induced voltage will be proportional to $\omega \cos \omega t$; that is, the output voltage will decrease as the frequency is reduced and will be zero for direct current. Response below 10 to 50 Hz is generally impractical.

Tape is erased by exposing the tape to a high magnetic field. An erase head similar to the recording head may be used, operated at a frequency higher than the maximum response frequency to avoid interference with the recorded signal. A high field magnetizes the particles to the maximum extent, thus wiping out the magnetization due to previous signals. Alter-natively the entire tape may be exposed to a strong 60-Hz field, which is slowly reduced by moving the tape away from the coil. As the tape passes from the high to zero field region, through slowly diminishing magnetic hystersis loops, the net magnetization approaches zero. Usually a moder-ate-amplitude high-frequency component or *ac bias*, is added to the recording-head current. It serves to linearize the current-magnetization response curve, especially for small signals. The bias signal is too high in frequency to be played back. For an audio recorder (20 Hz to 20 kHz) the bias frequency may be 100 kHz.

Fluctuations in tape speed, a common problem with low-quality recorders, results in a frequency variation in playback. Frequency variations are reflected in baseline variations with FM analog recording. It is a problem also with digital tone demodulation if the frequency shifts outside the bandwidth of the receiver, a distinct possibility. Amplitude variations, which might be caused by warped tape or an uneven distribution of magnetic particles in the tape, are always present with audio (AM) recordings. Assuming that the signal does not completely drop out, these variations do not adversely affect most analog and digital modulation methods because they are designed to be insensitive to amplitude. A test of a recorder capability is to record a square wave (100 to 1000 Hz). If the playback signal is sufficiently accurate in amplitude and frequency, direct recording is possible. It is difficult to obtain clean square waves from an AM recorder because of phase-shift and frequency limitations.

A standard audio playback circuit (Fig. 19-14) consists of a high-gain audio amplifier with a filter to compensate for the reduction in pickup-head voltage at lower frequencies. Basically the filter is a low-pass filter with the break frequenciy set equal to the lowest frequency to be played back, for example, $f_1 = 50$ Hz. All signals are attenuated in proportion to frequency, thus compensating for the linear increase in induced voltage with frequency. A typical playback voltage (for a constant-amplitude-current record) is given in Fig. 19-14a. The fast drop in amplitude at high frequencies is due largely to the record-head wavelength limitation discussed above. A second high-frequency filter is also added to reduce the effect of tape noise (hiss). While tape noise is fairly uniformly spread over the frequency band, it is more audible at higher frequencies. A filter is added to the record circuit to emphasize frequencies above a chosen point ($f_2 = 1.77$ kHz for the National Association of Broadcasters standard at a tape speed of $1\frac{7}{8}$ in/s). During playback a deemphasis filter is added to reduce the gain above the breakpoint. More precisely, during playback the compensation filter which reduces the signal in proportion to frequency is set so that it breaks to a horizontal slope above f_2 (see Fig. 19-14c).

Recording of analog signals, which may have a dc component, is done by frequency or pulse-width modulation, as discussed above. With the FM method the carrier frequency f_0 is often recorded on a second track to compensate for tape-speed variations, possibly by controlling motor speed.

Recording of digital signals may utilize either tone or pulse modulation with one of the formats discussed above. Standard telephone-tone fre-

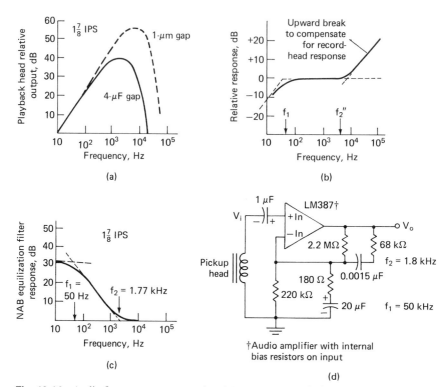

Fig. 19-14   Audio-frequency tape recorder: (*a*) uncompensated playback signal; (*b*) record characteristic; (*c*) playback characteristic; (*d*) playback circuit.

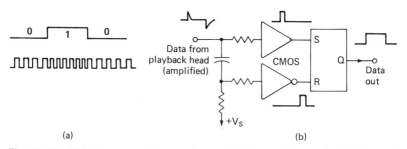

Fig. 19-15   Digital tape-recorder waveforms: (*a*) Kansas City Standard Code (300 b/s); (*b*) restoration of digital signal from NRZ pulses.

quencies (Table 19-1) can be recorded and reproduced with a high-quality recorder but not with most medium- or low-quality cassette recorders intended for home entertainment because speed control is usually inadequate. A more suitable method for low-cost cassette recorders is the *Kansas City Standard;* it uses two widely separated frequencies which need not be specified precisely. At a nominal data rate of 300 b/s, **0** consists of 4 cycles of 1200 Hz and **1** of 8 cycles of 2400 Hz (Fig. 19-15). The tones are conveniently generated by frequency division (divide by 2 or 4) from the same clock that drives the transmitter UART normally used with the tone encoder. In the receive mode, a 2400- or 4800-Hz clock can be derived from the recorder data itself, and this drives the receive UART. There is a similarity to Manchester and phase-encoding formats which also produce a double frequency for one logic level. If the data rate is 1200 b/s, the Kansas City Standard cycles (with the same pulse widths) become identical to the Manchester II code.

Often digital signals are recorded directly on tape by driving the tape to full magnetic saturation in one direction (north) for one logic level to full reverse saturation (south) for the other level. This method is implied by the term *nonreturn to zero* (NRZ), as distinct from allowing a zero or unmagnetized state on the tape. The limited low-frequency response of the tape can degrade the played-back square wave to positive- and negative-going pulses. As indicated in Fig. 19-15*b*, the original square wave can easily be restored by an *SR* flip-flop.

Chapter Twenty

Introduction to Microprocessors

A microprocessor is a simple but versatile computer in a single IC package. Its cost is low enough that even when accessory units such as memory must be added, it is practical in electronic instruments of modest size. Because of the diversity of the microprocessors, the variety of ways of connecting the microprocessor to the instrument through various support ICs (hardware), and the programming variations (software), it is possible to discuss here only a small fraction of useful microprocessor functions and techniques. Furthermore, only two specific microprocessors, the Intel 8080/8085 and the RCA 1802 (COSMAC), will be described. The 1802 was chosen because of its simplicity and adaptability to instrument control, in contrast to numerical data processing or "number crunching." All microprocessors have many features in common, and an understanding of one can be applied to others as well. In this sense, these are typical.

One purpose of this chapter is to provide an overall description of microprocessor operation for engineers and scientists who wish to communicate with engineers doing the detailed design of a microprocessor system. Another purpose is to provide the novice designer with a working knowledge of basic microprocessor hardware.

20-1   Microprocessor-System Organization

A block diagram of a small microprocessor system is shown in Fig. 20-1. The heart of the microprocessor is the central processing unit (CPU), which consists of an arithmetic logic unit (ALU), an internal control unit,

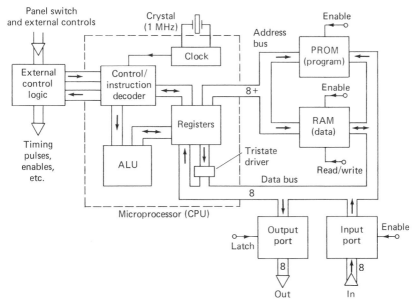

Fig. 20-1   Simplified block diagram of a typical microprocessor system.

and several registers. Arithmetic operations such as adding or comparing two numbers or bit manipulation (right or left shift) are done by the ALU. Included in the CPU group, loosely speaking, are the various timing circuits, drivers, and the clock. Often the clock is external and can be divided into subcycles or phases. The data bus of this 8-bit microprocessor consists of eight separate lines which can transmit data either in or out. The 8-bit binary number represented is termed 1 *byte*. Bidirectional transfer of data requires a tristate output driver which can be put into the high-impedance state, in particular during an input phase. Data can be transferred to or from memory and input/output (I/O) ports. All these units are connected to the same data bus but remain inactive until turned on (enabled, selected, or strobed) by the microprocessor control. Additionally, in the case of memory units, only the part of the memory corresponding to the address code is activated. Normally the strobe pulse is brief, about one clock cycle, and the various units connected to the bus are enabled sequentially. At no time can two units be enabled to drive the bus simultaneously or the bus data will be completely invalid. This time-multiplexing arrangement allows many units to use the same bus without confusion.

Most operations initiated by a single program instruction require a number of clock cycles (perhaps 8 or 16) to complete. A number of cycles is required because each operation takes several steps, e.g., sending the address to memory, strobing the memory (so that the instruction is picked up by the CPU from the data bus), decoding the instruction, performing the indicated operation, and storing the result. Timing diagrams supplied by the manufacturer detail the process. Timing pulses are sent by the microprocessor to the peripheral device to synchronize the operation.

Memory consists here of two parts, a PROM (programmable read-only memory), in which the program instructions are permanently stored, and a RAM (random-access memory), in which data is stored temporarily. Data stored in the RAM is lost when power is turned off, but that in a PROM remains. Standard nonprogrammable ROMs must be prepro-grammed by the manufacturer by a masking process during the manufac-ture of the IC, an approach practical only for volume production. For the average user, electrically programmable ROMs or PROMs are the only practical user-alterable ROMs. Alternatively, the RAM can replace the PROM in applications where (1) the program is loaded each time before each use (from tapes, perhaps), (2) the power is never turned off (battery standby), or (3) the program is frequently altered, as during initial trials. From the microprocessor's point of view, RAM and PROM are identical during read operations except for the address location which is specified (programmed and/or wired) by the user. Of course, only a RAM can respond to a *write* command. For the purposes of the following discussion it will be assumed that the program instructions are stored in PROM, which has been assigned the lower memory addresses (starting at 0). This is done because upon reset the microprocessor goes there for the first instruction.

Memory allocation, i.e., the selection of the size of PROM and RAM together with address, depends on the specific task. The PROM size must obviously equal or exceed the number of instructions required.

Control logic external to the microprocessor of some type must be supplied if only to reset the microprocessor to its starting state. Halt, single-step, and direct-memory-access (DMA) modes can also be imple-mented. State codes and timing pulsed to indicate the phase of an operation are provided to allow synchronization of external hardware.

Actually the term *control lines* can be applied to all lines which are not data or address lines. Some control lines originate or terminate in ports which normally transmit data but are used instead to carry *device-ready* or other control information. Often microprocessor diagrams show a bus labeled "control," but it must be recognized that it is a miscellaneous grouping rather than a homogeneous group, like the data and address buses.

Most operations require two distinct cycles, fetch and execute. During the fetch cycle, a coded program instruction (often 1 byte) is retrieved from PROM and internally decoded (within the CPU). The address of the next instruction is kept track of by an internal *program counter*, which normally increments by 1 at the end of the fetch cycle. During the execute cycle the operation set up by the previous fetch cycle is performed, e.g., the transfer of data from an input port to memory. Microprocessor operation consists mostly of a sequential series of regular fetch and execute cycles, occasionally broken up by operations with shorter or longer execute cycles or instructions which cause the program counter to jump to a different address.

20-2 Memory-Address Organization

Memory is usually divided into pages consisting of 256 bytes. This is necessary because a 1-byte address (8 bits) can specify at most 256 address locations (0 to 255). Nearly all microprocessors store a 2-byte address internally. Various microprocessors differ in the method of outputting the high address byte to the memory. Some have a 16-bit (2-byte) address bus, in which case eight of these output pins are used for the high byte of the address. The 1802 however has only an 8-bit address bus, so that the high and low address bytes must be transferred out at different phases of the fetch and execute cycles. The external hardware (latches) required in this case to restore the full address externally will be described subsequently. One way of looking at this organization is that the high address byte specifies the page number while the lower byte specifies the line within the page. Recognizing the division into pages is important in understanding certain data-transfer instructions which can only transfer within a page because the part of the instruction specifying an address is only 1 byte in size. The 8085 uses the data bus for the lower address byte.

A microprocessor has an instruction set which can do addition, subtraction, transfer to and from memory, branching, and the like. Most sets number between 80 and 256 (for an 8-bit machine), but only 20 to 40 instructions are really distinct. Many instructions are minor variations of each other, e.g., leaving the number $X - Y$ rather than $Y - X$ in the accumulator following subtraction. Others incorporate a register reference number as a part of the instruction code so that a single type of instruction has 8 or 16 variations.

During the first part of the fetch cycle, the coded instruction is fetched from PROM and interpreted. After execution, the next instruction is fetched from the next higher address location sequentially. This continues until an instruction causing the program to stop is encountered or, more commonly, some type of branch to another part of the program or loop occurs. Usually the byte representing an instruction or data as well as the addresses is represented by two hexadecimal numbers although binary, octal, and decimal representations are occasionally used (see Table 20-1).

TABLE 20-1 Decimal, Hexadecimal, Octal,and Binary Representation of Numbers 0 to 15

Hexadecimal	Binary	Decimal	Octal	Hexadecimal	Binary	Decimal	Octal
0	0000	0	0	8	1000	8	10
1	0001	1	1	9	1001	9	11
2	0010	2	2	A	1010	10	12
3	0011	3	3	B	1011	11	13
4	0100	4	4	C	1100	12	14
5	0101	5	5	D	1101	13	15
6	0110	6	6	E	1110	14	16
7	0111	7	7	F	1111	15	17

In Fig. 20-2 part of the memory and internal register organization of the 1802 is shown. Within the CPU the primary register is the accumulator (D register in COSMAC notation); a number of other internal registers are also available. Only 4 bytes each of RAM and PROM are shown. These registers are 16 bits (2 bytes) wide to allow addresses up to 16 bits (64 kilobytes = 65,536 bytes of data)† to be referenced and to facilitate double-precision-arithmetic operations. Since the data bus is 8 bits, it requires at least two instruction cycles to move the 2-byte word in or out of the CPU. A carry bit (DX) necessary for various arithmetic operations such as addition is present. It can be tested by various instructions or shifted to the accumulator but cannot be transferred into the data bus directly.

Within the CPU is a 16-bit-counter program register RP, which keeps track of the location in PROM of the next instruction. Normally it starts at location zero (or reset) and is automatically incremented to the next address (or by two or three addresses for 2- or 3-byte instructions), unless a branch or jump instruction causes it to do otherwise. Actually there are several registers which can be used as the program counter. More properly the counter register is designated $R(P)$, where P is a 4-bit number stored in yet another register. Initially, $P = 0$, and it remains so unless programmed to do otherwise. Registers 0, 1, and 2 are used for certain interrupt and direct-memory-access operations; setting $P = 3$ early in the

†By convention, 1 kilobyte = 1K = 2^{10} = 1024 bytes.

program is recommended. Here it will be assumed for simplicity that $R(P)$ is equivalent to a fixed register RP which is incremented automatically by the control unit.

Another important 16-bit register within the CPU is the data pointer register RX, which the RAM (or perhaps PROM) addresses of data soon to be needed are stored. The address in one of these registers appears on

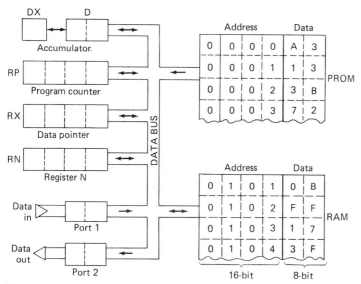

Fig. 20-2 Simplified memory and register layout for the RCA 1802.

the address lines (in 2 bytes, sequentially) during a RAM read or write operation. As a program proceeds, the address in RX frequently changes. Often a change in the upper byte (memory page number) is not required, but an appropriate instruction to change this number is, of course, available.

Like the program counter, the data pointer can be one of several registers and is more appropriately designated $R(X)$, or where X is a 4-bit number; for example, $X = 4$. This is initiated by software instructions, often done in the first steps of the program. Under these conditions the data pointer can be thought of as single register named RX.

The scheme by which a data-pointer register RX is used to designate a memory address at which the desired data is located is referred to as *indirect addressing*. Since several preliminary instructions must be first executed to set up the data-pointer register, it is less efficient than *direct addressing*, in which the address is stated as a part of the instruction. Indirect addressing, however, allows instruction subdivision into byte-sized pieces, allowing simplified device construction. It is a popular technique in lower-cost microprocessors. Actually there are a total of 16 2-byte registers. Those not utilized as a program counter or data pointer can be employed as counters or general storage registers, designated RN.

The structure of memory and the most essential internal registers of the Intel 8080 and 8085 are shown in Fig. 20-3. They are similar to those of

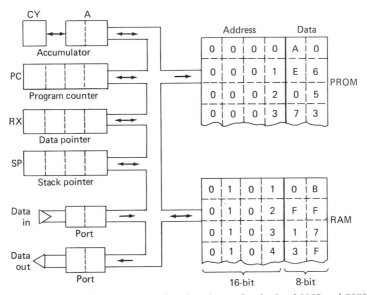

Fig. 20-3 Simplified memory and register layout for the Intel 8085 and 8080.

RCA although there are substantial differences in the type of permissible register data-transfer operations. Also the nomenclature differs. Three supplemental 2-byte registers are available, less than in the RCA, but the smaller number is compensated by easy register transfer instructions. Note that the accumulator and carry flag are named *A* and *CY* respectively.

20-3 Arithmetic, Logic, and Branching Operations

Instructions are specified by a hexadecimal operation code, by name, and by a mnemonic (abbreviation or shorthand name). These *op codes* are 1 byte and often are combined with one or two additional bytes to form the total instruction.

To add a binary number in the memory location specified by the address stored in the data-pointer register RX to the number in the accumulator $(D$ or $A)$ requires the following instruction:

Type	Mnemonic	Name	Operation	Op code
RCA	ADD	Add memory	D + M(RX) → D	F4
Intel	ADD M		A + M(RX) → A	86

The result is stored in the accumulator. If a carry is generated, the carry flag $(DF$ or $CY)$ is set. It is assumed that the desired 2-byte memory address has been placed in RX by some previous instruction and that valid data is in the accumulator.

A related instruction is *add immediate,* which is 2 bytes long. It differs in that the 1-byte address of the number to be added to the accumulator is obtained from PROM, specifically in the location immediately following the op code:

RCA	ADI	Add immediate	M(RP) + D → D	FC
Intel	ADI data		A + (byte 2) → A	C6

The "immediate" means that the data follows the instruction in PROM. It can be considered a part of the instruction. Note that a preliminary instruction which loads the data-pointer register is not needed. The notation "byte 2" means the second byte of the 2-byte instruction. The notation $M(RP)$ indicates that the address location RP is specified by the program counter, which, after retrieving the instruction op code, is incremented by 1 and therefore points to the location in PROM just after the op code. Note that RP is the address or register number of the address where the data is to be stored. After retrieving the address, the RP is again incremented by 1, ready for the next instruction. If the more general form of specifying the RCA program counter $R(P)$ is used, then P is the address of the address of the data to be stored (this is probably not confusing to a computer expert).

Among arithmetic instructions are *subtract, add with carry* (carry bit is added as needed for upper byte of double precision), and *subtract immediate.* These instructions are uncomplicated once the *add* and *add immediate* instructions are understood.

Logic-operation instructions are similar in form to arithmetic instructions. An example is the exclusive-OR (XOR), which combines a number in RAM (address in RX) with the number in the accumulator. The instruction is

RCA	XOR	Exclusive-OR memory	M(RX)· XOR· D→ D	F3
Intel	XRA data		A V M(RX) → A	AE

Each bit of the two operands is computed, and the result is placed in the corresponding bit position in the accumulator. This instruction is useful in testing two data types for equality, because if they are equal, the result is all zeros in the accumulator. A subsequent instruction can be then used to detect the zero accumulator condition. Other logic operations are *exclusive*-OR *immediate* and the standard and immediate forms of OR and AND operations.

An unconditional branch or jump to a new PROM address is done by the instruction

RCA	LBR	Unconditional long branch	M(RP, RP + 1)→ RP	C0
Intel	JMP	or jump	(byte 3)(byte 2) → PC	C3

This instruction breaks up the normal retrieval of an instruction from the next PROM address indicated by the program counter *RP*. The next instruction is obtained from the address stored at address *RP* and *RP* + 1 (2 bytes) and subsequent instructions from the new location of *RP*. In other words, the next 2 bytes in PROM following the op code are interpreted as an address rather than an instruction. RCA has a short branch instruction which has only 1 byte of address for transfer within a page. The main value of this instruction is a program organization tool similar to the GO TO instruction of Fortran.

A conditional branch instruction is similar to the unconditional branch instruction except it takes place only when a specified condition of the accumulator is true. If false, the next instruction is executed instead. When it is combined with arithmetic or logical operations, the result is similar to those performed by Fortran arithmetic IF statements. The conditional jump or branch instruction (3 bytes) which causes an address jump when the accumulator is zero is

RCA	LBZ	Long branch if D = 0	If D = 0: M(RP, RP + 1) → RP	C2
Intel	JZ		If z = 1: (byte 3)(byte 2) → PC	CA

Transfer of the program counter to the new 2-byte address, which is located immediately following the op code in PROM, occurs if the accumulator is zero. Otherwise it goes to the instruction in PROM which is the next position beyond.

Another useful branch instruction is quite similar to the instruction above except that transfer occurs if the carry bit (DF or CY) of the accumulator is zero

RCA	LBNF	Long branch if DF = 0	If DF = 0: M(RP, NP + 1) → RP	CB
		or		
Intel	JC	Jump if CY = 0	If CY = 0: (byte 3)(byte 2) → PC	D2

An example, illustrated here for the RCA 1802, is its application in a branch if a number A is less than B. Two instructions are required, assuming that A is in register D and B is in $M(RX)$:

	Mnemonic	Comment	Op code
	SD	Subtract A from B	F5
RCA		(DF = 1 if A < B)	
	LBDF	If DF = 1, go to M(RP)	C3

The Fortran equivalent is IF (A.LT.B) GO TO M(RP).

The feature of the subtraction instruction used here is that the carry bit DF is set equal to **0** during a borrow $D < M(RX)$; otherwise it is **1**.

Many additional short- and long-branch instructions are available. A short branch is restricted to changes of addresses within a page. The address high byte or page number remains the same. Included in the RCA instruction set are short branches which depend on the condition of four special 1-bit direct data-input ports called *external flags*, which are useful in sensing the status of external devices.

20-4 Register Transfer Operations

Transfer of data to and from registers within the CPU is an important software operation. Registers affected include the data pointer and program counter as well as general-purpose registers. Consider first the RCA 1802 instructions. Data can be transferred only from a particular register to the accumulator D or the reverse. There is no direct register-to-register or register-to-memory transfer; 16 registers are available. Instructions which transfer data from the accumulator D to register are

RCA	PLO	Put low R	D → RN(low)	AN
	PHI	Put high R	D → RN(high)	BN

The reverse transfer high- and low-byte instructions (RX → D) for RCA are GLO and GHI.

Instructions are available to set the 4-bit registers X and P, designating which registers are the data and program counter. The data-pointer register RX and program counter RP are set by the instructions

RCA	SEX	Set X	N → X	EN
	SEP	SEP	N → P	DN

Intel microprocessors transfer data from the data register to memory and the reverse. There is no direct data transfer from register to accumulator, but there is data transfer between registers. The memory-register transfer instructions are

Intel	MOV r,M	Move from memory	M(RX)→ RN	1K6†
	MOV M,r	Move to memory	RN→ M(RX)	7K†

†Expressed in octal.

Here K is a register code number [$K = 4$† for RX (high), $K = 5$ for RX (low)]. Transfer of data from one register r_1 to another r_2 is done by the instruction

Intel	MOV r_1, r_2	Move register	$(r_2) → (r_1)$	1KL†

†Expressed in octal.

Here K and L are the corresponding register octal code numbers.

Some instructions affect the individual bits within the accumulator data byte. An example is "shift left," which shifts bit D6 to D7, D5 to D6, etc. Bit 7 is transferred to the carry flag. Bit 0 is **0** (if shift without carry) or the contents of the previous carry bit (if shift with carry). The left shift without carry instructions are

RCA	SHL	Shift or	$D_N → D_{N+1}$	FE
		rotate left		
Intel	RLC	(without carry)	D7 → CY, **0** → DO	07

20-5 Input/Output (I/O) Ports

Data are usually brought in and out through ports. For the RCA 1802, data is transferred from memory to a port of the reverse (Fig. 20-4). There is no direct register- or accumulator-to-port transfer operation.

Control of ports is done by three N lines (port select) and, if needed, an $\overline{\text{MRD}}$ (memory read) line, which are brought out of the CPU. The N lines are at **0** when no I/O instructions are being executed. During an I/O instruction the lines specified by the instruction go to **1,** and these are used to turn on the port. The instruction which transfers data from the memory address in the data-pointer register RX to output port N is

RCA	OUT	Output N	M(RX) → Port N and RX → RX + 1	6N

where N is the port number ($N = 1$ to $N = 7$). During an output-port transfer the $\overline{\text{MRD}}$ line is at **1.**

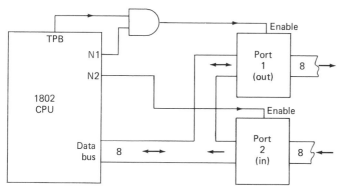

Fig. 20-4   Simplified input and output port connections for the RCA 1802.

Input instructions are similar to the output instructions except that $N' = N + 8$ replaces N. To transfer a number from port N to memory at the address stored in $RX,$ the instruction is

RCA	INP	Input N	Port N → M(RX) → D	6N'

During the output-port transfer, the $\overline{\text{MRD}}$ line is at **0.** Both an input and output can have the same number, but in this case the $\overline{\text{MRD}}$ line must be used to keep both ports from operating simultaneously. With this technique and a decoder (3-to-8-line) on the N-line output up to 14 ports can be controlled directly. Should larger numbers of ports be required, an indirect method involving the latching of memory address lines is available (over 10^5 ports are theoretically possible).

The Intel 8080/8085 transfers data between the accumulator (or a register) and I/O ports in much the same way as it transfers data to and from memory. The technique of treating ports as a part of memory is referred to as *memory-mapped I/O*. Two methods of separating RAM from ports are used; both involve turning on the appropriate enable line when the desired unit is to be activated. The first method (Fig. 20-5) is to turn

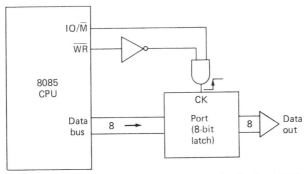

Fig. 20-5   Simplified output-port connections for the Intel 8085.

on an I/O status bit ($IO/\overline{M}$). The status flip-flop output is brought out for control purposes, in particular to switch off the memory and turn on the port. The 8080, by contrast, requires an external latch and decoder to generate port control lines. A specific port is turned on if the I/O bit and the address line are on. Two instructions, which input or output data from port N to the accumulator are

Intel	IN port	Input N	(data → A)	DB
	OUT port	Output N	A → (data)	D3

A second byte in the instruction is the port number N (8-bit).

An alternative method is to use the highest address bit A_{15} to switch between memory and I/O. It replaces the $IO/\overline{M}$ line. Presumably the required memory is small enough (under 32 kilobytes) for there to be no need for this bit in the RAM/PROM address. Intel has a number of chips which have ports as an accessory (see RAM, Fig. 20-13).

Circuit details of two types of input ports are shown in Fig. 20-6. The simplest port is a set of eight transmission gates, which are turned on simultaneously by the strobe or enable pulse as supplied by the microprocessor at the proper time. This type of port is satisfactory if the data remains unchanged until after the port is read. The data might originate from a set of data switches. If the data changes rapidly or the microprocessor may be delayed before reading the port, a latch type of port (Fig. 20-6b) is greatly superior. The latch pulse usually is supplied externally by the device generating the data, e.g., an A/D converter. Eight type-D flip-flops with tristate outputs are available in a single IC package (8212).

A ninth flip-flop functions as a data-ready flag. It can be connected to a

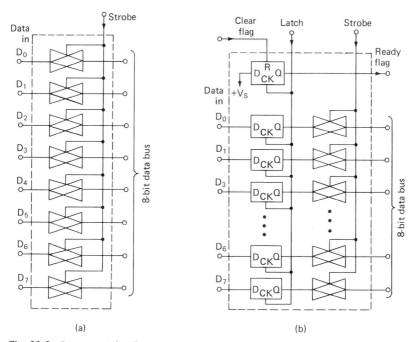

(a)   (b)

Fig. 20-6 Input-port details: (a) transmission-gate type, (b) latch type.

flag input or interrupt the microprocessor to signal the CPU that the port has data ready. After the input port has been read, the microprocessor must clear the ready flip-flop. One of the output ports can be utilized for this purpose.

A standard output port is shown in Fig. 20-7. It is essentially the same as the latched type-D flip-flop input port except that it lacks tristate output (although it could be included if needed externally). A ninth flip-flop can be included to indicate to the external device that output data is ready (the device must supply a clear pulse). The output port may be constructed from a pair of quad latches, for example, 4042, or single IC package (8212).

The RCA 1802 has four direct-input single-bit ports, or *external flags*, which make it possible to test the status of external signals, e.g., timing and data-ready signals. A group of branching operations is available which depend on the status of a particular flag. An example is branch if flag 1 (EF1) is at **0**.

20-6 Interrupt and Subroutine Transfers

The capacity of a microprocessor to exchange registers like the program counter adds considerably to the programming flexibility. Calling a subroutine and the return to the main program with Intel software requires one instruction for each

Intel	CALL	Call	$RP \to SP$ (byte 3)(byte 2) $\to$ PC	CD
	RET	Return	$SP \to RP$	C9

Here SP is a stack pointer (a 2-byte register). The address of the first instruction of the subroutine RX' is specified by bytes 3 and 2 of the

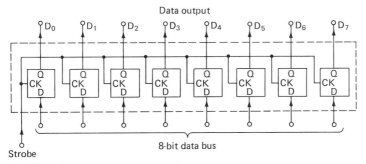

Fig. 20-7 Output-port circuit details.

instruction. Thus CALL puts the subroutine starting address into the program counter RP. Before doing so it saves the current contents of the program counter, which is later retrieved by the RETURN instruction, normally placed at the end of the subroutine program.

Subroutine calls for RCA can be done by changing the program counter (SEP instruction) after saving the current program-counter contents (MARK instruction). It has a similar return instruction.

Interrupt is a break in the main program initiated by an outside device to perform some high-priority operation. Interrupt is important, for example, in machine-control operations. It allows rapid response to sporadic data inputs, including alarm indicators. An interrupt process is initiated by a pulse applied to the interrupt (input) line of the microprocessor. After finishing its current instruction, a jump from the main program to an interrupt-service routine occurs. This is done by changing the program-counter register and is similar to a subroutine call. In the RCA 1802 this is done by changing P to 1 and X to 2. The starting address of the interrupt-service program in PROM must be placed in register 1 during program initialization. At the end of the interrupt program a RETURN instruction is executed and the program counter is restored to its previous value before interrupt. Also, the old value of the data-pointer register, which is also saved, is restored to that of the main program, which then proceeds as if the interrupt had not occurred.

An external circuit to provide and clear an interrupt signal when data is ready is indicated in Fig. 20-8. The external ports (here two) must supply

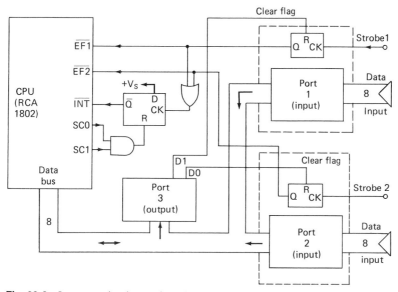

Fig. 20-8 Interrupt circuitry to the RCA 1802.

a strobe pulse to the appropriate data-ready flip-flop. Immediately the interrupt flip-flop is set, thus signaling the microprocessor that an interrupt is requested. An internal flip-flop is also set, and after finishing the current instruction the microprocessor will service the interrupt (transfer control to the address in program register 1, etc.). Transfer to the interrupt state is signaled by the state code lines (SC0 and SC1 = **1**), which causes the interrupt-ready flip-flop to reset. The data-ready flip-flops are not reset yet. The program (software) must now determine which port has caused the interrupt. It does so in this example by reading the status of the two flags (EF1 and EF2). If EF1 has been set, the data from port 1 is read, an output instruction port 3 causes the data-1-ready flip-flop to reset, and then the remaining interrupt program is executed.

Before returning to the main program, the flag EF2 is read to check for data ready on port 2 (as well as port 1) and the appropriate action taken. A priority of action in case of simultaneous interrupt must be established.

One method of interrupt for the 8085 is to apply a signal to the RST 6.5 input. This causes an immediate branch to address 3C, where the interrupt program presumably begins. Interrupt for the 8080 requires relatively complex external hardware, e.g., priority-interrupt control unit for implementation. It is equivalent to a hardware CALL instruction. Return to the main program is done by a RETURN instruction.

Systems with multiple interrupt often employ a *priority-encoder* IC to resolve the problem of simultaneous interrupts. Inputs to the encoder (perhaps eight lines) are the data-ready flip-flops of the various ports. They are wired in order of desired priority with input 1 highest. The output is the address (3-bit for eight inputs) of the highest-priority input line which is at **1.** The microprocessor is programmed to read from and service the port indicated by the encoder since it is the highest priority.

20-7 Timing Diagrams

The sequencing of the various microprocessor events is best understood by timing diagrams. Generally they are rather complex because of the many peripheral devices (memory, ports, etc.) which must be controlled. The CPU generates the timing pulses required for a particular operation. It is synchronized by the clock pulses (two-phase clock in the case of the 8080), to which the timing signals are referenced. The timing of various groups of operations differs.

The timing diagram for the input operations for the RCA 1802 is shown in Fig. 20-9. Both the fetch and execute states require eight clock cycles. Two external timing pulses (TPA and TPB) are provided for the address and port latches as well as for general timing. The sequencing of

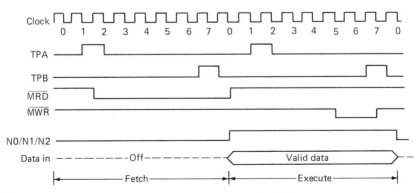

Fig. 20-9 Input-instruction timing diagram for the RCA 1802.

various operations discussed above (high- and low-address latching, fetch and execute cycles, memory and port enables) can be seen from these diagrams.

The number of clock cycles per operation for the Intel 8080/8085 is not constant, as indicated in Fig. 20-10. Machine cycle M_1 is an (instruction) fetch cycle which requires three clock cycles. M_2 is a also a memory-read

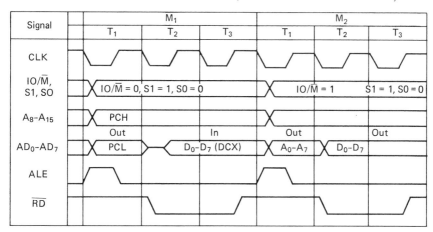

Fig. 20-10   Input-instruction timing diagram for the Intel 8085.

cycle which here obtains the port code (low byte) from PROM, and during M_3 the port code is transferred to the address lines while turning on the IO/$\overline{\text{M}}$ line. Other instructions have up to five cycles.

20-8   Memory (RAM and PROM) Citcuits

Of the numerous methods for storing digital data, the directly addressable semiconductor memory or random-access memory (RAM) is almost invariably used for microprocessor applications. Other memories such as magnetic core or serial storage may supplement the RAM but do not completely replace it, at least in small systems. Many types of RAMs are available which differ in total size, byte size or organization, speed, and clock requirements.

There is a basic division of memories into static and dynamic types. A dynamic RAM requires a clock (frequency about 10 kHz to 1 MHz) to continuously recycle or refresh the data internally. If the clock stops momentarily, the data is lost. Strobing of the output data ordinarily must by synchronized with the clock. Static RAMs have no clock or synchronization requirements. They are definitely easier to use, especially if the microprocessor is also static, like the RCA 1802. The discussion here will be restricted to static RAMs.

Memory size, or number of bits stored, is the RAM's most important characteristic. Invariably the number of bits is equal to some power of 2 in order to match the binary arithmetic of the microprocessor. A typical RAM is 1024 bits, or 1 kilobit in size. A large RAM might be 4 or 16 kilobytes while a small RAM might have only 256 bits. Bit or byte organization of RAMs differ. A RAM might be 1, 4, or 8 bits wide; i.e., each address would correspond to 1, 4, or 8 bits of data. For example, a 256 × 1 RAM would have 1 bit at 256 different addresses while a 32 × 8 RAM would have an 8-bit byte at 32 different addresses (both are 256-bit RAMs).

It is instructive to consider the internal block diagram of a RAM like Fig. 20-11. A special type of flip-flop holds each bit of data. When the local

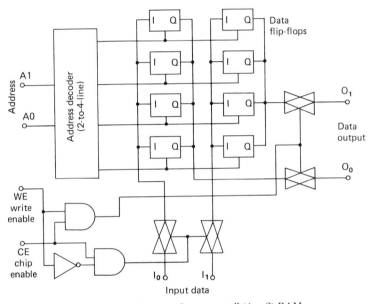

Fig. 20-11   Internal block diagram of a very small (4 × 2) RAM.

flip-flop enable line (from the decoder) is turned on, the flip-flop is connected to the data input and output lines through a transmission gate. Only when the input gate is turned on does the flip-flop follow the state (**0** or **1**) of the input. This occurs if the write-enable line (WE) is a logic **1** and the chip enable (CE) or strobe line is at logic **1**. The flip-flop is thus latched. The two enabled flip-flops corresponding to the two input data bits (I_0, I_1) are latched simultaneously. Note that the outputs are in the high-impedance state during this write process. If the WE line is low, the outputs of the enabled flip-flops appear at the RAM output (O_0, O_1) when the chip is enabled (CE = **1**). In this case the tristate drivers are not in the high-impedance state.

Usually in microprocessor applications the input and output lines of the corresponding data bits are tied together and connected to the data-bus lines. The write enable (WE) or read-write line from the microprocessor controls the direction of data flow along the bus. Note that in any case the logic shown will not allow simultaneous input and output data flow.

In memory-write operation, the address lines are turned on first because the address decoders exhibit appreciable delay. The time required for the address decoder to settle is referred to as the *memory-access time*, typically 0.1 to 1 μs. After the address has stabilized, the chip can be strobed by turning on the chip-enable (CE) line. During a write operation, if the CE line is turned on before the WE line, data from the RAM will appear on the data bus until WE is enabled. This is likely to foul up the data bus because the CPU is also dumping data onto the bus. Thus it is good practice to switch the WE line before or simultaneously with the CE line. During a read operation the CE line can be turned on (with WE off) before the address lines settle, but the data will not be stable and should not be taken off the bus until after the access time delay.

For many purposes the 8-kilobit RAM (1 kilobit $\times$ 8) consisting of 1024 bytes, or four pages, is adequate (Fig. 20-12). It is composed of eight 1024 $\times$ 1 static RAMs. The addressing decoding for the RCA 1802 is shown. A 2-bit latch for the higher address lines (page number) is provided. Latching of the high address bits occurs early in the fetch cycle (as timed by the

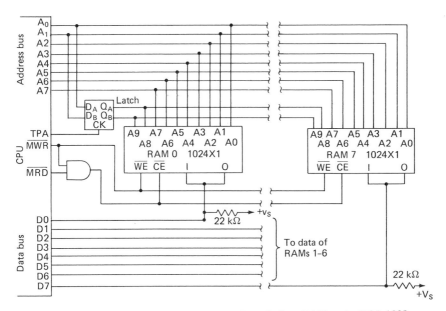

Fig. 20-12   Address and data-bus connections of a 1024-byte RAM to the RCA 1802.

TPA pulse) while the lower 8 bits are connected directly to the address pins of all the RAMs. The strobing and read-write lines ($\overline{\text{CE}}$ and $\overline{\text{WE}}$) are controlled by the $\overline{\text{MRD}}$ and $\overline{\text{MWR}}$ lines from the CPU.

The 8085 memory connections (Fig. 20-13) are similar. The main difference is that the low address byte is latched from the data bus at the proper time as determined by the address-latch enable line ALE (falling edge). The RAM chip shown (8156) also has an input/output port built in, a convenience which reduces the number of chips and the wiring required in a system. Address connection to the 8080 is easier to understand since all 16 address lines are brought out, but two other chips are required to support the CPU, a clock generator and a system controller. The connection of a ROM or PROM to the microprocessor is essentially the same as that of a RAM except that there is no write-enable line. In other words the reading from RAM or ROM (or PROM) is practically identical. Some erasable PROMs require an additional negative power supply, a small inconvenience.

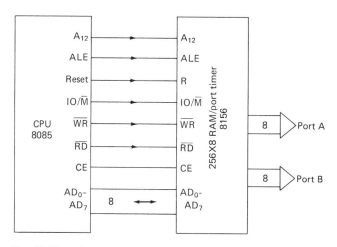

Fig. 20-13   Address and data-bus connections of a 256-byte RAM to the Intel 8085.

Programming or writing into PROM is a comparatively difficult procedure which must be done with care. In all electrically programmable ROMs the bits are set or "burned in" by applying a higher-voltage pulse of short duration. This is usually done by a PROM programmer. Voltages and procedures are different for the various types and makes and therefore a general-purpose programmer must be rather complex, especially if high-speed and accuracy-checking circuits are included.

Perhaps the simplest PROM is the fusible-link type. Each bit is connected to a small Nichrome or silicon fuse, which can be blown, or melted, by the relatively high programming voltage (12 V instead of the normal 5-V supply). The address of the bit to be blown is supplied as usual. When not blown, the bits are at **0.** A programming voltage pulse is applied to the data output line if a **1** is desired for a particular bit.

An obvious disadvantage of the fusible-link PROM is the lack of provision for correcting errors or altering the program. It is low in cost, however, and quite satisfactory for fully debugged programs, especially when programmed from RAM automatically.

Erasable, electrically programmable read-only memories (EPROMs) are perhaps the most generally useful PROMs. They are composed of FETs with long-term, practically permanent charge storage on the gates. As with any FET, a voltage or charge on the gate can turn off the transistor. By trapping the charge on the gate in an electrically insulating region, the transistor can be turned off until the charge leaks off. With existing PROMs it is estimated that the time constant in the dark is decades; i.e., the transistor (1 bit) can be turned off practically permanently. When exposed to ultraviolet light, the insulator becomes slightly conducting, and under intense radiation the charge leaks off within a few minutes. All bits are thus returned to the **0** state. To charge the gate to the **1** state in the first instance requires the application of a voltage high enough (25^+ V) to cause an avalanch breakdown of the insulating region. By this process a bit (**1**) is burned in. If a mistake is made, the entire PROM must be erased by exposure to ultraviolet light through a window on the top of the IC package. The programming pulses must be carefully timed to prevent damage to the PROM and perhaps repeated if verification indicates that the bits have not been changed to **1** on the first try. An automatic programmer is almost essential. Electrically erasable PROMs, termed electrically alterable read-only memories (EAROM), are also available.

20-9 Programming Methods

The programming for a microprocessor is similar to programming in a higher-level language such as Fortran but is much more tedious because the small details of machine operation must be accounted for. Initially an overall flowchart is made which lists in order the tasks to be performed. Major branches and loops are included, but details are ignored. Next the numbers and functions of registers and ports are assigned. To a large extent the choice is arbitrary, but of course once hardware is fixed, i.e., the circuit is wired, the software, e.g., port numbers, must correspond. Next the memory (PROM and RAM) addresses are assigned. For example, PROM might be allotted four pages (1 kilobyte) of memory from 0000 to 03FF while RAM might be allotted one page, from 0400 to 04FF.

It is assumed that the actual memory address lines are properly wired according to the chosen assignment; i.e., the software and hardware coding must agree. When this is done, a detailed flowchart can be prepared which references specific ports and memory locations by assigned number.

Computer-aided programming is desirable, if not practically necessary, for all but the smallest programs. If large central computer facilities (IBM, CDC) are available, one approach is to obtain a cross-compiler program intended for the specific microprocessor selected. The compiler accepts commands in mnemonic form, converts to op code, and allows branch locations to be specified as a symbolic name. Absolute hexadecimal addresses are calculated by the compiler, a great convenience, especially when additions or corrections to the program are required. Various error checks are made, and the compiled program is listed (printed) in a convenient form with space for comments. Most programs are intended for keyboard input from a time-sharing terminal.

Often a simulation program is associated with the cross-compilation program. It carries out the program in PROM in the same manner as the microprocessor but with a supervisory program to monitor the process. Should a problem develop, the steps in the calculation can be examined and traced easily. The simulator is intended to simulate the CPU and memory sections rather than a complete microprocessor system since it does not model the response of peripheral hardware. Successful simulation does not mean the microprocessor system will work but if simulation is unsuccessful, the system certainly will not work.

A microprocessor system can be programmed to compile and simulate another microprocessor system, in particular the system under development. These development systems are costly but desirable alternatives to cross-compilation and simulation by central computer.

Few small microprocessors have multiply or divide commands in the instruction set, but these operations can be done by combination of standard instructions such as addition and shift. Usually these operations are written as subroutines (perhaps 30 to 60 bytes long) to allow call from several parts of the program. These standard routines should be supplied by the manufacturer as a list of instructions or perhaps as a preprogrammed ROM. Other standard routines include binary-to-BCD conversion (and the reverse) and double-precision addition and subtraction.

20-10 Microprocessor Types

The versatility of microprocessors is such that with the proper peripheral devices nearly any arithmetic or control operation can be performed with any microprocessor. There is a great difference, however, in calculation time, number of instructions, and external hardware required to perform a particular task. Some microprocessors, e.g., the RCA 1802 or Fairchild F-8, are more instrument-control-oriented. They can move input and output data through ports with minimal hardware. Software instructions related to branching, as required by many control functions, are also relatively simple. Other microprocessors, e.g., the Zilog Z-80 or Texas Instruments TMS 9900, are more computation-oriented. They can transfer blocks of data and perform arithmetic and logic operations efficiently. Programming is easier and execution is faster.

The tendency to view the microprocessor as a low-cost or personal computer when the application is instrument control should be avoided. Even simple machines respond rapidly in typical dedicated-control applications, so that all but a small fraction of the time is spent idle. Speed and a powerful instruction set may be quite unnecessary. Availability of software is important, however, and may override other considerations. One reason for the popularity of the Intel 8080 and software-compatible machines is the large number of programs and subroutines available.

Microprocessors are classified basically by data-bus (or word) size. Most are 1-byte (8-bit) machines but 16-, 4-, and even 1-bit devices are available. Generally a larger word size implies higher accuracy or speed and a more powerful instruction set, approaching that of a minicomputer. Some 16-bit microprocessors have multiply and divide instructions, a definite convenience. The cost and circuit-board complexity also increase with word size.

A few manufacturers offer bit-slice microprocessor ICs, which are actually subunits of microprocessors. Each is equivalent to 2 bits (usually), and a 12-word machine can be made with six slices. Hardware design for a bit-slice machine is much more difficult than for standard microprocessors and is best left to computer engineers.

As indicated above, a number of microprocessors have software compatible with the Intel 8080. These include the discussed, Intel 8085, and the Zilog Z-80, all of which have built-in clocks and are generally more compact than the 8080. The Z-80 has additional instructions which simplify programming. Other popular 8-bit microprocessors are the Mostek 3870 and Motorola 6800, which have features roughly comparable with those of Intel.

DATA SWITCHING, CONTROL, AND READOUT DESIGN EXAMPLE

Example 8   Digitally Controlled Timer

DESCRIPTION: A timed pulse selected by a three-digit thumbwheel switch is produced when a switch is pressed.

SPECIFICATIONS

Resolution: three digits (±0.01 s)

Interval: 0 to 9.99 s†

Output: TTL-compatible (0/3.6 V)

Trigger: Push-button switch (or TTL pulse)

DESIGN CONSIDERATIONS: The design is based on a three-stage decade down counter (see Figs. 17-13 and 17-14) connected to a clock generator (555). A clock produces a stream of 10-ms pulses. The counter is preset to N (12-bit BCD number) by a brief pulse (PE) when the start switch is pressed. The **0** output of the 4522 is high when the counter (all four Q outputs) is zero provided the CF input connected to the **0** of a previous stage is high. Upon preset, **0** therefore becomes low and T_{out} becomes high. During the count process the **0** output of the most significant digit goes high first, thus allowing, through the cascade connection (CF), the next most significant **0** to go high when it reaches a zero count (otherwise it resets to 9, as usual for a decade counter). After the least significant digit reaches zero after all other digits are zero, its 0 goes high and T_{out} returns to zero, ending the timed interval.

Other interval times can be obtained by changing the clock frequency. Interval accuracy depends on the frequency stability. For the 555 timer it is the 0.1 to 1 percent range. A 60-Hz timer with a six-digit counter might be preferred for longer times (0.1- or 1-s resolution).

DATA SWITCHING, CONTROL, AND READOUT DESIGN PROBLEMS

47 A one-shot multivibrator that will deliver a 5-V, 50-μs pulse (±10 μs) is required. It is to be triggered by the positive edge of a pulse (5 V) with a width longer than 1 ms.

48 Design a push-button switch activated timer which will turn on a 115-V light (100 W) for 3 s. A 12-V relay (coil resistance of 300 Ω) with 2-A contacts is available. One contact of the switch is grounded.

49 Design a circuit which will deliver an output pulse approximately 1 μs wide after counting 100 input pulses. Assume that TTL-compatible logic is used and that a clear pulse occurs before the input pulses start. Further input pulses must not result in another output pulse (unless a clear pulse occurs).

50 Give a circuit for a PLL type of frequency multiplier with a multiplication factor of 200. Assume a square-wave input f_i in the range 0.3 to 1.5 Hz. Use a 565, 4046, or similar IC.

51 Give a circuit for a 4-to-16-line decoder made from two 4-to-10-line decoders (and other gates).

52 Design a three-channel analog time-multiplex system which requires only one line (one wire plus ground). The input signal is assumed limited to ±5 V full scale. A fourth channel may be utilized for synchronization, e.g., higher than full-scale voltage as detected by a comparator.

53 A three-digit frequency counter using the 60-Hz line as a time base is

†Proportional to clock frequency, which is easily changed.

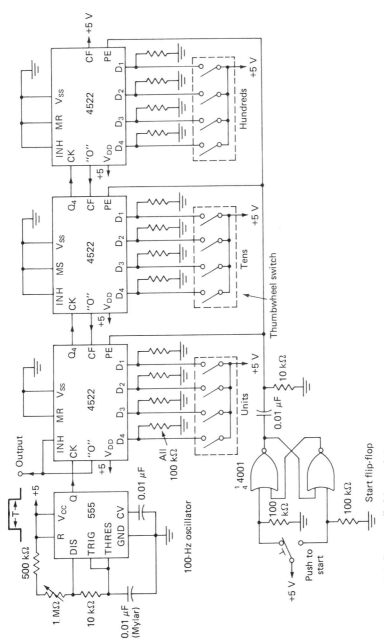

Fig. E-8 Digitally controlled interval timer.

required. The maximum frequency is 9.99 kHz, and a new reading (display update) must occur every 0.2 s or faster. A latched output is suggested.

54 A simple two-digit frequency counter using a single 555 timer as the time base is required. The timer pulse width is the count time and the period is the display update time (timer in astable mode).

55 Design a digitally controlled interval timer based on a presettable down (or up) counter. It is triggered by a brief (~0.1-ms) start pulse. Assume a clock input frequency of 1 kHz and a maximum time of 99 ms (resolution 1 ms). The time should be set by a two-digit thumbwheel switch.

56 Data from a thumbwheel (4-bit) is to be transmitted along a serial data line to a seven-segment decoder, where it is displayed. Use parallel-serial and serial-parallel registers. For more of a challenge, do this without the clock and reset lines.

57 Give a complete circuit for a pulse counter with latch and multiplexed four-digit display.

58 Design a tone decoder that will trip when frequencies of 700 *and* 1000 Hz are present.

59 Design a tone encoder which will produce the two tones 700 and 1000 Hz when a pulse (5 V, CMOS-compatible) is applied to the input.

60 A tape-recorder digital encode-decode circuit is required which will plug into a standard cassette recorder. Test a recorder to find its pulse response and then design the appropriate circuits which use the NRZ or pulse-ratio formats.

61 Devise the circuits to transmit and receive a digital signal over a single narrow-band line or link. The frequency response is 100 to 3000 Hz, and the attenuation varies between −3 and −10 dB. Data rates will not exceed 10 b/s.

62 A tape-recorder-controlled slide projector is required. When the slide is to be advanced, a 1-kHz tone 1 s long appears in the recording, which otherwise consists of speech (describing the slides). To advance a slide, a relay contact must be closed for 0.2 s minimum. Devise a circuit which can be attached to the tape-recorder speaker output (1 V peak to peak minimum for tone) to accomplish this.

63 A remote light switch (relay or triac) controlled by a 30-kHz signal transmitted on the power line is required. Assume that the signal will be attenuated by a factor of 10 and that at this frequency the line shunt impedance is over 3 Ω and interference (noise) under 20 mV peak to peak. Design a system to accomplish this task.

64 Design an output port (8-bit) from a microprocessor to drive a seven-segment digital display. The port is a pair of CMOS quad latches (CD4042), and the display is a pair of hexadecimal displays with TTL inputs.

65 Design a 16-bit pulse counter to be controlled and read through an 8-bit microprocessor port. The port is bidirectional or, alternatively, consists of both an input and output port. The output port is for control (D0 = counter clear, D1 = counter enable, D2 = byte select).

66 Give a circuit for a port that will read two BCD digits from a thumbwheel. Next expand the circuit so that it will read in six digits.

67 Look up the characteristics of a commercial D/A converter. Design a circuit based on this which will deliver a dc voltage under the control of a microprocessor through an 8-bit port.

Part Five
Power Circuits

Chapter Twenty-One
Power Supplies

Power supplies are often taken for granted, and with good reason; most supplies generally deliver a pure, regulated dc voltage as expected and without trouble. For many or most instruments the convenience and modest cost of a commercial modular power supply make it the best choice. Those who make this choice will consult this chapter only to clarify the terminology involved in power-supply specification. However, power supplies are not difficult to construct and often cost less when put together by the user, especially if multiple output voltages or other nonstandard features are needed.

21-1 Power-Supply Characteristics

A single power supply is a constant dc voltage source and is equivalent to a battery with an output voltage of V_S. Either the positive or negative terminal may be grounded, but for most applications discussed in this text, especially logic circuits, the ground is negative, as indicated in Fig. 21-1a. A dual supply is the equivalent of two batteries in series with a common connection (Fig. 21-1b). Almost invariably the common connection is grounded when used to power op-amps.

It should be noted that terminal markings are similar on single and dual supplies (+, −, gnd/com) but have different meanings. A ±15-V dual supply for op-amps is really a 30-V center-tapped supply. Because of the center tap, or common, both positive and negative signal excursions with respect to common are possible. The terminal labeled ground is case-

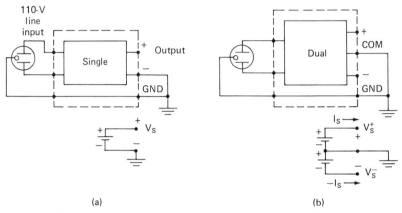

Fig. 21-1 Conventional line-operated power supplies: (*a*) single; (*b*) dual.

ground, which may or may not be connected to common (dual supplies) or V_S^- (single supplies). It is also well to know, especially when connecting test instruments, whether the circuit "ground" is in fact connected to a true earth ground through the third (ground) wire of the line cord either directly or indirectly such as through the ground lead of an oscilloscope or other test instrument.

The output voltage V_S of a supply is its most important parameter. Fixed voltage supplies, that is, ±15, +10, or +5 V, are most convenient and least costly for instrument power although variable voltage supplies, for example, 0 to 20 V, are often convenient for test purposes.

Maximum output current is the next most important parameter. Presumably the set output voltage will change insignificantly under load, i.e., as current is drawn, until the current limit is exceeded. A low-cost, unregulated supply will overheat or blow a fuse if excess current is drawn, but nearly all regulated supplies now are protected by a circuit which limits the current to some safe value, perhaps 10 to 50 percent above the stated maximum. The current limit is effective for practically any regulated supply when short-circuited for brief intervals, but not all supplies can withstand a continuous short without overheating and damage. Better supplies have a thermal shutdown in addition to overcurrent protection. For long life a supply should not be operated continuously more than 70 to 90 percent of its current capacity. Because the cost and bulk of power supplies increases with their current capacity, a supply with excess capacity is usually not desirable, but it will generally function well even when lightly loaded.

Good voltage regulation is always desirable and often essential. Actually most op-amp and logic circuits are rather insensitive to power-supply fluctuations, but most designers do not wish to bother with an analysis of power-supply fluctuations on circuit performance (it is easier to specify a regulated supply and forget the problem). This is an attractive solution since the cost of a regulator is modest, perhaps even less than a conventionally filtered supply, at least at lower power levels. Parameters characterizing regulator performance are regulation against load, regulation against line, ripple, and precision.

Regulation against load refers to the change (reduction) in output voltage ΔV_S as current I_S is drawn. It is often specified as the fractional change in output voltage when loaded. A good but not exceptional supply will change no more than 0.2 percent as the current is varied between zero and maximum output. Another way of specifying this is the effective internal (Thevenin) resistance of the supply. A 15.0-V supply which drops to 14.7 V when a 100-mA load is drawn, for example, has an internal resistance ($R_o = \Delta V/\Delta I$) of 3 Ω. Regulation against line is the percentage change in output voltage for a given percentage change in line voltage (nominally 110 or 120 V rms, usually). A good but not exceptional supply will change no more than 0.2 percent for a 10 percent change in line voltage.

Ripple and/or noise refers to a presumably small ac component which is generally present in addition to the desired dc component. Ripple has frequency components which are multiples of line frequency (mostly 60 and 120 Hz). A good but not exceptional regulator will have a ripple content no higher than 2 mV rms when measured at full load.

Precision refers to absolute accuracy or difference between the stated voltage and the actual voltage. A 10.00-V supply with a precision of 0.2 percent will have an output voltage between 9.98 and 10.02 V under all conditions of line fluctuations and load changes. By contrast a supply with 0.2 percent regulation against load may have an actual output voltage of 9.80 V no load, which will drop at most to 9.78 V under load. For most applications, in fact, it is important that the supply voltage remain constant, but the exact value is not critical. Precision supplies are needed, however, where a reference voltage is assumed, as in some digital voltmeters and A/D converters.

21-2 Rectifier-Capacitor Input Section

Nearly all line-operated instrument power supplies use a stepdown transformer to provide the proper voltage and the necessary isolation between the instrument ground and the power line. Current passes through the diode (rectifier) to charge the capacitor (Fig. 21-2). The peak voltage to which the capacitor is charged is equal to the peak voltage of the transformer secondary less the voltage drop across the rectifier (~0.7 V). Since transformers are rated in rms volts at full load, a standard 12-V transformer will deliver 16.8 V peak and more if lightly loaded. Transformer

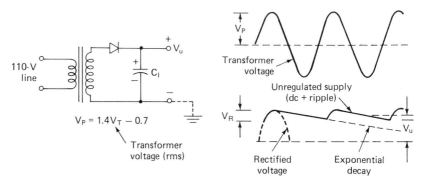

Fig. 21-2 Power-supply input sections: (*left*) half-wave rectification; (*top right*) voltage input and output; (*bottom right*) ripple analysis.

current ratings are also stated in rms but apply to a resistive load and must be derated because the current is drawn from the transformer in pulses. Power loss is proportional to the current squared, and therefore the transformer heating will be substantially higher for a given average current if the current is pulsating. In practice, the maximum direct current available is about 40 to 80 percent of the transformer rating.

It should be pointed out that power-supply parameters such as ripple are difficult to calculate accurately because the exact characteristics of the individual components, e.g., transformers, are usually not known. Thus calculations of the precise magnitude and waveshape of ripple under particular conditions is of little value. All that is really wanted are estimates or limits of supply ripple or regulation. Instead of going into the details of power-supply analysis, reliance will be made here on practical rules and graphs.

Full-wave rectification (Fig. 21-3) is desirable except for small load currents. The two main advantages are that the input capacitor C_I need only be half as large because the discharge time is half as long (frequency twice as high) and the transformer loss is less when current is drawn for both the positive and negative cycles. Transformer loss is higher for half-wave rectification not only because the peak current is higher but also because a net direct current which flows through the secondary can cause magnetic saturation of the transformer core. These losses are not very important if the transformer is lightly loaded. Another advantage of full-wave rectification is that the high-voltage surge which often occurs as the transformer is switched off is harmlessly discharged through the rectifiers to the capacitors (for half-wave rectifiers the surge across the diode may exceed the inverse breakdown voltage and destroy the diode).

The two methods of obtaining full-wave rectification are shown in Fig. 21-3. Where a suitable transformer is available, the center-tapped circuit is preferred because it can be used in dual supplies (see Fig. 21-6).

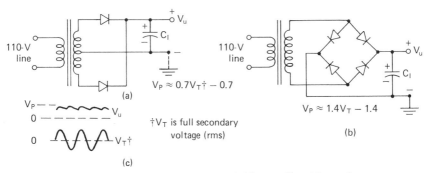

Fig. 21-3 Full-wave rectifiers: (*a*) center tap; (*b*) bridge rectifier; (*c*) waveform.

Current I_L is drawn from the supply by the load, which is represented here by an equivalent resistor R_L. It is the discharge of the capacitor C_I through R_L which causes the voltage drop between cycles seen on an oscilloscope as a sawtooth or ripple. Normally the power-supply input-section voltages are calculated when maximum current is drawn, since this corresponds to the worst case (maximum ripple). To a fair approximation the maximum direct current drawn from the complete supply I_L, even with regulation, is equal to that drawn from the input section, and therefore the effective load resistance R_L is equal to V_u/I_L where V_u is the unregulated voltage (dc or average value). Between cycles (half-cycles for full wave) the capacitor discharges exponentially with a time constant of $R_L C_I$ (see Fig. 21-1). Between cycles the unregulated voltage V_u is roughly

$$V_u = V_p e^{-t/R_L C_I} \approx V_p \left(1 - \frac{t}{R_L C_I} \right) \tag{21-1}$$

where the binomial approximation has been used to obtain the second expression. It is valid for a small voltage drop between cycles, i.e., for large values of C_I. In terms of this equation, the average or dc unregulated voltage is

$$V_u \approx V_P \left(1 - \frac{T_c}{2 R_L C_I} \right) \tag{21-2}$$

where T_c is the period between cycles ($T_c = \frac{1}{60}$ for half wave and $T_c = \frac{1}{120}$ for full wave).

The peak-to-peak ripple v_r from this equation is

$$v_r \approx \frac{V_P T_c}{2R_L C_I} \qquad (21\text{-}3)$$

The input capacitor C_I must be large to reduce the ripple to an acceptable level but should not be larger than necessary because these capacitors are costly and bulky. An acceptable compromise is to choose a capacitor such that the ripple (peak to peak) is between 5 and 20 percent under full load. Selection of a capacitor is simplified by the graph and

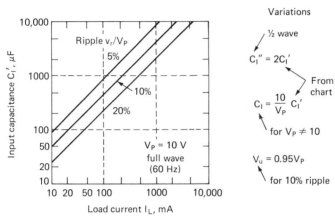

Fig. 21-4   Input-capacitor selection graph. Note scale factor for peak voltages V_P other than 10 V.

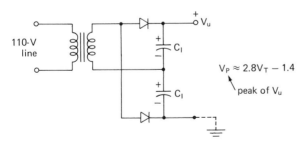

Fig. 21-5   Voltage-doubler circuit.

formulas of Fig. 21-4. To use this graph, the maximum load current I_L and peak voltage V_P must be estimated. Note that usually V_P is only slightly higher (2 to 10 percent) than the unregulated voltage V_u and only slightly lower (5 to 15 percent) than the peak transformer voltage. For example a 12.6-V rms transformer into a full-wave bridge rectifier will produce a dc voltage V_u of 16 V (10 percent ripple). A load current of 50 mA would require a 200-μF capacitor for a 10 percent ripple.

Occasionally a voltage-doubler rectifier circuit (Fig. 21-5) is used if a transformer of sufficiently high secondary voltage is not available. The disadvantages of half-wave rectification are magnified, and the circuit is suited primarily for light current loads. The capacitors C_I are 4 times the value calculated by the graph of Fig. 21-4.

Inductor input-filter supplies are unpopular because of the additional cost and weight of the inductor. They are more efficient, an advantage for high-power applications, but otherwise have little to recommend them.

21-3 Unregulated Supplies

An unregulated supply can consist of just the transformer and input section. Sometimes a ripple filter is added. In Fig. 21-6 a complete unregulated dual ($\pm$20-V) supply is given. It is designed by the methods dis-

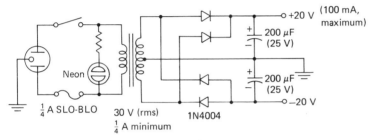

Fig. 21-6 Dual unregulated power supply.

cussed above. Such a supply is suitable where unregulated power is required (lamps, relays) or as an input to a voltage regulator.

Sometimes a passive RC filter section is added to reduce ripple. It should be pointed out, however, that ripple reduction is accomplished more effectively with IC voltage regulators, and RC filters are not routinely useful. An appropriate filter is shown in Fig. 21-7. It is simply a low-pass filter intended to reduce the dc voltage only slightly (V_u to V_u') while

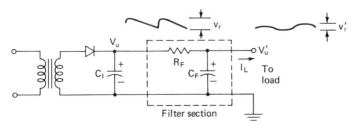

Fig. 21-7 RC power-supply filters.

reducing the ripple greatly (v_r to v_r'). The reduction in dc voltage $V_u -$ V_u' due to the load current I_L is calculated from Ohm's law. Note that the voltage is load-dependent; i.e., the load regulation is poor. The ripple-reduction factor $A = v_r'/v_r$ is estimated from the low-pass-filter attenuation factor $A \approx X_F/R_F$, where $X_F = 1/2\pi fC$ is the impedance of the capacitor ($X_F = 13\ \Omega$ for a 100-μF capacitor at 120 Hz or full wave). This approximation treats the ripple as a sine wave, assuming $X_F \ll R_F$ (true for any effective filter), and ignores phase shift as irrelevant. It is difficult to obtain an attenuation factor of more than 5 or 10 at even modest load currents without large capacitors or poor regulation.

21-4   Diode Voltage Regulators

Zener or voltage-regulator diodes are the basis of nearly all regulated supplies. The simplest regulator circuit consists of a Zener diode and limiting resistor (Fig. 21-8). As stated in Chap. 2, a Zener diode is normally

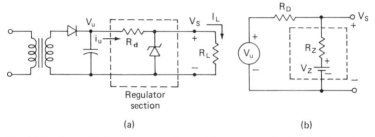

(a)   (b)

Fig. 21-8   Zener diode voltage regulator: (*a*) power-supply circuit; (*b*) equivalent circuit.

reverse-biased into the breakdown region, which occurs at a particular voltage V_Z. The equivalent circuit of the Zener diode biased in this region is a battery V_Z with a small series resistance R_Z. Current from the unregulated supply i_u which flows through R_D divides between the Zener I_Z and external load I_L. This current i_u is roughly constant, equal to $(V_u - V_Z)/R_d$. As the load current changes, the Zener takes up the difference. Load current must be limited or the diode current will drop to zero and stop regulating.

As an example, suppose a 10-V Zener ($R_Z = 20\ \Omega$) has been selected, and the load current is not to exceed 15 mA. A safe minimum current for the diode is 10 mA, so that the unregulated current i_u should be 25 mA. For a 18-V unregulated voltage V_u, R_s should be about 320 Ω. To calculate the line and load regulation, the equivalent circuit (Fig. 21-8*b*) is convenient. At full load ($I_Z = 10$ mA) the voltage drop across R_Z is 0.2 V, while for no load ($I_Z = 25$ mA) it is 0.5 V. The regulation against load is therefore (0.5 − 0.2)/10 or 3 percent. If the line voltage were to drop by

10 percent (V_u = 16.2 V), I_V would drop to 18 mA, I_Z would be reduced by 7 mA, and the output voltage would drop by 0.14 V. This corresponds to a regulation of 1.4 percent for a 10 percent line change.

The Zener-current change for a given line-voltage change is rather large. Care must be taken in the circuit design to ensure that the Zener current never becomes excessive or drops to zero for any normal combination of line-voltage changes, load-current changes, and component tolerance (±5 to ±20 percent for most Zeners). Note that these variations are magnified if the difference between the unregulated voltage and regulated voltage is small.

Zener diodes have maximum power dissipations $I_Z V_Z$ ranging from ¼ W to over 3 W. Their life expectancy is enhanced considerably by operating below half rated power. Higher-power regulator diodes are not popular since regulators with transistors or ICs cost less and perform better.

A variation on the diode regulator is the Zener-diode–transistor regulator shown in Fig. 21-9. The voltage across the Zener is calculated as above.

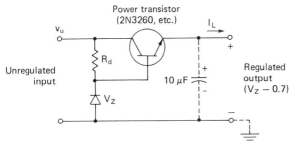

Fig. 21-9 Emitter-follower–Zener voltage regulator. Current overload not provided.

It is applied to the base of the transistor connected as an emitter follower, so that the output (emitter voltage) is equal to the Zener voltage less the base-emitter drop (0.7 V). Because the base current is relatively small (I_L/β), the Zener current does not change as much as the circuit of Fig. 21-8. A further advantage is that the power is dissipated in the power transistor, a less costly device than a power Zener diode. A disadvantage of this circuit is a lack of short-circuit-current limit. While adequate for many purposes, it is nowhere near as good as the IC regulators discussed below.

21-5 Op-Amp Regulators

Although not as convenient as the specialized ICs, voltage regulators based on op-amps perform well. Another reason for describing them here is to understand the usual series-regulator circuit operation. One such regulator is shown in Fig. 21-10. A Zener-diode regulator provides the voltage reference V_Z. Any voltage difference between this and the potentiometer output V', a fraction α of the output voltage V_S, is amplified by IC1. The IC output V_0' drives the base of Q1. Normally Q2 is cut off and can be ignored. The series transistor must have high power and current

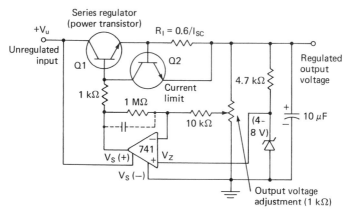

Fig. 21-10   Series voltage regulator, op-amp version.

characteristics. Often Q1 is replaced by a Darlington transistor for higher current gain. Since Q1 is effectively an emitter follower (and the drop across R_I is small), the regulated output voltage V_S will equal the base voltage less the usual base-emitter diode drop (0.7 V). Negative feedback adjusts V_0, and thus V_S, so that the IC1 input voltage difference ($V_Z - \alpha V_S$) is small; that is, $V_S - V_Z/\alpha \approx 0$. This is the basis of the voltage-regulator action. Note that α is less than unity for this circuit, which means that the Zener voltage must be less than the output voltage.

As with other active voltage regulators, good ac as well as dc regulation is provided. The reduction in ripple is much more effective than with passive RC filters. Care must be taken to ensure an adequate voltage difference (1 to 2 V) across Q1 at all times, particularly during the ripple-cycle minimum, or the transistor will saturate. This problem is often seen as a negative notch in the output, or dropout of regulation, synchronized with the line frequency when the regulator is heavily loaded. Should this occur, the transformer voltage or size of the input capacitor must be increased.

The output capacitor (electrolyte or tantalum) provides a low impedance at higher frequencies where the regulator effectiveness decreases. Logic and high-frequency (megahertz) circuits require another parallel ceramic or paper capacitor (0.01 μF to 0.1 μF) near the circuit (not at the power supply) for adequate bypassing.

Short-circuit or current-overload protection is provided by Q2. Normally the voltage drop across the low-value resistor R_I in series with the load is so small that Q2 is off. Only when a high load current flows will the voltage drop exceed the emitter-base diode voltage (0.6 V). In this case the collector current will flow through Q2 and Q1, causing the output voltage to be reduced. Under short-circuit conditions ($I_L = I_{SC}$), IC1 loses control, the voltage drop across R_I is constant at 0.6 to 0.7 V, and the current is limited. The current limit protects Q1 and perhaps the external circuit as well. Maximum power is dissipated in Q1 during a short circuit, and adequate cooling must be provided to allow for this possibility.

21-6   Overvoltage Protection

Sometimes overvoltage protection is desirable. ICs and other semiconductors are damaged by excess voltage. Application for a short time of a voltage a bit (10 percent) above the maximum ratings is fairly safe, but a factor of 2 overvoltage is likely to destroy the device. All the ICs in an instrument can be wiped out by a transient voltage surge, a worrisome possibility, especially in large and expensive instruments. To understand how this might happen, it must first be realized that the usual series regulator (IC or op-amp versions) responds to overvoltage by turning off the series transistor, that is, Q1 of Fig. 21-10. Of course, the overvoltage must be supplied by an external source since the output voltage would normally drop to zero with the series transistor off. Since series regulations do not act as current sinks, they passively allow the output voltage to rise unchecked. In instruments which have several supplies (say +5 and + 15 V) an accidental cross connection will cause the output of lower voltage supply to rise to that of the higher voltage supply, at least if the current limit of the higher voltage supply is sufficiently high. Another way of introducing an overvoltage is through an external input connection. Suppose that an IC input lead is accidentally connected to the 110-V line. The IC will immediately blow and may short the input through the destroyed IC to the power supply. Without overvoltage protection, all the remaining ICs connected to the supply and the supply itself may be destroyed.

A simple and effective overvoltage protection circuit is the SCR "crowbar" circuit of Fig. 21-11. Normally the SCR is off and has no effect on the regulated supply. Should a voltage surge exceed the Zener voltage, current will flow through the Zener into the SCR gate and turn the SCR on, thus shorting the power supply. The SCR remains on until the power supply is switched off. The SCR current rating should exceed the maximum surge current expected. Negative surges are limited by the reversed diode. Negative supplies can be protected by the same circuit by reversing the leads.

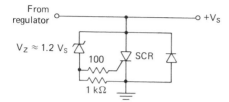

Fig. 21-11 Overvoltage protection circuit.

21-7 Voltage-Regulator ICs

Voltage regulators are effective, convenient, inexpensive, and therefore popular in electronic-instrument power supplies. Basically they are equivalent to the op-amp voltage regulators described above except that the circuit is largely incorporated into one package. Only a few of the many units available will be discussed here.

The simplest units are the three-terminal fixed-voltage regulators. Popular sizes are 5 and 15 V. All that is needed is to connect the input and common to an unregulated supply, and the regulated voltage is taken from the output. Most regulators require that the input voltage be at least 2 to 3 V higher than the output. Current limit (short-circuit protection) is provided. When operated without heat sinks, as is usually the case, the power dissipation $I_L(V_u - V_S)$ is limited, typically to about 1 W. These units are therefore most satisfactory for low-current applications. The unit shown is a positive regulator (input and output voltages are positive with respect to the common terminal). A series of negative three-terminal voltage regulators is also available (for example, 7905).

Conversion of these fixed-voltage regulators into variable-voltage or odd-value regulators requires only the addition of two resistors (Fig. 21-12b). Since the regulator holds the voltage between the output and common terminals V_R constant, division of the voltage V_S by the resistors results in a regulated output determined approximately by the resistor ratio. A correction for the regulator quiescent current I_o (typically 3 to 10 mA) flowing through R_2 is required. Because I_o varies somewhat with load, regulation of the variable-voltage configuration is poorer than the fixed-voltage regulator itself.

A convenient dual-supply IC is shown in Fig. 21-13. In addition to output-current limit, this unit has thermal shutdown to protect it against overheating due to a long-term short circuit or an inadequate heat sink. For full current output at higher input voltages, a heat sink is desirable. Another feature is that the negative voltage follows or tracks the positive voltage. When this is used with op-amps, the signal output is affected less when the $+V_S$ and $-V_S$ supplies change together. A variable-output-voltage version (4194) utilizing an external potentiometer is also available.

One of the most versatile regulators is the 723 chip. It is especially useful as a regulator with an external power (series pass) transistor, as indicated in Fig. 21-14. Separate power transistors are relatively inexpensive and easy to cool with a heat sink. They are practically indispensable for higher current supplies. The main disadvantage is the additional cir-

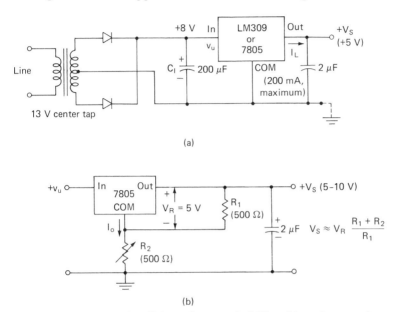

(a)

(b)

Fig. 21-12 Power supply utilizing a three-terminal IC positive voltage regulator: (*a*) fixed and (*b*) variable.

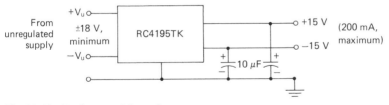

Fig. 21-13 Dual-output IC regulator.

cuit connections and components required. Other regulator circuits useful for special applications have been developed and are described by the manufacturers.

21-8 Single-to-Dual Supplies

Occasionally only a single supply is available to power a circuit which normally requires a dual supply. An example is an op-amp powered by an automobile battery. One solution is to split the power-supply voltage in half, each half acting as a section of the dual supply (Fig. 21-15). In this case the power-supply tap is the same as the dual-supply common, which is normally connected to the circuit ground. If the battery is grounded, obviously the circuit cannot be also, but the common can be thought of as a *floating ground* for convenience in circuit analysis. If little current flows in

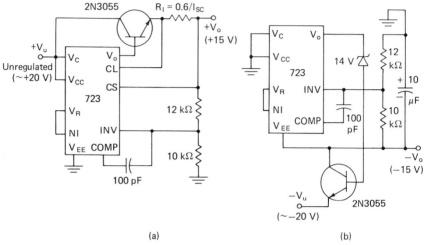

(a)   (b)

Fig. 21-14 External-transistor IC voltage regulators: (*a*) positive output; (*b*) negative output; resistor values shown are for ±15 V output.

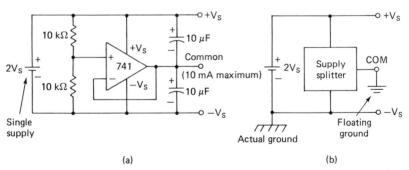

(a)   (b)

Fig. 21-15 Split single supply as light-duty dual supply. The commons or grounds of the two supplies differ.

the common return, i.e., the currents in the positive $+V_S$ and negative $-V_S$ supplies are nearly equal, only a center-tapped resistor would be needed to split the supply. Usually, however, the current drains from the negative and positive supplies differ, and the supply splitter must be able to take up the difference. A unity-gain amplifier (Fig. 21-15) works well for this purpose at modest current levels. If larger unbalanced (common) currents occur than the op-amp can handle (10 to 20 mA for the 741), an output dual-transistor current booster (Chap. 22) may be added. Voltage regulators may be added to one side of the dual supply as discussed above if the input/output voltage difference is sufficiently high (2 to 3 V).

21-9   DC-to-DC Converters

A dc-to-dc converter is a power supply with a dc input and dc output. Usually the output is higher in voltage than the input although sometimes it is of equal voltage but opposite polarity. Generally the input is a battery. These converters generate a square wave which is rectified and filtered after passing through a capacitor or step-up transformer. In Fig. 21-16

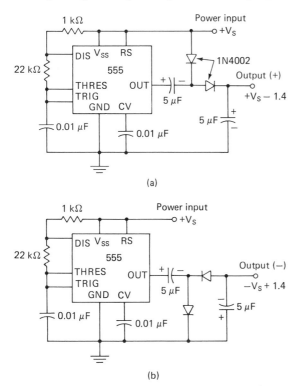

Fig. 21-16 Lower-current dc-to-dc supplies: (*a*) voltage doubler; (*b*) negative output.

two simple relatively low-current converters are shown. The square wave is generated by an astable multivibrator (555 timer). In Fig. 21-16*a* the rectified (half-wave) and filtered voltage is added to the existing supply to produce an approximate voltage doubling. Because of the diode drops and ripple the voltage actually will be somewhat less. In Fig. 21-16*b* the diodes are reversed, and the output voltage is of the opposite polarity.

This converts a single battery into a dual-polarity supply, as may be needed for op-amp circuits. The current capacity is somewhat low (~50 mA) because of the IC limitations but is adequate for many applications.

Where higher output current and voltage are required a transformer driven by power-switching transistors can be used. A circuit utilizing widely available components is shown in Fig. 21-17. It operates at a

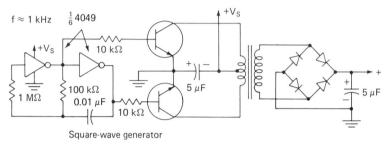

Fig. 21-17 Transformer-coupled dc-to-dc-converter power supply.

frequency which is easy to generate and filter (1 to 10 kHz). Rectifier design considerations have been previously discussed (see Figs. 15-1, 21-2, and 21-4). The square wave, generated by a CMOS astable multivibrator, switches the two power transistors in a push-pull configuration. The output voltage V_c on the capacitor is $V_c = 2NV_S - 1.4$, assuming negligible ripple, where N is the transformer turns ratio (full primary to secondary). For example, if the transformer is a standard 12.6-V center-tap filament transformer (rated 110 V 60 Hz primary) which has a turn ratio of about 9, the output voltage will be about 90 V when connected as a step-up transformer. Since the transformer is operated other than intended (higher frequency, primary and secondary reversed), the performance is difficult to predict accurately and some experimentation may be required. Transformers which match specific transistors with specific input and output voltages are manufactured but are not widely available. They have the advantages of higher efficiency and simpler driving circuits, often because of an additional feedback winding. If a suitable transformer (and circuit) is available, it may be preferable to the more general circuit of Fig. 21-17.

21-10   Batteries and Battery Charges

Portable battery-powered instruments are increasingly popular as a consequence of the development of smaller low-power ICs. A list of widely available nonrechargeable batteries is given in Table 21-1. In general the capacity of a particular type, expressed in amperehours (Ah), increases

TABLE 21-1 Nonrechargeable Battery Characteristics

Class	Type or model	No. of cells	Voltage	Capacity, Ah	Recommended maximum current, mA	Weight, g	Popular use
Carbon-zinc	AA	1	1.5	0.25–1	80	17	Penlight
	C	1	1.5	0.5–2	80	39	Flashlight
	F	4	6	2–10	250	640	Lantern
	2U6†	6	9	0.05–0.3	15	33	Transistor radio
Alkaline	VS1325‡	6	9	0.1–0.5	30	40	Replacement for 2U6
Mercury	VS1‡	1	1.4	0.3–0.5	80	12	Transistor devices
	M25	6	8.4	0.2–0.5	30	46	Transistor radio
Silver oxide	S5	1	1.6	0.07	3	1.2	Hearing aid

†RCA number.
‡Burgess number.

with the battery weight at a given voltage. It is therefore possible to estimate the capacity of a battery by comparing its weight (or volume) with those given in the table, taking into account the number of cells. It should be observed that while the capacity differences between equal-sized batteries of different types is not extremely large, silver oxide and, to a lesser extent, mercury cells are small for a given capacity and are best suited for highly miniaturized instruments. The shelf life of batteries is quite variable but usually is over 1 year at room temperature and much longer at lower temperatures (refrigerated). Capacity varies with manufacturing technique, shelf life, and desired endpoint voltage; the values in the table should be considered only estimates.

Battery capacity depends also on current magnitude (and cycling if intermittent). It will decrease by roughly 15 percent for each factor-of-2 increase in current beyond the suggested maximum. A carbon-zinc cell (often called the *Leclanche cell*) has a voltage variation of 20 to 50 percent over its useful life. A mercury-cell voltage will decline much less, about 2 to 10 percent until close to the end of its life, when the voltage drops rapidly. Alkaline (zinc–manganese oxide) cells are similar to the carbon-zinc but have about 50 percent greater capacity and can deliver more current without a loss in capacity.

No 5-V batteries are readily available. An inexpensive substitute is a 6.3-V lantern battery with two forward-conducting diodes in series to drop the voltage to 5.0 ± 0.5 V. If a 5-V regulated supply is desired, a battery of 8 V or above is needed because most regulators have a 2- or 3-V drop.

Most of the comments concerning nonrechargeable cells, in particular the relation of capacity to weight, also apply to rechargeable cells (Table 21-2). Battery life, expressed as the number of times a battery can be charged and discharged, is maximized by keeping the current drain within the suggested maximum and avoiding complete discharge. However, battery exercise consisting of cycling from full charge to 90 percent discharge is beneficial.

Battery charging may be done at a slow rate (trickle charge at a roughly constant current), which does not require a voltage control, or at a fast rate, which does. A cell can be damaged (or even made to explode) by a high current, especially when fully charged; therefore either the charging current must be low or the state of charge, as measured by the terminal voltage, must be monitored.

Two trickle-charge circuits are shown in Fig. 21-18. Charging current I_c, as set by the series resistor $I_c = (V_S - V_B)/R_S$, must not exceed the specified trickle-charge current when the battery is charged. In Fig. 21-18b the battery is attached to the normal supply as a standby source of power in case of line power-supply failure. Although the voltage will drop from its normal regulated value of V_S down to V_B, many circuits will

TABLE 21-2   Rechargeable Battery Characteristics

Class	Type model	No. of cells	Voltage		Capacity, Ah	Trickle charge rate, mA	Recommended maximum current	Popular use
			Nominal	Charging				
Nickel-cadmium	CD14†	1	1.2	1.40	0.8	5	400 mA	Type C replacement
	CD25†	8	9.6	11.2	0.2	2	65 mA	Transistor devices
Alkaline	C	1	1.5	1.75	1	25	200 mA	Type C replacement
Gel	D	1	2	2.35	2.5	5	500 mA	Portable instruments
	X	3	6	7.05	5	10	2 A	Portable instruments
Lead (oxide)	......	6	12	14.4	50	200	10 A	Automobile

†Burgess number.

operate satisfactorily at a lower voltage, or at least stored digital data will be preserved.

Fast chargers (Fig. 21-19) simply consist of a current-limited power supply which turns on when the battery voltage drops below its fully charged value. A standard voltage-regulated power supply (Fig. 21-14)

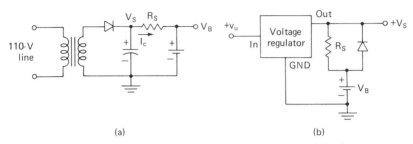

(a)   (b)

Fig. 21-18   Battery trickle-charge circuits: (*a*) standard; (*b*) battery standby.

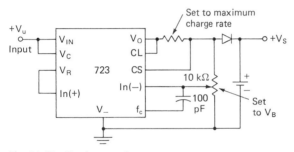

Fig. 21-19   Fast battery charger.

works well. The current limit of the regulator must be set at or below the maximum charging current to avoid battery damage. If several batteries are involved, they must be in series so that they charge and discharge together. Care must be taken to set the regulator voltage to the battery's fully charged value (within 2 to 5 percent). If the voltage is set too high, the charging current will not reduce to its trickle value when the battery is fully charged, while if it is set too low, the battery will not charge fully.

Chapter Twenty-Two

Power Amplification and Control Circuits

Occasionally an electronic instrument must deliver a relatively high power into an external load. Standard op-amps and ICs are low-power devices which are rather limited in the current and voltage they can deliver. Usually a power amplifier as a final stage provides the solution. Several types are described here. Also considered in this chapter are triac and SCR controls for loads such as motors or heaters placed across the power line.

22-1 Power-Output Transistor Configurations

Power dissipation is an important and often limiting factor in power-amplifier design. Discrete power transistors can dissipate considerably more power than ICs, especially when they are attached to a heat sink; for this reason they are often used. Efficient configurations are desirable both to increase the maximum available output power for a given set of transistors and to eliminate unnecessary drain from the power supply. Dissipation is minimized by keeping the power-transistor current close to zero when no output voltage is required. With the push-pull configurations (class B amplifiers) two transistors are required. One transistor controls the positive voltage-output swing and the other the negative swing. Two types of configurations are shown in simplified form in Fig. 22-1. Bias is adjusted so that both transistors are slightly conducting when the input signal v_i is zero. On positive signal swings the current through Q1, and

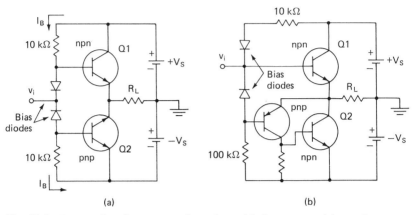

(a)   (b)

Fig. 22-1   Basic push-pull output configurations with dc response: (*a*) complementary symmetry; (*b*) totem pole with driver.

therefore the load R_L, increases while Q2 turns off. On negative swings Q2 is on and Q1 is off. The bias diodes compensate for the transistor base-emitter voltage. When properly adjusted, collector currents (Q1 and Q2) are small (1 percent of peak value) with zero signal. Some means of bias equalization between transistors is desirable, e.g., by the addition of small emitter resistances. Sometimes separate driving stages are provided for Q1 and Q2. A close match between transistors is difficult to achieve, especially at the crossover point where one transistor is turning on while the other is turning off. As a consequence of this mismatch a notch is often observed in the signal near zero.

Care must be taken to avoid turning Q1 and Q2 on simultaneously, as might occur with improper bias, since in effect the power supplies are then shorted through the transistors. When hot, transistors draw additional bias current, especially when biased from a fixed voltage source. Transistor power dissipation is the average (over 10 to 100 seconds) of $I_C V_{EC}$, where I_C is the collector current and V_{EC} is the voltage drop across the transistor. Under some conditions the increase in power dissipation with bias current can heat the transistors enough for the transistors to be destroyed, a condition known as *thermal runaway*. Current limit of some type is essential.

Where efficiency or power dissipation is unimportant, the simple transistor emitter-follower configuration may suffice (Fig. 22-2). The emitter resistor R_E determines both the maximum (negative for npn) signal current I_n and the zero-signal transistor dissipation P_D; that is, when v_o = 0, $I_n = V_S/R_E$ and $P_D = I_n V_S$. Current limit is provided by R_S. Dual-polarity high-current power supplies are needed with the configurations shown; a disadvantage is that a high-current dc output is required. For ac

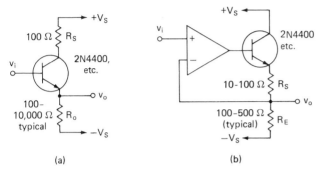

(a)   (b)

Fig. 22-2   Emitter-follower current amplifiers: (a) standard amplifier; (b) op-amp with current booster.

output, a single power supply can be used if the load is coupled through a capacitor (and the input signal is properly biased). Practical versions of this circuit are described in Fig. 22-6. A transformer-coupled ac amplifier is discussed below (Fig. 22-5).

Another type of power amplifier utilizes a switching transistor (Fig. 22-3). The driver Q2 turns the switching transistor Q2 on and off rapidly.

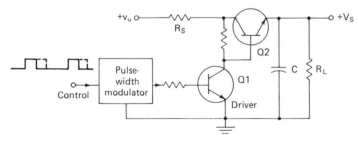

Fig. 22-3   Switching-transistor principle.

When on (transistor in saturation) the capacitor charges (through R_S) to replace current drain through the load R_L. The ratio of on to off time (duty cycle or pulse width) is automatically varied so that the capacitor voltage has the desired value. Control is accomplished by a pulse-width modulator. Efficiency is high because the transistor is either off (no power dissipation) or saturated (only a fraction of a volt transistor drop). The rate of switching must be at least one or two orders of magnitude higher than the capacitor time constant $R_L C$. Many higher-current power supplies are based on this method.

22-2 Op-amp Output Booster

Higher current can be drawn from the output of an op-amp booster. The booster of Fig. 22-4 utilizes the complementary-symmetry configuration.

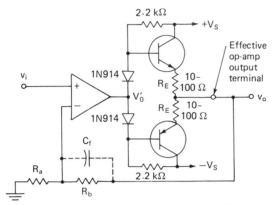

Fig. 22-4 Op-amp output-current booster. Noninverting configuration is illustrated.

Notice that the transistor amplifier output v_o is the effective op-amp output (rather than v'_o). Any op-amp feedback configuration will work, e.g., the noninverting configuration. The emitter resistors R_E limit the output current as well as provide bias equilization between transistors. Feedback reduces the crossover notch to a negligible level. However with some transistors a high-frequency oscillation may occur near the crossover point as a consequence of an enhanced phase shift or poorer frequency response near transistor cutoff. This problem usually can be eliminated by the addition of a feedback capacitor C_f or application of another op-amp feedback-compensation method.

22-3 AC Amplifiers

While the amplifiers described above can amplify alternating as well as direct current, they have the inconvenience of requiring a high-current negative supply as well as a positive supply. The ac amplifiers described have only a single supply. Transformer or capacitor output coupling is involved. Where distortion is not critical, such as driving a servo motor, a transformer-coupled (class B) push-pull amplifier (Fig. 22-5) may be suitable. Here Q1 conducts on the positive signal cycles and Q2 on the negative. One driver (IC1) and diode provides the proper bias while the other driver (IC2) inverts the phase for Q2. The transformer combines the transistor currents and steps the voltage up (or down). If we take into account the voltage drop across R_E, the peak-to-peak output v_o is approximately $0.9V_S n$, where n is the turns ratio between the total primary and the secondary.

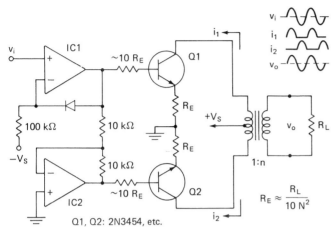

Fig. 22-5 Transformer-coupled ac amplifier.

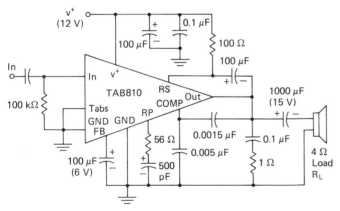

Fig. 22-6 An IC power ac amplifier.

Complete IC ac amplifiers capable of delivering modest power into a low-impedance load R_L, such as a 4-Ω speaker, have been developed for the consumer electronics industry. An example is given in Fig. 22-6. The low-frequency response is determined by the external capacitors, especially the output capacitors C_o. Internally most circuits are similar to those of Fig. 22-1. Normally a single, unregulated supply $+V_S$ is all that is required. Metal strips (tabs), which act as heat sinks, often protrude from the IC package. If the strips are connected to larger heat sinks, this unit may deliver 1 to 5 W.

22-4 SCR and Triac Drivers

Both the SCR and triac are high-current high-voltage semiconductor switches, often termed *thyristors*. The SCR will pass current in one direction, but the triac will also pass alternating current. Applications of the SCR include high-voltage power supplies and dc motor control. Triacs are popular controls for ac motors and lighting loads operating from the 110- to 440-V ac lines.

An SCR is turned on by applying a voltage between the gate and cathode sufficient to raise the gate current to a threshold value (Fig. 22-7).

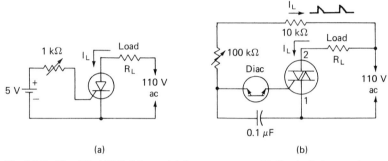

(a)   (b)

Fig. 22-7 Simplified SCR drivers: (*a*) dc gate current; (*b*) diac (pulse) control.

Switched on, the SCR is equivalent to two forward-conducting diodes in series (voltage drop 1.2 to 1.6 V). It remains on until the cathode-to-anode voltage becomes less (more negative) than a holding voltage (about +1.2 V), which for an ac anode voltage occurs at the end of the positive half of the cycle. Gate-current requirements vary widely, ranging from below 1 mA for a "sensitive gate" SCR to well over 100 mA for a standard-sensitivity medium- to high-anode-current (5 to 50 A) SCR. Many require higher gate current than can be delivered by the usual logic or op-amp ICs. However, the gate current need be applied only briefly (typically 2 to 20 μs), and therefore most SCR drivers are designed to deliver short pulses which have a high peak but a low average current (to reduce power requirements).

A popular SCR-gate pulse generator, when the control is derived from a higher voltage source, is the diac or trigger diode (Fig. 22-7*b*). It has the property of remaining nonconducting (off) until a threshold voltage is reached (typically 20 to 40 V), at which point it suddenly becomes highly conducting until the voltage drops to a low value (~1.2 V), when it again switches off. Almost invariably the diac is connected between a charged capacitor and the gate so that when the capacitor voltage reaches the diac threshold value, it discharges through the SCR gate. Current is limited only by the low SCR and diac internal resistance and therefore has a high peak value. Without the diac in Fig. 22-7*b*, the value of the resistance needed to turn the SCR on would be relatively low and the power dissipated in the resistor would be excessive.

Gate-current requirements for the triac are similar to those of the SCR if the gate polarity is proper and/or the triac is of the proper type. Terminals 1 and 2 of triacs correspond to the cathode and anode of an SCR, respectively. Under conditions of ac gate drive, where the gate-to-terminal-1 voltage and terminal-2-to-terminal-1 voltage have the same polarity (quadrants I and III), the gate turn-on current (magnitude) is comparable to that of the SCR. In the circuit of Fig. 22-7*b*, the triac can be substituted for the SCR since the diac driver will deliver a trigger pulse during both the positive and negative cycles. Some types of triacs can also be triggered by a positive gate voltage when terminal 2 is negative (quadrant IV operation). In this case any drive circuit which works for the SCR will work for the triac (such as Fig. 22-8). Care must be taken to determine

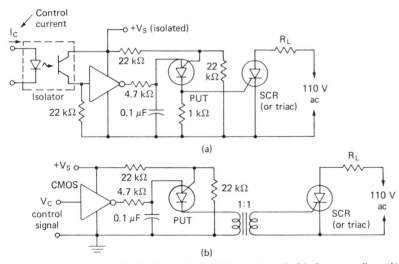

Fig. 22-8 PUT-type isolated SCR and triac drivers; (*a*) optical isolator coupling; (*b*) transformer coupling.

whether a triac will operate in quadrant IV if a positive drive is chosen since some types of triacs have an extremely poor sensitivity or can be damaged if operated in this quadrant. In contrast other types of triacs may require such a low gate current that they can be turned on by a CMOS driver.

A convenient isolated driver for the SCR or triac shown in Fig. 22-8 utilizes a PUT (programmable unijunction transistor). It will deliver short high-current pulses (5 to 50 μs, 100 mA typically), as required by the SCR or triac. If an isolated power supply is available, the driver circuit of Fig. 22-8a can be used. A trigger pulse is applied through the optical isolator which lowers the threshold voltage of the PUT and causes the capacitor to discharge into the SCR or triac gate. One or more gate pulses are produced for each trigger pulse (positive edge). The second PUT driver circuit (Fig. 22-8b) produces a train of pulses as long as the input is positive. The pulses are coupled to the SCR or triac through a pulse transformer intended for SCR trigger applications. The transformer provides the required isolation between the control circuit and the power line. The SCR or triac is turned on by the first pulse in the train after the anode becomes positive (SCR) or terminal 2 voltage crosses zero (triac). Subsequent pulses have no further effect.

22-5 Small Motors and Drivers

Switching or speed control of small motors is occasionally required for instrumentation applications. The types of motors likely to be encountered are reversible dc motors, reversible ac motors, three-phase servomotors, synchronous ac motors, and stepping motors. A dc motor with a permanent magnetic field, e.g., the small motors used in toys, can be driven with a range of dc voltages. Direction of rotation reverses with polarity reversal. Speed increases with voltage and decreases with load. Speed is not constant without a velocity sensor and feedback control. A power amplifier with a dc output is a suitable driver, as indicated previously (Fig. 22-4). The magnitude of the supply voltage V_S should be equal to the maximum or full-speed voltage. Of course, the power supply and power amplifier must be able to deliver the current required by the motor under full-load conditions. Typically 6 to 30 V at 0.2 to 5 A is required. It is desirable to set the power-supply current limit just beyond the full-load current point to provide protection in case the motor stalls and draws excess current. Control of motor speed and direction is accomplished by varying the amplifier input voltage and polarity.

A common reversible ac servomotor has both field and control windings. The field winding is connected to the line (usually 110 V 60 Hz). The control winding is connected to an ac source which is phase-reversible and of variable amplitude. The motor speed increases with control voltage up to its maximum, where the control voltage is equal to the line voltage. Motor direction reversal is accomplished by reversing the control-winding phase (from approximately 0 to 180° with respect to the field-winding phase). A transformer-coupled ac power amplifier (Fig. 22-5) is a suitable driver. In most cases a 60-Hz control or filament-type transformer operated in reverse is a satisfactory step-up transformer. Motor speed and direction are determined by the amplifier ac input voltage and phase. Control by a dc voltage may be done through an ac bridge and amplifier with an FET as a variable resistor as the control element. When the bridge is in balance, the motor is stopped. Increasing the dc control voltage increases the motor speed in one direction, while a decrease (more negative) voltage increases the speed in the reverse direction. In addition to being the same frequency, the phase of the bridge "generator" voltage must be approximately in phase (or 180° out of phase) with respect to the field windings.

Three-phase servomotors are sometimes employed for remote positioning, especially where relatively rapid response is required. In principle, and occasionally in practice, the connection between the transmitter and receiver servos is direct (Fig. 22-9). The transmitting servo can be thought

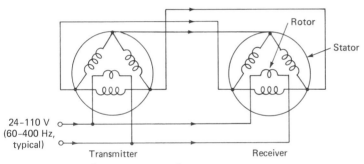

Fig. 22-9 Three-phase servo connection.

of as a transformer with a primary (rotor) which can be rotated continuously over 0 to 360°. There are three secondaries (stators) mechanically fixed in position at 120° angles. As the rotor is moved, the amplitude and phase of the voltages induced in the three windings vary and are a unique function of angular orientation. The receiver servo has the same construction, and the stator voltages are the same as the transmitter. Correspondingly the magnetic fields in the receiver match the transmitter at any instant, and therefore the receiver rotor moves to match the position of the transmitter. More powerful servos require an ac amplifier on each phase. Actually it is the winding phase angles which determine the receiver position, and the transmitter and a servo can be replaced with a circuit which generates signals with the appropriate phases.

The speed of synchronous motors is locked to the line frequency. Since the line frequency (60 Hz) is constant, especially over the long run, they are good clock motors. Control over speed over a modest range at reduced output torque can be accomplished by a variable-frequency drive.

22-6 Stepping-Motor Drivers

A stepping motor is a pulse-controlled motor which rotates in discrete steps, typically 100 or 200 per revolution. One step of a 100-step motor, for example, corresponds to an angular rotation of 3.6°. It has a reproducible angular position, which is very desirable in many instrumentation applications, although the speed and torque are mediocre.

The operating principle of the stepping motor is shown in Fig. 22-10

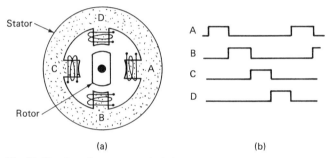

(a)   (b)

Fig. 22-10 Stepping-motor principle: (*a*) four-step motor (hypothetical); (*b*) winding step sequence (sam for 100- and 200-step motors).

for a hypothetical four-step motor. When no coils are energized, the permanent-magnet rotor remains fixed in position because it is attracted to, and held by, the closest stator pole. It can be moved to a neighboring pole by applying sufficient current through the electromagnetic coil of that pole. Suppose, for example, that the north pole is at position *D*. If pole *A* is now magnetized as a south pole, the rotor north pole will swing to *A*. It can be seen that if the poles are pulsed in an *ABCD* sequence, the rotor will continuously rotate clockwise and if pulsed in an *ADCB* sequence, the rotation will be counterclockwise. Practical stepping motors have many more poles on the stator (and rotor) but they are usually organized in sets of four, so that the same sequence of four pulses into four windings is required. The motion can be stopped or reversed after any step.

A practical drive circuit for a stepping motor is given in Fig. 22-11. The step sequence given is controlled by a 2-bit counter and decoder. Greater motor torque is obtained by overlapping the pulses and doubling their pulse length. Each input pulse produces one step. The direction of the counter and motor is controlled by the counter up-down control

In order to conserve power a gate connected to the transistor drivers turns off the motor current during prolonged periods where no movement is required.

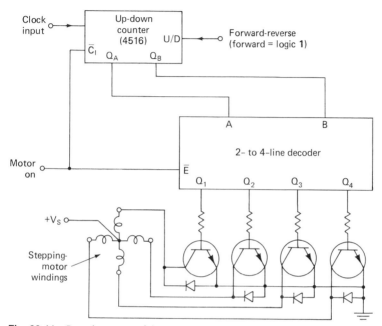

Fig. 22-11 Stepping-motor driver.

POWER-CIRCUIT DESIGN EXAMPLE

Example 9 Dual 15-V Supply

DESCRIPTION: This standard +15/−15 V regulated supply will deliver 70–150 mA and is suited for powering op-amp circuits.

SPECIFICATIONS

Power line: 115 V, 60 Hz (15 W)

Output voltages: ±15.0 ±0.3 V

Regulation (line or load): 1 percent

Maximum current: 70 mA (average, both sides)

120 mA (short-term)

Protection: short-circuit and thermal overload

DESIGN CONSIDERATIONS: A dual IC regulator with internal current and thermal limits was selected. The input voltage must be 3 to 5 V higher than the regulated output, so that the unregulated voltages should be +19 and −19 V. Allowing for diode drop and assuming a 10 percent ripple, the transformer should deliver 20.7 V peak (14.7 V rms) minimum across each half secondary. A 30 V rms center-tapped transformer was selected. From Fig. 21-4 the input capacitors should be 350 μF (or more) for a 120-mA load (10 percent ripple). Power dissipation for a 120-mA load is 1 W.

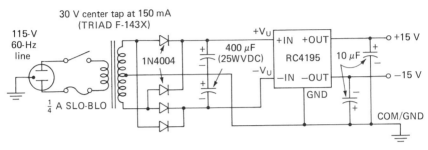

Fig. E-9   Dual power supply for op-amps.

POWER-CIRCUIT DESIGN PROBLEMS

68 Design a regulated power supply that will deliver ±12 V (dual) at 100 mA *and* +5 V at 500 mA using only one transformer (26 V rms center-tapped at 1 A). Calculate the power dissipation in the regulators and choose a proper heat sink if needed. A resistor in series with the 5-V regulator (chosen so the regulator input voltage is 8 V at full load) will minimize heating.

69 A power supply for a microprocessor requires a battery backup. The battery is 8.8 V at full charge (3-Ah gel type). Voltages and currents required are + 8 V at 1 A, +15 V at 50 mA, and −15 V at −50 mA. Regulation is not necessary (these are the specifications of the S-100 bus many microprocessors use). Design such a supply, including the battery charger.

70 Design a combination power supply and op-amp (voltage gain of 10) which will operate on a single 9-V battery. The quiescent (no-signal) battery current should be under 0.1 mA. Maximum output voltage required is ±2 V (ac or dc) into a 2-kΩ load. A micropower op-amp (776 or LM4250) is suggested.

71 Design a triac circuit that will control the speed of a motor (110 V 10 A maximum) from a microprocessor port. Isolation of the power line from the port ground is necessary.

72 Design a power amplifier (voltage gain of 10) that will deliver an output (ac or dc) of ±30 V at 1 A (30 W peak). A "high-voltage" op-amp and discrete transistors are suggested. For more of a challenge, redesign for an output of ±100 V maximum.

73 Design a stepping-motor control in which the speed (unidirectional) is set by thumbwheel (two-digit). The motor has 100 steps per revolution, and the maximum speed is 99 r/min (corresponding to 99 on the thumbwheel dial).

Index